Textbook on Cloning, Expression and Purification of Recombinant Proteins

Kakoli Bose
Editor

Textbook on Cloning, Expression and Purification of Recombinant Proteins

Springer

Editor
Kakoli Bose
Integrated Biophysics and Structural Biology Lab, Advanced Centre for Treatment, Research and Education in Cancer (ACTREC)
Tata Memorial Centre
Mumbai, India

Homi Bhabha National Institute
BARC Training School Complex
Mumbai, India

ISBN 978-981-16-4986-8 ISBN 978-981-16-4987-5 (eBook)
https://doi.org/10.1007/978-981-16-4987-5

This Springer imprint is published by the registered company Springer Nature Singapore Pte Ltd.
The registered company address is: 152 Beach Road, #21-01/04 Gateway East, Singapore 189721, Singapore

I dedicate this work to Dr. Lalith Kumar Chaganti who conceptualized this book being by my side—a beloved student, later an extraordinary postdoctoral associate, and an amazing human being whom we lost to complications arising from tuberculosis.

This work is also dedicated to a dear friend, Mr. Arindam Basu and millions of other brave hearts across the globe who finally succumbed to the atrocities of ongoing COVID-19 pandemic.

These lost souls will remind the world to challenge future pandemics and deadly diseases with tireless efforts toward path-breaking scientific discoveries.

Preface

The twentieth century had been a torchbearer of several seminal scientific discoveries that revolutionized the world through cogent advances in technology and biomedical sciences. It is simply intriguing to learn that identification of a small plasmid DNA in the early 1950s led to a series of scientific explorations with the consequential revelation of many important biomolecules, including restriction endonucleases and ligases. These paramount discoveries helped germinate the concept of engineering a gene simply inside a test tube. Introduction of genes that encode "*proteins of interest*" into plasmid DNAs, followed by their expression in any microorganism such as *E. coli*, and subsequent purification in vitro for research and commercial purposes, gave birth to the field of recombinant DNA technology. This steady journey since mid-twentieth century soon burgeoned exponentially over the years to bring forth radical changes in the fields of agricultural, food, and biomedical research. Furthermore, this technology has been successfully employed to develop therapeutics such as antibodies and vaccines, thus uplifting the overall quality of life.

This textbook has been envisaged to provide thorough information on this important topic of biotechnology focusing on cloning, expression, and purification of heterologous target proteins primarily in the bacterial system. This book is divided into eleven chapters where each and every aspect of recombinant protein expression and purification has been chronologically and lucidly elaborated. Apart from providing a broad overview, the book annotates the recent developments in the field along with vivid protocols and methodologies to enable researchers to apply the knowledge in their own laboratory experiments. Furthermore, each chapter provides easy understanding of various concepts through user-friendly flowcharts, illustrations, troubleshooting tips, and problems. This textbook also aims at providing up-to-date information with thorough discussions on current and potential clinical applications.

I am indebted to all the contributing authors for their hard work and for sharing their thoughtful insights on protein expression and purification. I also thank my other lab members and friends who spent considerable amount of time proofreading and giving invaluable inputs.

Mumbai, India Kakoli Bose

Contents

Editor and Contributors

About the Editor

Kakoli Bose is currently working as Scientific Officer "G" at Advanced Centre for Treatment, Research and Education in Cancer (ACTREC), Mumbai, India. She is known for her expertise in a diverse range of fields from protein biochemistry, enzymology, structural biology, biophysics to cell biology and computational biology. Her research activities are focused on elucidation of the role of macromolecules that are involved in pathways of apoptosis and their associated diseases. She is an active member of many elite scientific societies and associations, including Biochemical Society, UK, Indian Biophysical Society, Indian Association for Cancer Research, and American Chemical Society. Dr. Bose is also a core member of the Editorial Board of coveted international journals such as Bioscience Reports and the Biochemical Journal, Portland Press, UK. She is a recipient of many prestigious awards, including the National Women Bioscientist Award from Govt. of India. She has a teaching experience of more than 10 years in the fields of biochemistry, biophysics, and structural biology. She has several publications in reputed international journals to her credit. Furthermore, her publication involving a four-state equilibrium unfolding process remains a landmark and pioneering work till date.

Contributors

Saujanya Acharya Integrated Biophysics and Structural Biology Lab, Advanced Centre for Treatment, Research and Education in Cancer (ACTREC), Tata Memorial Centre, Mumbai, India
Homi Bhabha National Institute, BARC Training School Complex, Mumbai, India

Kakoli Bose Integrated Biophysics and Structural Biology Lab, Advanced Centre for Treatment, Research and Education in Cancer (ACTREC), Tata Memorial Centre, Mumbai, India
Homi Bhabha National Institute, BARC Training School Complex, Mumbai, India

Roshnee Bose Integrated Biophysics and Structural Biology Lab, Advanced Centre for Treatment, Research and Education in Cancer (ACTREC), Tata Memorial Centre, Mumbai, India

Lalith K. Chaganti Integrated Biophysics and Structural Biology Lab, Advanced Centre for Treatment, Research and Education in Cancer (ACTREC), Tata Memorial Centre, Mumbai, India

Ayon Chakraborty Integrated Biophysics and Structural Biology Lab, Advanced Centre for Treatment, Research and Education in Cancer (ACTREC), Tata Memorial Centre, Mumbai, India

Shubham Deshmukh Integrated Biophysics and Structural Biology Lab, Advanced Centre for Treatment, Research and Education in Cancer (ACTREC), Tata Memorial Centre, Mumbai, India
Homi Bhabha National Institute, BARC Training School Complex, Mumbai, India

Shubhankar Dutta Integrated Biophysics and Structural Biology Lab, Advanced Centre for Treatment, Research and Education in Cancer (ACTREC), Tata Memorial Centre, Mumbai, India

Rucha Kulkarni Integrated Biophysics and Structural Biology Lab, Advanced Centre for Treatment, Research and Education in Cancer (ACTREC), Tata Memorial Centre, Mumbai, India
Homi Bhabha National Institute, BARC Training School Complex, Mumbai, India

Raghupathi Kummari Integrated Biophysics and Structural Biology Lab, Advanced Centre for Treatment, Research and Education in Cancer (ACTREC), Tata Memorial Centre, Mumbai, India
Homi Bhabha National Institute, BARC Training School Complex, Mumbai, India

Aasna L. Parui Integrated Biophysics and Structural Biology Lab, Advanced Centre for Treatment, Research and Education in Cancer (ACTREC), Tata Memorial Centre, Mumbai, India
Homi Bhabha National Institute, BARC Training School Complex, Mumbai, India

Rashmi Puja Integrated Biophysics and Structural Biology Lab, Advanced Centre for Treatment, Research and Education in Cancer (ACTREC), Tata Memorial Centre, Mumbai, India
Homi Bhabha National Institute, BARC Training School Complex, Mumbai, India

Nitu Singh Regional Centre for Biotechnology, NCR Biotech Science Cluster, Faridabad, Haryana, India

A Brief Introduction to Recombinant DNA Technology

1

Roshnee Bose and Kakoli Bose

Abstract

The discovery of plasmid DNA in the early 1950s marks the beginning of recombinant DNA (rDNA) technology although there remained a big hiatus between these two events. The versatility of the circular DNA plasmids enabled them to carry foreign genetic material and multiply them in the bacterial systems. Furthermore, Nobel winning discovery of restriction enzymes facilitated the quantum leap in rDNA technology that aided in engineering of plasmids so as to insert foreign genes encoding human proteins. The significant technological advances in this field that eventually led to *in vitro* expression and production of human proteins and their variants have taken biomedical science several steps ahead with innumerable applications in basic and industrial research as well as drug development. This chapter introduces rDNA technology with a brief overview of its historical landmarks, important discoveries, and path-breaking applications.

Keywords

Recombinant DNA technology · Plasmid · Restriction enzymes · Protein purification · Foreign genetic material · Drug development · Industrial research

R. Bose
Integrated Biophysics and Structural Biology Lab, Advanced Centre for Treatment, Research and Education in Cancer (ACTREC), Tata Memorial Centre, Mumbai, India

K. Bose (✉)
Integrated Biophysics and Structural Biology Lab, Advanced Centre for Treatment, Research and Education in Cancer (ACTREC), Tata Memorial Centre, Mumbai, India

Homi Bhabha National Institute, BARC Training School Complex, Mumbai, India
e-mail: kbose@actrec.gov.in

K. Bose (ed.), *Textbook on Cloning, Expression and Purification of Recombinant Proteins*, https://doi.org/10.1007/978-981-16-4987-5_1

1.1 Overview: Recombinant DNA Technology

Recombinant DNA technology is a paradigm-shift discovery in the field of molecular biology and protein biochemistry [1]. This technique uses genetic engineering to alter the chemistry of genetic material where two or more DNA molecules from different organisms are adhered together and thereafter incorporated into the genome of the host organism. These modified DNA molecules that are prepared using laboratory methods bring about changes in the phenotype of the host and therefore is called recombinant DNA (rDNA). The techniques involved in the formation of rDNA and its integration into the host genome are collectively referred to as recombinant DNA technology. This elegant process involves a plethora of biotechnical methods such as creation of rDNA, molecular gene cloning, and gene transfer that have been elaborated in the subsequent chapters.

Developing an rDNA involves a series of sequential steps, which are briefly discussed below:

- Isolation of genetic material from the organism.
- Cutting of vector DNA (plasmid DNA) and the gene of interest (foreign gene) using the same restriction enzyme.
- Amplification of gene of interest using Polymerase Chain Reaction (PCR) [2].
- Adhering the foreign gene fragment with the cut vector using the enzyme *ligase*. This leads to the formation of recombinant vector through a process called *ligation* [3].
- Making host cell competent and introducing the recombinant vector into the host through a process called *transformation.*
- Segregation of transformants and non-transformants through a screening process, and consequently obtaining the desirable gene product.

1.2 Brief History of Recombinant DNA Technology

The idea of developing organisms with desired characteristics or simply the creation of genetically modified organisms (GMO) has baffled mankind since the inception of bioengineering and discovery of genetic material. From the early seventeenth century, scientists started exploring the possibility of manipulating the genome of an organism. They also pondered whether it is possible to insert traits of our choice into an organism. In the 1960s, with the discovery of restriction enzymes, this idea slowly turned into reality [4]. Further research and development in the subsequent years led to the invention of other associated techniques pertaining to biotechnological and molecular biology research, including gel electrophoresis, PCR, mutagenesis, and microbiological culture [2, 5, 6]. With pioneering research and major breakthroughs, today we cannot imagine our lives without this transilience in the field of biotechnology.

Soon after the discovery of restriction enzymes, leading-edge research began in the field of genetic engineering. In the late 1960s, enormous progress was made in understanding the architecture of cell, its genetic component, difference in its composition between prokaryotic and eukaryotic systems, DNA replication, translation, and so forth. For the said purpose, the bacterium *Escherichia coli* has been extensively studied [7]. This facilitated development of the process of transduction that enabled transfer of infection from one bacterial strain to another by certain viruses called bacteriophages that are capable of infecting and replicating within bacteria [8–10].

The brief history of discovery of different methodologies pertaining to recombinant DNA technology and its applications has been illustrated chronologically in Fig. 1.1.

1.3 Tools in rDNA Technology

With the advent of rDNA technology, enormous efforts were directed toward expansion in the fields of bioengineering and molecular biology. Several molecular biology techniques were developed to cater to the increasing demands of production of recombinant human proteins and enzymes in bacterial system both for industrial use and for research purposes. A brief description of these important tools and their applications are provided below.

1.3.1 Restriction Enzymes in Cloning

Restriction enzymes play a pivotal role in rDNA technology. Due to their specificity and pre-defined cleaving patterns, restriction enzymes can be considered the backbone of rDNA technology. Their enzymatic properties are utilized to cut a plasmid DNA at a particular site and insert the gene of interest for further expression and characterization [4].

They are primarily divided into three categories:

- *Exonucleases*—Restriction enzymes that cleave nucleotides from 3′ or 5′ terminal ends of the DNA strand.
- *Endonucleases*—Restriction enzymes that cut double-stranded DNA (dsDNA) at any point but the terminal ends.
- *Restriction Endonucleases*—These enzymes recognize a particular sequence of the DNA molecule and cleave only at specific positions. Endonucleases are known as molecular scissors that are naturally present in bacteria as a part of their defense mechanism. They were first discovered by Werner Arber in the late 1960s in *E. coli* [11]. Since, the endonucleases lacked specificity, they could not be used in biotechnological applications until restriction endonuclease *HindII* (from *Haemophilus influenzae* strain R_d) was discovered in 1970 [12].

Fig. 1.1 An illustration depicting the important events chronologically in the history of recombinant DNA technological advancements [10, 37–53]

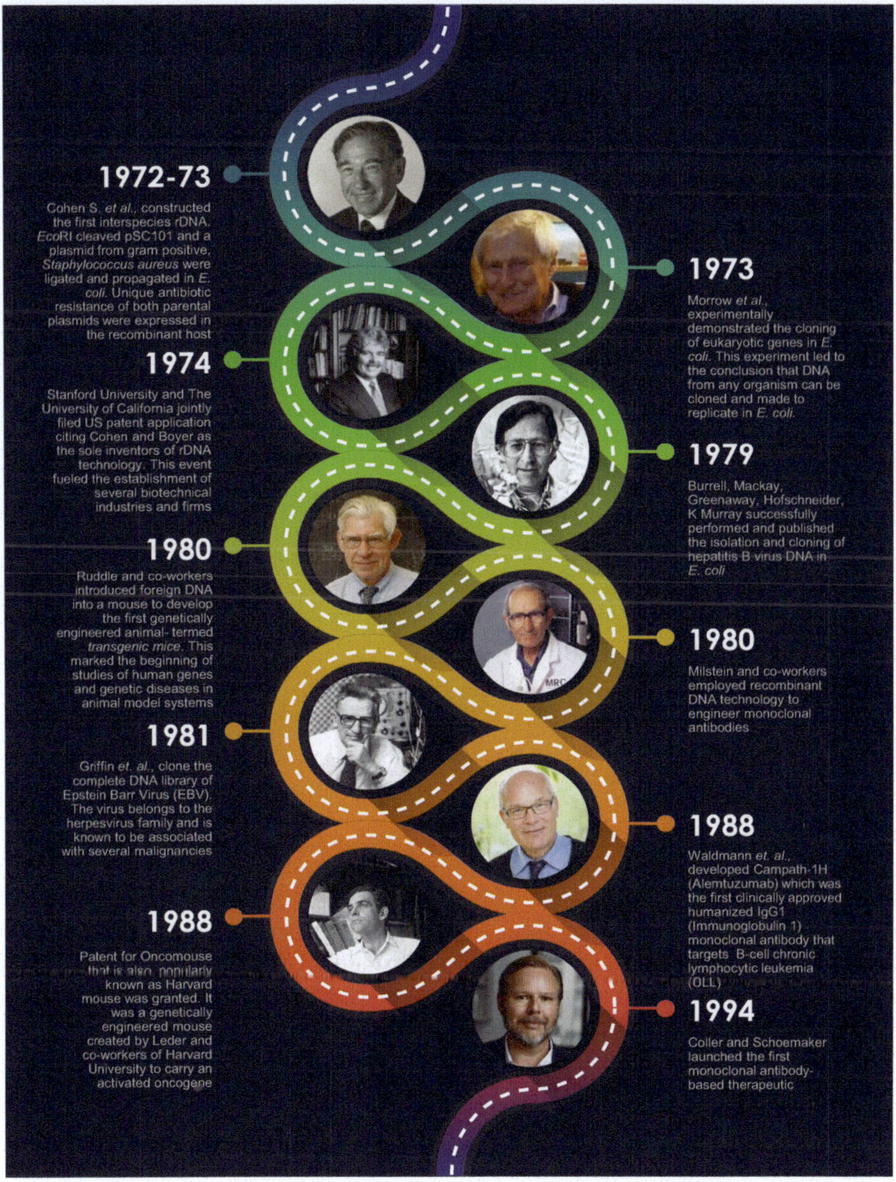

Fig. 1.1 (continued)

Naming of restriction enzymes follows a defined nomenclature as discussed below:

- First three letters of their name is derived from Latin name of the organism that produces them.
- First letter of their name is the first letter of the genus while next two letters are derived from the first two letters of the species (e.g., *Hin* from ***Haemophilus influenzae***).
- In some cases, the strain is also included in the nomenclature (here, "d" from strain R_d).
- The Roman numeral at the end denotes the number sequence at which the restriction enzyme was discovered, therefore *Hind**II*** was the second one to be identified.

Specific sequences recognized by individual restriction endonucleases are known as "*recognition sequences*." These are palindromic sequences, i.e., they read the same forward and backward. The recognition sequence for *HindII* is given below:

↓
5'----- GTPyPuAC ----- 3'

3'----- CAPuPyTG ----- 5'
↑

When *HindII* finds the above sequence in a DNA molecule, it cleaves the covalent bonds between pyrimidine (Py) (either T or C) and purine (Pu) (either A or G). This particular type of cutting in the middle results in blunt ends with no overhangs that cannot base pair with other nucleotide sequences [13]. Luckily for biologists, all restriction endonucleases do not form blunt ends; many of them produce sticky ends as well. Sticky ends are produced when single-stranded overhangs are left in the cut recognition sequences that are capable of base pairing. For example, *HindIII* makes staggered cuts to produce sticky ends as shown below:

↓
5'-----A¦A G C T T----- 3'

3'-----T T C G A¦A----- 5'
↑

When cut using the same restriction enzyme, the DNA fragment of interest and vector DNA produce identical sticky ends that can be glued together with the help of DNA ligase to form a rDNA [3].

1.3.2 Vectors

The vectors are engineered bacterial plasmid DNAs that are capable of self-replication and have very high copy numbers [14–17]. With an origin of replication (*ori*), a multiple cloning site (MCS), and an antibiotic resistance gene (e.g., Ampicillin resistant, Amp^R that can be used as a selectable marker), they act as a vehicle carrying a foreign gene (passenger DNA). This introduction of foreign gene occurs at MCS with the help of a restriction endonuclease. The first artificial vector, pBR322 vector was constructed in 1977 by Boliver and coworkers [18]. The features required to facilitate cloning into a vector are discussed below:

- **Origin of Replication (ori)**—Origin of Replication is the region in a DNA molecule from where DNA replication process is initiated. *Ori* also controls the copy number of the passenger DNA.
- **Selectable Marker**—A selectable marker is a gene present in the vector that helps in identifying transformants and selectively eliminating non-transformants. The most common selectable marker is antibiotic resistance gene such as Ampicillin resistant, Amp^R.
- **Restriction or Cloning Sites**—Restriction site codes for specific recognition sequences. At this site in a cloning vector, the corresponding restriction endonucleases make a cut enabling the introduction of passenger DNA. A cloning vector may have several cloning sites. When an engineered vector contains 20 or more restriction sites, they are collectively called Multiple Cloning Sites (MCS) or polylinker. pBR322 vector has MCS encoding for over 40 major restriction sites.
- **Replication of Plasmid (ROP)**—ROP codes for proteins that assist replication of cloning vector.

Currently, different types of vectors are commercially available (for cloning and expression) and are discussed at length in Chap. 3.

1.3.3 Competent Host Organism

These are the organisms (e.g., *E. coli*) into which the rDNA is inserted for further replication and expression. The host takes in the plasmid vector with the desired gene insert through a method called "transformation" [19]. However, it is to be noted that the host cell does not readily take up the recombinant vectors unless it is made *competent*. Depending on the cell type, various methods have been developed to accomplish this task, such as gene gun, heat shock, and micro-injection [20, 21].

Post transformation, only the transformed bacteria survive on plates harboring the same antibiotic. Thus, these antibiotic-resistant bacteria will form colonies on the agar plates. These colonies are grown in appropriate growth medium for overexpression and production of proteins of interest from the genes inserted in the plasmid [22, 23].

1.4 Protein Expression and Production in Bacterial Systems

The importance of the robustness of bacterial systems that aids in the production of a large number of plasmid vectors is manifold. One of the most popular and routine uses of this technology is overexpression and production of proteins of interest that has revolutionized biomedical research and industries alike. Therefore, the bacterial systems are aptly termed as "protein factories." The steps for in vitro protein expression and purification are briefly provided stepwise [22, 24, 25]. Furthermore, a schematic representing the entire workflow of cloning of recombinant genes to protein expression and purification is shown in Fig. 1.2.

- The bacterial colonies harboring the foreign gene-incorporated plasmid DNA is grown in a suitable medium (Luria-Bertani or LB) at 37 °C in a temperature-controlled shaker incubator.
- Once the log phase is reached, i.e., when the optical density at 600 nm (OD_{600}) of the culture is 0.6–0.7, the expression of the protein of interest is induced (e.g., using Isopropyl ß-D-1-thiogalactopyranoside or IPTG).
- The culture is grown for next 4–5 h or overnight at 30 °C followed by pelleting down through centrifugation and cell lysis.
- The soluble slurry that is obtained post second-round of centrifugation contains over-expressed recombinant protein as well as several other bacterial proteins. It is to be noted that sometimes the recombinant protein is insoluble and requires distinct protocol for its recovery as described later in the book in detail.

Therefore, it becomes imperative to purify the recombinant protein (>95% purity) prior to its large-scale production commercially or for structural and functional characterizations in a research laboratory. Several protein purification techniques (ion-exchange, gel filtration, and affinity chromatography) that utilize different inherent properties of the proteins have been developed in the past few decades to fulfill the growing need of the recombinant proteins in industry and in biomedical research.

1.5 Important Applications and Future Perspectives of rDNA Technology

From unraveling the human genome to creating transgenic mice, rDNA technology has established itself to be an integral part of biomedical research [26–28]. It has also created a niche in the agricultural industry [29] through development of genetically modified crops that are resistant to pests as well as adverse weather and soil conditions [30]. This technology has been used in diagnostics as well such as

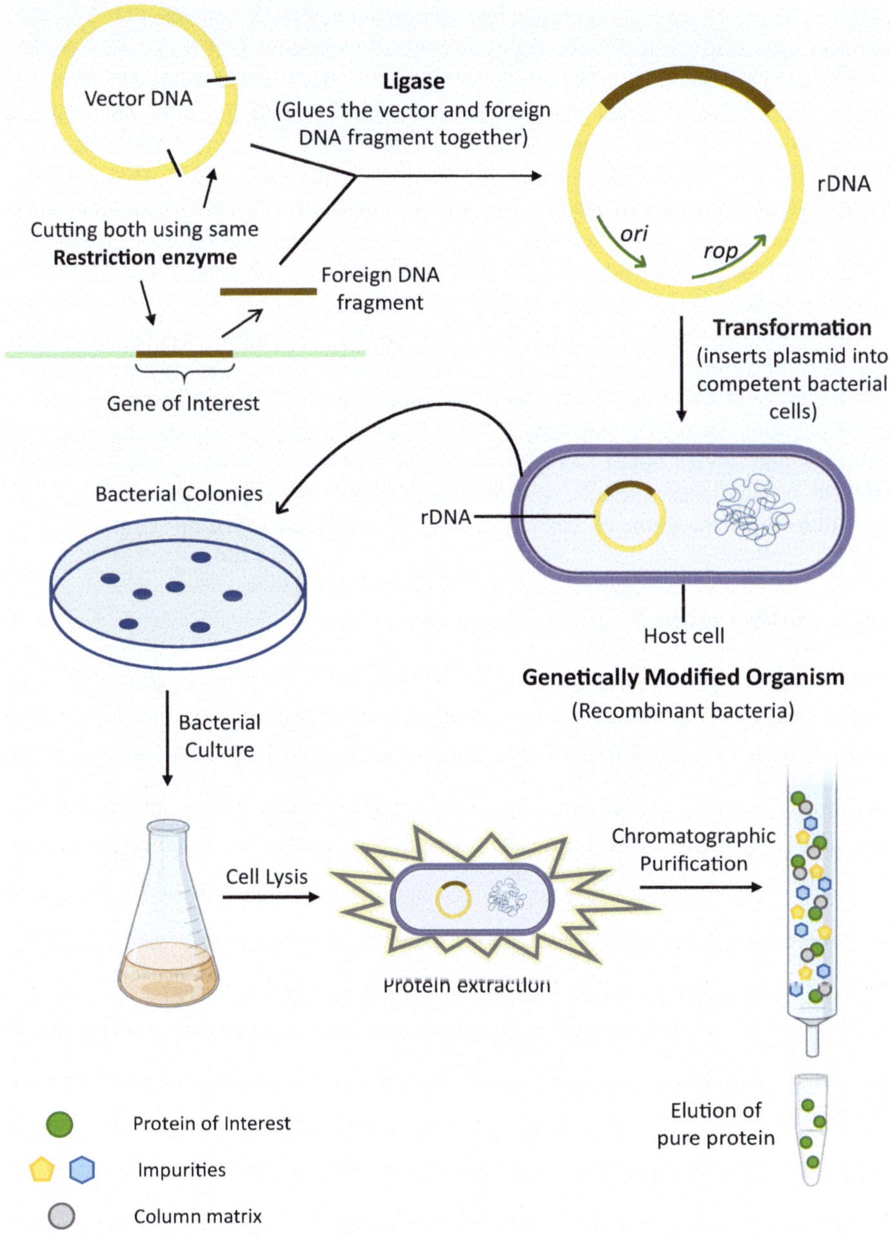

Fig. 1.2 Schematic representation of the overall workflow involving cloning, *in vitro* expression, and purification of recombinant proteins. The figure is self-explanatory

identification of hepatitis and human immunodeficiency (HIV) viruses [31, 32]. Furthermore, plasmid vectors have been successfully employed to transfer a specific gene to lung tissues in a clinical trial to study its effect in the treatment of cystic fibrosis (CF). CF, which is a genetic disease that occurs due to a single mutation in a gene named cystic fibrosis trans-membrane conductance regulator (CFTR), severely affects lungs and digestive systems of the affected individual [33]. However, the greatest breakthrough is associated with the development of protein-based therapeutics such as vaccines and medicines (e.g., insulin for patients suffering from type 1 diabetes) [34, 35] as well as production of recombinant proteins for research and industrial benefit.

Furthermore, introduction of CRISPR (Clustered Regularly Interspaces Short Palindromic Repeats)-Cas9 technology in the past decade has given a new dimension to the field of biomedicine through better understanding of complex disease processes such as AIDS and cancer [36]. With its ability to dissect the intricate pathways and devise novel therapeutic strategies of life-threatening diseases using CRISPR-based disease models, recombinant DNA technology and genetic engineering will become the game changers in medical biotechnology of this century.

1.6 Conclusions

This chapter is a prelude to the following sections of the book on protein expression and purification. The book aims at providing a basic understanding of the techniques and protocols involved in recombinant gene cloning in bacterial expression systems so as to obtain pure proteins in optimum quantity and in their native conformations for further research and desired applications. It also discusses the real-life difficulties experienced by researchers in each and every step, and ways to overcome them in an elaborate manner with specific examples.

Acknowledgments The authors thank Advanced Centre for Treatment, Research and Education in Cancer (ACTREC) for providing necessary infrastructure and resources for successful completion of the chapter. The authors acknowledge Ms. Chanda Baisane, Bose Lab, ACTREC for formatting the manuscript.

References

1. Morrow JF. [1] Recombinant DNA techniques. In: Methods in enzymology. New York: Academic Press; 1979. p. 3–24.
2. Green MR, Sambrook J. The basic polymerase chain reaction (PCR). Cold Spring Harb Protoc. 2018;2018(5):pdb.prot095117.
3. Patrick Higgins N, Cozzarelli NR. [4] DNA-joining enzymes: a review. In: Methods in enzymology. New York: Academic Press; 1979. p. 50–71.
4. Boyer HW. DNA restriction and modification mechanisms in bacteria. Annu Rev Microbiol. 1971;25:153–76.
5. Green MR, Sambrook J. Analysis of DNA by agarose gel electrophoresis. Cold Spring Harb Protoc. 2019;2019(1):pdb.top100388.

6. Voytas D. Agarose Gel Electrophoresis. Curr Protoc Immunol. 1992;2(1):10.4.1–8.
7. Hacker DL, Wurm FM. 2.29 - Recombinant technology. In: Moo-Young M, editor. Comprehensive biotechnology. 2nd ed. Burlington: Academic Press; 2011. p. 401–6.
8. Fedoroff NV, Zinder ND. Properties of the phage f2 replicase: I. Optimal conditions for replicase activity and analysis of the polynucleotide product synthesized in vitro. J Biol Chem. 1972;247(14):4577–85.
9. Fedoroff NV, Zinder ND. Properties of the phage f2 Replicase: II. Comparative studies on the ribonucleic acid-dependent and poly(c)-dependent activities of the replicase. J Biol Chem. 1972;247(14):4586–92.
10. Jackson DA, Symons RH, Berg P. Biochemical method for inserting new genetic information into DNA of simian virus 40: circular SV40 DNA molecules containing lambda phage genes and the galactose operon of *Escherichia coli*. Proc Natl Acad Sci U S A. 1972;69(10):2904–9.
11. Lehman IR, Roussos GG, Pratt EA. The deoxyribo-nucleases of *Escherichia coli*. III. Studies on the nature of the inhibition of endonuclease by ribonucleic acid. J Biol Chem. 1962;237:829–33.
12. Roszczyk E, Goodgal S. Methylase activities from *Haemophilus influenzae* that protect Haemophilus parainfluenzae transforming deoxyribonucleic acid from inactivation by *Haemophilus influenzae* endonuclease R. J Bacteriol. 1975;123(1):287.
13. Roberts RJ. [2] Directory of restriction endonucleases. In: Methods in enzymology. New York: Academic Press; 1979. p. 27–41.
14. Balbás P, Bolívar F. Back to basics. In: Balbás P, Lorence A, editors. Recombinant gene expression: reviews and protocols. Totowa, NJ: Humana Press; 2004. p. 77–90.
15. Tolmachov O. Designing plasmid vectors. In: Walther W, Stein US, editors. Gene therapy of cancer: methods and protocols. Totowa, NJ: Humana Press; 2009. p. 117–29.
16. Friehs K. Plasmid copy number and plasmid stability. In: Scheper T, editor. New trends and developments in biochemical engineering. Berlin, Heidelberg: Springer; 2004. p. 47–82.
17. Lederberg J. Plasmid (1952–1997). Plasmid. 1998;39(1):1–9.
18. Bolivar F, et al. Construction and characterization of new cloning vehicle. II. A multipurpose cloning system. Gene. 1977;2(2):95–113.
19. Inoue H, Nojima H, Okayama H. High efficiency transformation of *Escherichia coli* with plasmids. Gene. 1990;96(1):23–8.
20. Chan W-T, et al. A comparison and optimization of methods and factors affecting the transformation of *Escherichia coli*. Biosci Rep. 2013;33(6):e00086.
21. Hanahan D, Jessee J, Bloom FR. [4] Plasmid transformation of *Escherichia coli* and other bacteria. In: Methods in enzymology. New York: Academic Press; 1991. p. 63–113.
22. Zerbs S, Giuliani S, Collart F. Chapter 11 - Small-scale expression of proteins in *E. coli*. In: Lorsch J, editor. Methods in enzymology. New York: Academic Press; 2014. p. 117–31.
23. Chaganti LK, Kuppili RR, Bose K. Intricate structural coordination and domain plasticity regulate activity of serine protease HtrA2. FASEB J. 2013;27(8):3054–66.
24. Singh N, et al. Dual regulatory switch confers tighter control on HtrA2 proteolytic activity. FEBS J. 2014;281(10):2456–70.
25. Acharya S, Dutta S, Bose K. A distinct concerted mechanism of structural dynamism defines activity of human serine protease HtrA3. Biochem J. 2020;477(2):407–29.
26. Nimmo HG, Cohen PTW. Applications of recombinant DNA technology to studies of metabolic regulation. Biochem J. 1987;247(1):1–13.
27. Tait RC. The application of molecular biology. Curr Issues Mol Biol. 1999;1:1.
28. Kosuri S, Church GM. Large-scale de novo DNA synthesis: technologies and applications. Nat Methods. 2014;11(5):499–507.
29. Godfray HCJ, et al. Food security: the challenge of feeding 9 billion people. Science. 2010;327(5967):812.
30. Georges F, Ray H. Genome editing of crops: a renewed opportunity for food security. GM Crops Food. 2017;8(1):1–12.

31. Xiao-Jie L, et al. CRISPR-Cas9: a new and promising player in gene therapy. J Med Genet. 2015;52(5):289.
32. Xiao Q, Guo D, Chen S. Application of CRISPR/Cas9-based gene editing in HIV-1/AIDS therapy. Front Cell Infect Microbiol. 2019;9:69.
33. Ranasinha C, et al. Efficacy and safety of short-term administration of aerosolised recombinant human DNase I in adults with stable stage cystic fibrosis. Lancet. 1993;342(8865):199–202.
34. Ladisch MR, Kohlmann KL. Recombinant human insulin. Biotechnol Prog. 1992;8(6):469–78.
35. Mao R, et al. Insulin and its single-chain analogue. Appl Microbiol Biotechnol. 2019;103 (21):8737–51.
36. Lu X-J, et al., editors. CRISPR-Cas9 technology: applications and human disease modelling. Brief Funct Genomics. 2017;16(1):4–12.
37. Arber W, Linn S. DNA modification and restriction. Annu Rev Biochem. 1969;38(1):467–500.
38. Gottesman MM, Hicks ML, Gellert M. Genetics and function of DNA ligase in *Escherichia coli*. J Mol Biol. 1973;77(4):531–47.
39. Berg P. Moments of discovery. Annu Rev Biochem. 2008;77(1):15–44.
40. Khorana HG. Total synthesis of the gene for an alanine transfer ribonucleic acid from yeast. Pure Appl Chem. 1971;25(1):91–118.
41. Berg DE. Genes of phage λ essential for λd_v plasmids. Virology. 1974;62(1):224–33.
42. Pollack R. Biologists in pinstripes. Sciences. 1982;22(7):35–7.
43. Danna K, Nathans D. Specific cleavage of simian virus 40 DNA by restriction endonuclease of Hemophilus influenzae. Proc Natl Acad Sci. 1971;68(12):2913.
44. Mertz JE, Davis RW. Cleavage of DNA by R1 restriction endonuclease generates cohesive ends. Proc Natl Acad Sci. 1972;69(11):3370.
45. van Embden J, Cohen SN. Molecular and genetic studies of an R factor system consisting of independent transfer and drug resistance plasmids. J Bacteriol. 1973;116(2):699–709.
46. Morrow JF, et al. Replication and transcription of eukaryotic DNA in *Escherichia coli*. Proc Natl Acad Sci U S A. 1974;71(5):1743–7.
47. Burrell CJ, et al. Expression in *Escherichia coli* of hepatitis B virus DNA sequences cloned in plasmid pBR322. Nature. 1979;279(5708):43–7.
48. McMichael AJ, et al. A human thymocyte antigen defined by a hybrid myeloma monoclonal antibody. Eur J Immunol. 1979;9(3):205–10.
49. Gordon JW, et al. Genetic transformation of mouse embryos by microinjection of purified DNA. Proc Natl Acad Sci. 1980;77(12):7380.
50. Arrand JR, et al. Molecular cloning of the complete Epstein-Barr virus genome as a set of overlapping restriction endonuclease fragments. Nucleic Acids Res. 1981;9(13):2999–3014.
51. Riechmann L, et al. Reshaping human antibodies for therapy. Nature. 1988;332(6162):323–7.
52. Leder P, Hill C, Stewart TA. Transgenic non-human mammals. US Patent 4736866; 1988.
53. Coller BS. The role of platelets in arterial thrombosis and the rationale for blockade of platelet GPIIb/IIIa receptors as antithrombotic therapy. Eur Heart J. 1995;16(Suppl_L):11–5.

Cloning and Gene Manipulation

2

Aasna L. Parui, Lalith K. Chaganti, Rucha Kulkarni, and Kakoli Bose

Abstract

Gene cloning has continually been one of the many important discoveries in the areas of molecular and cell biology. The advent of recombinant DNA technology has revolutionized the field of biology in many ways, as it allowed the scientific community to understand the functional and cellular aspects for a wide array of essential genes. In brief, gene cloning involves inserting the DNA fragment of interest in a suitable vector and amplifying this recombinant molecule in host cells such as *E. coli*. However, the basic procedure also involves numerous steps such as restriction enzyme digestion and ligation reactions to be carried out in order to obtain a perfect recombinant DNA molecule. The purpose of this chapter is to present an overview of the cloning process, with the details of various steps involved and the techniques used in the identification of the recombinant clones.

Keywords

Cloning · Recombinant DNA · Restriction digestion · Ligation · PCR

A. L. Parui · R. Kulkarni · K. Bose (✉)
Integrated Biophysics and Structural Biology Lab, Advanced Centre for Treatment, Research and Education in Cancer (ACTREC), Tata Memorial Centre, Mumbai, India

Homi Bhabha National Institute, BARC Training School Complex, Mumbai, India
e-mail: kbose@actrec.gov.in

L. K. Chaganti
Integrated Biophysics and Structural Biology Lab, Advanced Centre for Treatment, Research and Education in Cancer (ACTREC), Tata Memorial Centre, Mumbai, India

K. Bose (ed.), *Textbook on Cloning, Expression and Purification of Recombinant Proteins*, https://doi.org/10.1007/978-981-16-4987-5_2

2.1 Introduction

The word "cloning" refers to the asexual reproduction required to obtain organisms that are genetically identical to one another and to their parents. This process is in contrast to the sexual reproduction where offspring are not identical. The cloning method involves generation of large population of cells with identical DNA molecules obtained from a single living cell by a process of replication of the original single DNA molecule. The word "cloning" is also applied to genes, which is an extension of this concept in molecular biology [1].

Gene or DNA cloning is a common practice used by researchers to create exact copies (clones) of a particular gene or a DNA sequence using recombinant genetic engineering techniques. Major breakthrough in the cloning experiments was obtained by Herbert Boyer, Stanley Cohen, Paul Berg, and their colleagues in the early 1970s [2]. The traditional technique for gene cloning involves transfer of a target DNA or gene fragment from one organism to different cloning vectors (described in Chap. 3) or autonomously replicating genetic element, such as bacterial plasmids (small, circular piece of extra chromosomal DNA) and bacteriophages, which serve as mediums to propagate the cloned DNA within the cell. Apart from bacteria, plasmids are also naturally present in archaea and eukaryotes such as yeast and plants. They provide additional benefits to the organisms by conferring properties such as antibiotic resistance, virulence, and degradative abilities. The plasmids also contain an origin of replication (*ori*), which helps in controlling the host range and copy number of the plasmid within the host. For cloning purposes, the plasmids have been designed artificially in such a way that DNA of interest can be easily inserted into these vectors and propagated substantially for various purposes. Apart from the *ori* site and antibiotic resistance gene, the lab-engineered plasmids have a Multiple Cloning Site (MCS—short segment containing various restriction enzyme sites required for easy insertion of the gene of interest), a promoter region (allows transcription of the downstream gene, present especially in expression plasmids), and a selectable marker gene (Fig. 2.1). These segments in a plasmid help ease the task of modifying the plasmids based on different experimental requirements, thus making them an attractive tool for molecular cloning.

Further, the recombinant vector is transferred into suitable host cells such as *E. coli* for production of multiple DNA copies. Plasmids that have an antibiotic resistance gene are typically employed in DNA cloning or bacterial transformation. The presence of the antibiotic resistance gene allows for the selection of bacteria harboring the desired plasmid. Thus, bacteria with the recombinant plasmid will thrive in an antibiotic-containing media while plasmid deficient bacteria will not be resistant to the antibiotic and fail to survive (Fig. 2.2). This method helps the cells containing the desired recombinant DNA to be distinguished from the others, and their selection becomes possible [3].

Gene cloning technique is used for several downstream applications, such as DNA sequencing, mutagenesis, genotyping, or heterologous protein expression. However, precision in the development of the basic cloning steps plays key roles behind all these applications. The recent technological advancements in molecular

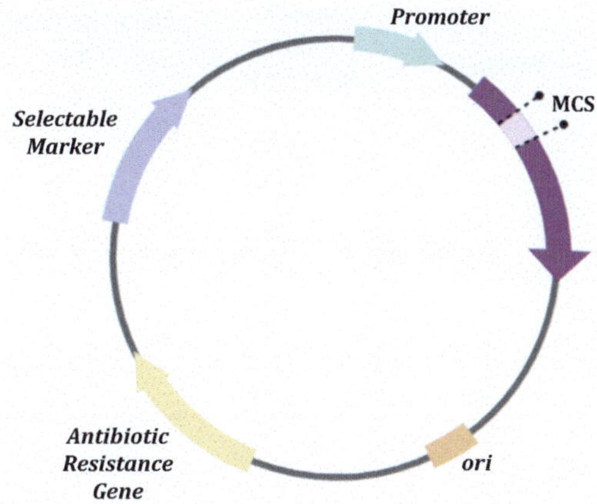

Fig. 2.1 Generalized plasmid map. A plasmid contains an *ori* (origin of replication) site, an antibiotic resistance gene (required for selective propagation of only plasmid-containing cells), and a selectable marker. The promoter region is present upstream of the MCS (multiple cloning site) where the gene of interest is inserted

genetics have allowed scientists to study, develop, and explore various modifications in the genomes for a wide range of organisms. Foreign DNA may now be introduced in bacterial plasmids with the use of restriction endonucleases (described in Sect. 2.4.1) and can be replicated further. Bacterial cells transformed with this foreign DNA can now express the genetic information and make suitable products encoded by the desired genes. Thus, by molecular cloning, we can learn a lot about the structure and *modus operandi* of different genes. Moreover, production of bulk amount of specific gene products including unique as well as rare proteins has become industrially feasible. We can also use such plasmids to transform the genetic constitution of other organisms. In this chapter, we will discuss the general strategies and principles of gene cloning as well as genetic engineering tools that can be used for a wide range of research purposes with a focus on their applications in recombinant DNA technology.

2.2 DNA Libraries

Acquiring the genetic information has become a major step for any field of biological sciences. This requires navigation through the complete genomic sequence of a specific organism, for either understanding the function of a particular gene of interest or the relevance of the entire genome. A DNA or gene library is a compilation of cloned DNA fragments that collectively represent the genes isolated from a particular organism. This DNA fragment collection may then be utilized for the identification of specific genes and other DNA sequences of interest, which is analogous to selecting desired books from any conventional library. Specific DNA fragments are generated by digesting the genome or genes with the help of specific restriction enzymes. The generated fragments are further cloned into specific plasmid vectors, and then transformed into suitable host cells [4]. The total number of all

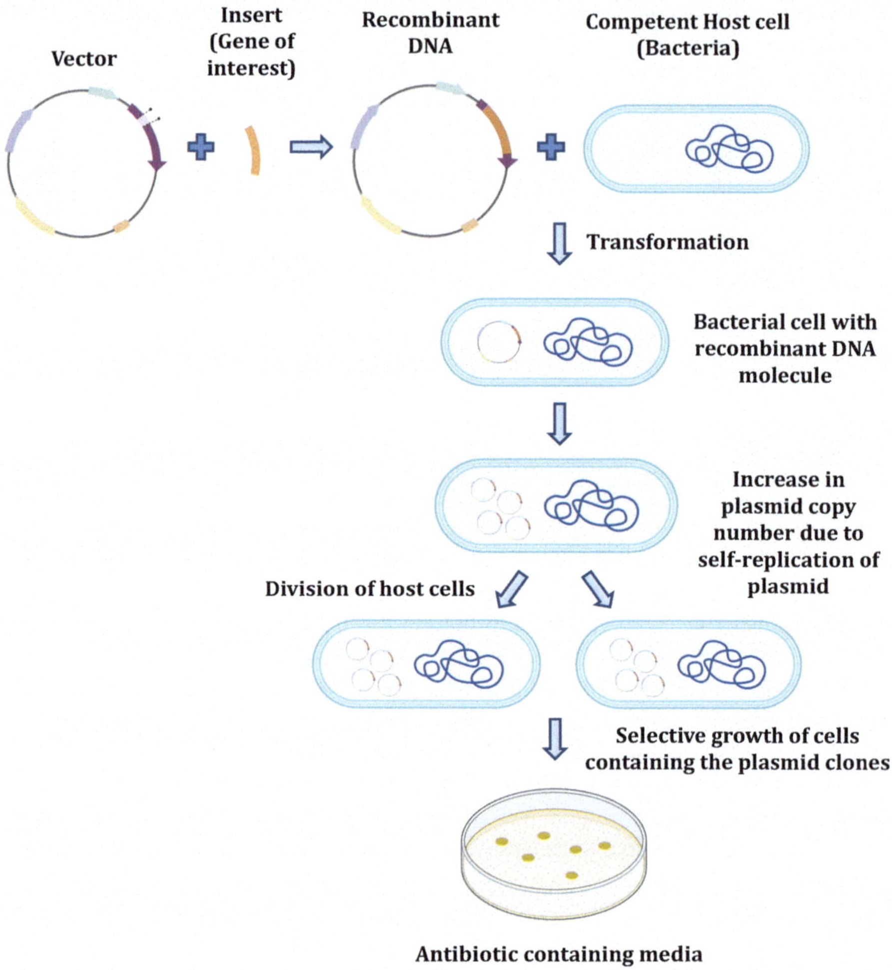

Fig. 2.2 Steps in gene cloning. The gene of interest is inserted into a suitable vector at the MCS site and this recombinant DNA molecule is then transformed into a compatible bacterial host. The *ori* site promotes the replication of recombinant plasmid inside the host. After numerous divisions of the transformed cells, the recombinant clones are obtained on the selective media supplemented with the antibiotic. Cells transformed with the recombinant plasmid will only grow on the selective medium

DNA molecules of a particular genome/field of interest makes up that particular library. The target or specific DNA from the library is further screened with a molecular probe. Once prepared, the library can be propagated indefinitely in the host cells and can be readily retrieved whenever a new probe is available to seek out a particular fragment from the entire library.

There are two types of DNA libraries that can be used to isolate specific DNAs: (1) genomic library and (2) cDNA library. The choice of the particular type of gene library depends on factors such as protein production from a specific gene and studying genetic architecture.

2.2.1 Genomic Library

A genomic library is a collection of clones that contain DNA fragments representing the total genomic DNA of a specific organism of interest. Depending upon the organism and size of its genome, this library can be prepared either in bacterial plasmids, phage vectors, cosmids, bacterial artificial chromosome (BAC), or a yeast artificial chromosome (YAC). Chapter 3 of this book elaborates on different vectors and discusses the importance of choosing them for distinct cloning purposes. A detailed outline of the construction of genomic library is schematically represented in Fig. 2.3. The steps involved in construction of a genomic library are:

1. Isolation and purification of genomic DNA:
   The first step in construction of any genomic library requires isolation of complete genomic DNA from the organism of interest (bacteria, virus, plants, or animals). Depending on the type of organism, the procedures engaged in the isolation of genomic DNA vary widely. In eukaryotes, genomic DNA can be prepared either from nuclear DNA or any organelle-specific DNA. Nuclear genomic library is prepared by specifically isolating the DNA from the nucleus. The eukaryotic cell nuclei are purified by digestion with protease and organic (phenol-chloroform) extraction. In case of organelle genomic library, first the

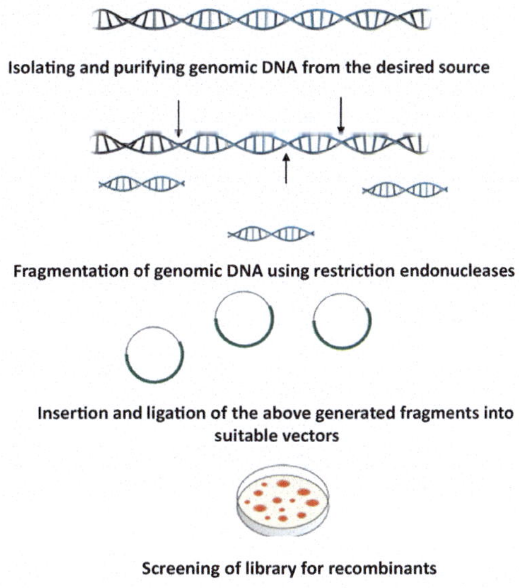

Fig. 2.3 Preparation of a Genomic library. Genomic DNA is isolated from the organism of interest using methods such as Phenol-Chloroform extraction [7]. It is further subjected to random fragmentation using either physical (sonication) or enzymatic (restriction endonucleases) methods. The fragmented DNAs are cloned into suitable vector and the transformed recombinants are then selected under appropriate selection pressure conditions. The target DNA is screened from the recombinant clones using methods such as autoradiography and PCR

respective organelle is purified and then the DNA is isolated only from that particular purified organelle. Organelle separation procedures vary for different organelles.

2. Fragmentation and restriction digestion of genomic DNA:
The isolated genomic DNA is very long and needs to be cut into fragments of ideal sizes. This can be achieved either by fragmentation or by enzymatic digestion. Physical methods include pipetting the DNA molecule or applying intensified ultrasound waves (sonication), whereas the enzymatic method involves the use of restriction enzymes as described in Sect. 2.4. Generation of DNA fragments of various sizes depends on the distribution probability of specific restriction enzyme site within a gene. Therefore, complete digestion of the genomic DNA generates very short fragments of variable sizes depending on the presence of the restriction enzyme site in the entire sequence. As a consequence, the desired gene of interest in its complete form might not be represented within a library. Therefore, partial restriction enzyme digestion is usually employed to generate overlapping fragments containing one or more gene [4]. The generated fragments are then purified by either gel electrophoresis or density gradient centrifugation techniques, which are then further cloned into a suitable vector.

3. Ligation of the DNA fragments:
The third step is to insert the generated DNA fragments into a suitable vector as shown in Fig. 2.3 below. Different vectors such as plasmids, λ phage, YAC, and BAC (described in Chap. 3) are used for cloning the DNA fragments. YAC (up to 2000 kb) and BAC (up to 300 kb) are considered suitable vectors for cloning larger DNA molecules [5]. However, it is difficult to clone a large insert into these vectors; therefore, bacteriophage λ or cosmid vectors are usually employed for generating genomic libraries. Since a larger insert size (up to 40 kb) can be accommodated by these vectors compared to plasmids (~10 kb), there is a greater chance of cloning a gene sequence with both the coding sequence and regulatory elements in a single clone. T4 DNA ligase is typically used for ligating the selected DNA sequences into the vectors. Details about the steps involved in ligation are discussed in the Sect. 2.5.
Selection of the number of clones required for construction of a genomic library is the most important step. One should ensure that the constructed library is a representative of the entire genome. However, since any genomic insert generated by a particular restriction enzyme has an equal chance of being in the library compared to any other insert, the number of clones to be pooled depends on the size of the organism's genome "*f*" and the average insert size. The probability (*P*) of including any DNA sequence in a random library of *(N)* independent recombinants is represented by Eq. ((2.1):

$$N = \ln\,(1 - P)/\ln\,(1 - f) \tag{2.1}$$

where

$f =$ total genome length/average insert size.

$N =$ necessary number of recombinants.

$P =$ desired probability that any fragment in the genome will be present.

$f =$ fractional proportion of the genome in a single recombinant.

Thus, bigger the library, greater is the chance of finding a gene in that particular library. On the contrary, increasing the insert size would allow fewer clones that are needed to represent a genome.

4. Library screening:

   A common method employed to screen the library is colony hybridization. Each transformed host cell of a library will have just one vector with one insert of DNA. First colonies of host cells carrying the plasmid or phage libraries are plated onto an agar plate with a suitable antibiotic such as ampicillin. This will ensure growth of only those cells that are transformed with vectors containing antibiotic resistance gene. The colonies are then transferred onto a nitro cellulose membrane for further processing. Once the cells are attached to the membrane, they are lysed, deproteinized (to avoid protein contamination), and the released DNA is denatured by alkaline treatment. Later, hybridization is performed between the target DNA and labelled DNA probe (complementary sequence to the target DNA). The target DNA can then be identified by autoradiography. Polymerase chain reaction (PCR) and immunological screening can also be used as alternatives to colony hybridization [6].

PCR screening is generally used to identify uncommon DNA sequences among diverse cocktails of molecular clones by increasing the quantity of a specific sequence. The library is plated as plaques or colonies on agar plates and individually these colonies are inoculated into the wells of the multi-well plate. PCR reactions are performed with primers flanked by a unique target sequence to identify the clone of interest. This method is applicable upon availability of detailed gene sequence for designing of typical primers.

Immunological screening includes the use of antibodies that identify antigenic determinants on polypeptide specifically. It does not depend on the function of the foreign protein produced, instead requires a protein-specific antibody. This screening technique is similar to colony hybridization; however, instead of using labelled DNA probe, antibodies are used to specifically detect the target protein (Table 2.1).

2.2.1.1 Applications

Genomic libraries can be used for many purposes:

- The whole genomic sequence of an organism can be produced.
- Serves as a repository for genomic sequences for the development of transgenic animals.

Table 2.1 List of vectors used for generation of DNA libraries

Vectors	Origin	Insert size (kb)	Properties and applications
Plasmid	Naturally occurring multicopy Circular DNA	10	2–6 kb size, commonly used for molecular cloning, contains origin of replication, reporter and antibiotic resistance genes
Phage	Bacteriophage λ	10–20	Genome size is 47 kb, contains double-stranded DNA (dsDNA) and infects *E. coli*, efficient packaging system, used to study individual genes
Cosmid	Plasmid containing a bacteriophage λ cos site	40	Contains a λ phage cos site, which allows packaging, transmits as plasmids in *E. coli*, helpful for subcloning YAC, BAC, PAC, etc.
BAC	*E. coli* F factor plasmid	300	Circular DNA molecules, up to 7 kb length, have F-plasmid, high-capacity vectors used for investigation of bulky genomes
YAC	Yeast (*Saccharomyces cerevisiae*) centromere, telomere, and autonomously replicating sequence	2000	Linear DNA, have properties of a yeast chromosome, contains selection markers for identification of successful transformants that are used for analysis of large genomes

- The structure of a given chromosome can be investigated.
- Genomic libraries from higher eukaryotes are important to study untranslated regions (regulatory elements) of a gene, including promoters or introns.
- In prokaryotes, genomic libraries are used to clone relatively smaller gene fragments.

2.2.2 cDNA Library

cDNA library is a collection of complementary DNA (cDNA) fragments which have been cloned individually into separate vector molecules. In cDNA libraries, DNA copies complementary to the transcribed RNA sequences (usually the mature mRNA) of an organism are produced by the reverse transcription of RNA by the reverse transcriptase enzyme [8, 9]. Thus, these cDNA libraries contain only the coding sequences generated from the fully transcribed and spliced mRNA produced from the expressed genes (exons). Unlike the genomic library, these cDNA libraries lack repetitive sequences, introns (non-coding regions), regulatory regions, and enhancers of the gene. Hence, cDNA libraries are prepared primarily from the higher eukaryotes and not from the lower eukaryotes or prokaryotes, which lack these regulatory elements.

2.2.2.1 Construction of a cDNA Library

A detailed outline for construction of a cDNA library is described below:

(a) Initial extraction and purification of mRNA:
This step involves the isolation of total mRNA from the cells. Eukaryotic mRNA consists of 50–250 adenylate residues (poly-A tail) at the 3′ end, which facilitates simple separation of mRNA through affinity chromatography using oligo(dT). Chromatographic column or Magnetic beads coupled with oligo(dT) are usually used to purify mRNA from the much more prevalent rRNAs and tRNAs in a cell lysate. The poly-A tail at the 3′ end of the mRNA enables its efficient binding to the oligo(dT) beads. After providing sufficient washes to remove the impurities, these mRNA can then be eluted using strong magnetic force or low salt buffer; the bound mRNA is isolated from the total RNA content (Fig. 2.4). The recovered mRNA is then analyzed by agarose gel electrophoresis, before using it as a template for cDNA synthesis [10].

(b) Production of cDNA:
Once mRNA is extracted, the complementary DNA strand is synthesized using reverse transcriptase enzyme to make mRNA:DNA duplex. Herein, a short oligo (dT) primer with free 3'-OH is annealed to a poly-A tail of mRNA and the primer is extended by reverse transcriptase to generate the complementary DNA strand (Fig. 2.5). Now, the mRNA template from the mRNA:DNA hybrid is removed by alkaline hydrolysis using an RNAse H enzyme and this generates a single-stranded cDNA (ss-cDNA). By producing a short hairpin loop at its 3′

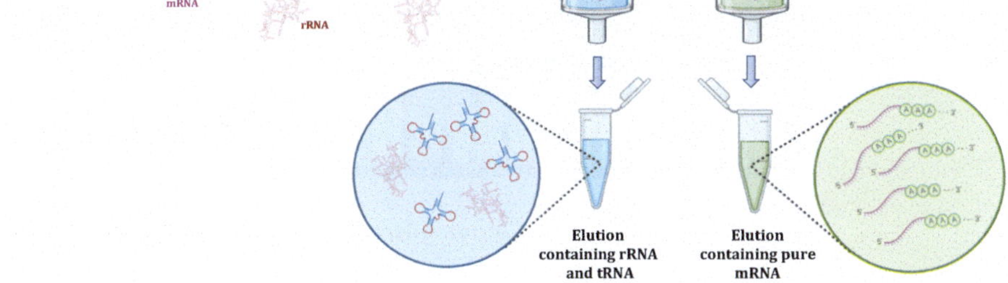

Fig. 2.4 Isolation of mRNA from total RNA content. The poly(A) region at the 3′ end of the eukaryotic mRNA allows its selective isolation from total cellular RNA content. It is loaded on an oligo(dT) affinity chromatography column under high salt conditions that promotes hybridization between the 3′ poly(A) tails of the mRNA and the oligo(dT)-coupled matrix. The rRNAs and tRNAs are washed out of the column after hybridization and the mRNA is then eluted with a low salt buffer

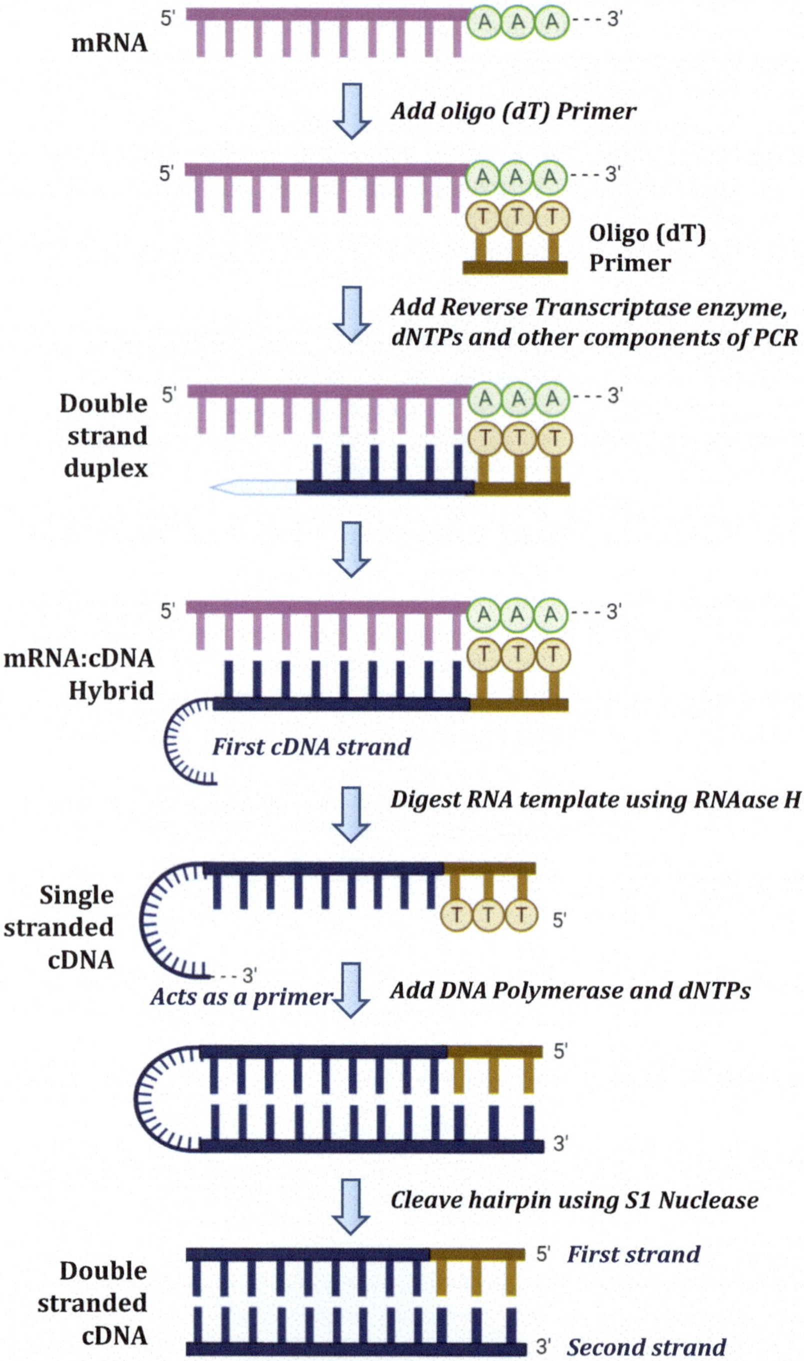

Fig. 2.5 cDNA synthesis. In the presence of dNTPs, the first strand of cDNA is synthesized by reverse transcriptase and oligo (dT) primer. A hybrid mRNA-cDNA is generated, followed by digestion of the mRNA template by alkaline hydrolysis and the enzyme ribonuclease H. The natural

end, ss-cDNA acts as its own primer. Due to the hydrophobicity of the bases, ss-nucleic acid molecules have a tendency to form such secondary structures. The free 3′-OH in the hairpin loop is essential for the generation of the complementary DNA strand. Thus, the ss-cDNA is converted into a double-stranded (ds) cDNA with the help of DNA polymerase. The generated ds-cDNA initially has a hairpin loop at one end. This is then removed by S1 nuclease treatment and the final product is blunt-ended ds-cDNA molecule [4].

(c) Ligation of cDNA into the vector:
The generated cDNAs are cloned into plasmid and bacteriophage vectors; however, plasmids are extensively used for cloning and isolation of the desired cDNAs. These ds-cDNAs are ligated into appropriate vector either by using a blunt end ligation or by adding linkers to ds-cDNA ends. Due to the inefficiency of blunt-end ligation, small restriction-site linkers are initially ligated to both ends before cloning into any suitable vector. In this method, 10–12 base pair (bp) long hybridizing complementary oligonucleotide linkers with a restriction enzyme site is ligated to the ds-cDNA ends using T4 DNA bacteriophage ligase. The resulting ds-cDNAs with linkers at both ends are digested with the respective restriction enzyme to generate cDNAs with sticky ends. Restriction digestion of cDNAs with internal restriction site can be overcome by efficient modification of ds-cDNA with the methylases before adding linkers. This methylation step ensures the protection from the action of restriction enzymes.

(d) Library screening:
Screening of colonies with cDNA is similar to genomic library. The most common methods used are hybridization methods and immunological assays, which are elaborated in the previous sections.

2.2.2.2 Applications of cDNA Library

cDNA libraries can be used for many purposes:

- Unlike the genomic DNA libraries, cDNA can be directly expressed in prokaryotic organisms.
- Discovery of Novel genes.
- Storage of less information as a result of elimination of the non-coding regions.
- cDNAs are used for in vitro study of gene function.
- A cDNA library is useful for isolating genes that code for specific mRNAs.
- cDNA libraries are also useful to identify the tissue-specific mRNAs, where certain genes are expressed only in one cell type but not in the other.
- cDNA libraries are important in reverse genetics, where more genomic information obtained from genomic libraries is of less use.

Fig. 2.5 (continued) hairpin of the first cDNA strand acts as a primer for the synthesis of the second strand. Using a self-priming method, DNA polymerase I catalyzes synthesis of the second strand and further the hairpin is cleaved using S1 nuclease. Double-strand cDNAs corresponding to the many different mRNAs extracted from the cell are formed at the end of this reaction

2.2.3 Difference Between Genomic and cDNA Library (Table 2.2)

Table 2.2 Difference between genomic and cDNA library

Genomic library	cDNA library
Prepared directly from the genomic DNA	Prepared using mRNA as a template
Represents entire genome of the organism	Represents only those genes of genome which express under specific conditions or tissues
Restriction endonucleases and ligases are important for its construction	Reverse transcriptase enzyme plays an important role in its construction
They carry introns, non-coding regions, and other regulatory elements	They lack introns and contain only the coding regions
Represent the DNA of both eukaryotic and prokaryotic organisms	Represent the DNA of only eukaryotic organisms
They are not capable of expression in prokaryotes (such as bacteria) because they carry introns, and prokaryotes do not have machinery to process non-coding regions	They are capable of expression in bacteria because they do not contain the non-coding regions

2.3 Polymerase Chain Reaction (PCR)

2.3.1 Background

PCR is a comparatively straightforward technology that amplifies the DNA template for producing specific DNA fragments in vitro. Practically, the conventional ways to clone a DNA sequence into a vector and to replicate it can involve days or weeks, while amplifying the DNA sequences using PCR just takes hours. While a large volume of biological materials are required for most of the biochemical analysis including nucleic acid detection with radioisotopes, the PCR method takes relatively less number of reagents and effort. In a reduced amount of time, PCR is able to accomplish higher sensitivity for detection and amplification levels of particular sequences. The technical characteristics make it highly helpful for use in fundamental as well as commercial research and also in genetic identification testing, forensics, industrial quality control, and in vitro diagnostics. Basic PCRs are widely employed in many molecular biology laboratories where DNA fragments are amplified and DNA or RNA sequences are detected from a cell or a particular environment sample. Furthermore, PCR has expanded well ahead of basic amplification and detection, and several extensions were recently made to the original PCR method [4].

2.3.2 Components of PCR

DNA template: The double-stranded DNA (dsDNA) sample containing the specific target sequence for amplification.

DNA polymerase: It is an enzyme that synthesizes new strands of DNA complementary to the target sequence. Of the different types of DNA polymerase enzymes, the first and most commonly used are *Taq* DNA polymerase (from *Thermis aquaticus*) and *Pfu* DNA polymerase (from *Pyrococcus furiosus*). The latter is currently being used widely because of its higher fidelity in copying DNA. These enzymes may be slightly different, yet each possesses two key features that put them pertinent for PCR:

1. They can amplify new DNA strands from a DNA template using specific primers.
2. They are susceptible to higher temperatures.

Primers: These are small ss-DNA sequences that are complementary to the target sequence. The polymerase begins synthesizing new DNA from the 3′ free hydroxyl group of the primer.

Nucleotides (dNTPs or deoxynucleotide triphosphates): The four single units of the nucleotide bases, viz., A (Adenine), T (Thymine), G (Guanine), and C (Cytosine) that are essentially the "building blocks" of new DNA strands.

2.3.3 PCR Protocol

Following steps are to be included in a particular PCR experiment (Fig. 2.6):

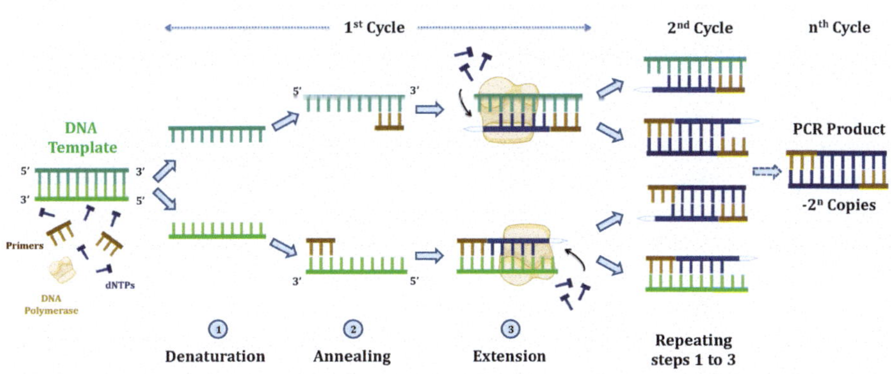

Fig. 2.6 Basic steps of PCR. The dsDNA is denatured into two ssDNA and the respective primers bind at their 3′ ends in the annealing step. Extension of the new strand occurs with the help of DNA polymerase and dNTPs. The resulting DNA fragments are again denatured in the next cycle and the three steps are repeated for specific number of cycles to obtain the amplified PCR product

1. *Initial denaturation*: The initial step of PCR includes denaturation of the target DNA by heating it to 95 °C for 5 min. It involves separation of the two intertwined strands of DNA to produce the essential single-stranded DNA (ssDNA) templates.
2. *Annealing*: In the second step of PCR, the reaction temperature is decreased to ~40–60 °C for 15–60 s so that the oligonucleotide primers can bind to the denatured specific target DNA by forming stable and specific associations. Further, these primers serve as the docking site for the DNA polymerase.
3. *Extension*: During this step of PCR, the DNA polymerase synthesizes new complementary DNA strands by binding to the primer. The temperature in extension is usually raised to 72 °C as this is an optimum temperature for most of the DNA polymerases such as *Taq* or *Pfu* that is present in the reaction mixture. Instead of two, a total number of four DNA stands are obtained after the extension step.
4. *Amplification*: During this step, the temperature is increased to 95 °C again. Each of the ds-DNA molecules, comprising one strand of the original molecule and one newly synthesized strand of DNA that were obtained from the previous step, again get denatured into single strands. This begins the second cycle of denaturation–annealing–extension, at the end of which there are eight DNA strands that are obtained.

However, it is to be noted that for every template and primer permutation, each step of the cycle should be optimized individually. If the temperature is comparable between annealing and extension, these two steps can be merged in one step in which both primer annealing and extension can be done. The amplified products may be evaluated for sizes, quantities, and sequences after 20–40 cycles and subsequently employed in other experimental methods.

2.4 Restriction Digestion

Gene cloning requires the recombinant DNA molecule to be cut in a very precise manner such that insertion of the new DNA fragment is only at one particular site. Restriction digestion is a procedure where DNA is cut in appropriate sites using restriction endonucleases [3]. These sites are present only at a particular region in the entire vector (called MCS region) to avoid any unnecessary cuts that would generate various fragments of the same DNA molecule. The target DNA molecule is mixed under specific reaction conditions with restriction enzymes for digestion. These enzymes distinguish and attach to the particular DNA sequences, then cleave at specified nucleotides sequence. Restriction digestion may lead to formation of blunt ends (ends of a DNA molecule that finish in a base pair) or sticky ends (ends of a DNA molecule that have a nucleotide overhang) (Sect. 2.4.1). Restriction digestion is usually the step preceding insertion of a foreign gene into a vector via a process called ligation. The results of a restriction digestion can be analyzed by gel electrophoresis, a process wherein the digested products are separated on the basis of their

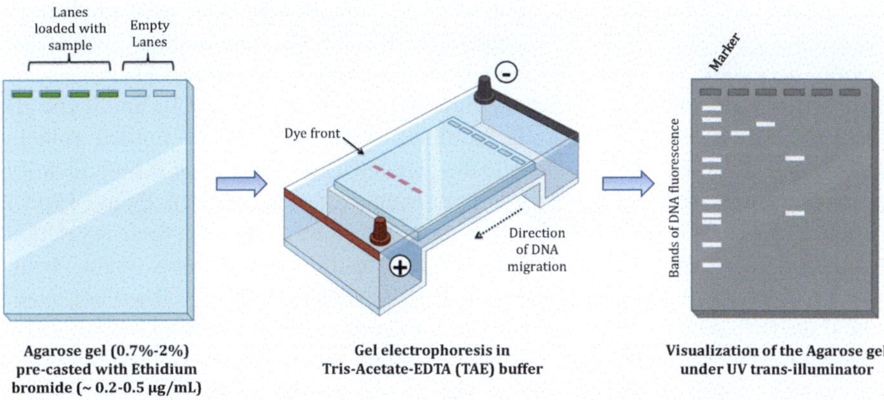

Fig. 2.7 Separation of DNA by Agarose gel electrophoresis. An agarose gel matrix (depending on the DNA size to be separated) containing ethidium bromide (EtBr) is pre-casted on a plastic tray. The DNA samples are mixed with the tracking dye (to determine the extent of DNA migration) and loaded into the wells of the gel. When visualized under UV transilluminator, the intercalated ethidium bromide fluoresces and the molecular weight can be determined from extend of migration

molecule length in a polymer-based gel (agarose). The gel is run against an electric field, where the negatively charged DNA molecules are allowed to travel from the anode to cathode and thus the separation occurs (Fig. 2.7). Visualization of the DNA is done with the help of a fluorescent dye such as ethidium bromide (EtBr) that intercalates into the DNA major grooves and fluoresces under UV light.

The constituents required for a restriction digestion are a DNA template, suitable restriction enzyme, a digestion buffer, and at times bovine serum albumin (BSA) to avoid sticking of enzymes to the tube surfaces and for stabilizing enzymes in overnight reactions [11]. At a certain temperature the reaction is incubated for the optimal activity of the restriction enzyme and after desired time the reaction is stopped by heat deactivation.

2.4.1 Restriction Enzymes (Endonucleases)

Many molecular biology methods are ingrained upon the skill to digest DNA molecules in a precise and predictable way (also known as "cutting" or "cleaving"). The advancement of this technology relies upon the discovery of bacterial restriction enzymes or endonucleases. Bacterial species contain restrictive enzymes that detect "nucleotide" patterns of DNA called palindromic (inverted repeat) sites of restriction [12]. Restriction sites are usually 4 to 8 base pairs (bp) long. The enzymes recognize and cleave at this site, generating a 5′ phosphate and a 3′ hydroxyl group at cleavage point. The restriction enzymes are usually named after the bacteria of which they are isolated. The initial letter of the genus is used, followed by the first two letters of the species. The type of strain or sub-strain sometimes follows the species designation in the name. Roman numerals are usually used to show if the specific enzyme was the first, the second, the third, etc. For example, the first enzyme extracted from *Escherichia coli* strain RY13 is named EcoRl. So far, hundreds of restrictive enzymes accessible commercially have been discovered and isolated [12].

Restriction enzymes are of mainly three types—Type I, Type II, and Type III [12, 13], Type I and III being the complex ones having only limited role in recombinant DNA technology. Despite the fact that Type I and Type II both identify particular restriction sites, there is a significant variation between them. Type I restriction enzymes cleave ds-DNA at random locations away from their restriction recognition sites, resulting in indistinguishable restriction fragments. As a result, Type I restriction enzymes are of no use in molecular genetics. On the other hand, Type II restriction enzymes produce distinct and predictable restriction fragments by digesting the ds-DNA inside (or very near to) their restriction sites. Type II restriction enzymes can be further classified based on the type of cuts they make in the DNA leading to generation of either a *sticky* or a *blunt* end [13].

Some restriction enzymes digest DNA asymmetrically along their recognition sequence, leading to a single-stranded overhang on the digested end of the DNA segment. These overhangs, called "sticky ends," consist of unpaired nucleotides that are produced at both the 5'and 3′ ends. Cohesive ends are the ones produced by longer overhangs. The sticky overhangs are usually palindromic sequences, those that read the same from both 5′ to 3′ and 3′ to 5′ directions. The sticky ends make it possible for the vector and the insert to bind together. When the sticky ends are compatible, i.e., when the base pairs are complementary on the vector and the insert, the two parts of DNA are joined by a ligation process. Another advantage of sticky end generating enzymes is that less amount of enzyme is required when ligating the vector and the insert while cloning [14]. EcoRI, for example, identifies the sequence 5'GAATTC 3′ and makes a staggered cut, resulting in sticky ends with base pair overhangs. The formation of sticky overhangs is schematically explained in Fig. 2.8a below.

The second class of Type II restriction endonucleases (depending on the type of cut) includes the "blunt end" (also termed as the non-cohesive ends) generating restriction enzymes. These types of ends are generated when the enzyme gives a straight cut, thus terminating both strands into base pairs. This means there are no unpaired DNA strands or overhangs generated at the ends. Also, more amount of ligase enzyme as well as DNA is required for ligating the blunt-ended DNA molecules efficiently since there are no complementary ends produced [15]. For example, enzyme SmaI recognizes the sequence 5'GGGCCC 3′ and cuts both strands of the DNA between the same nucleotide pairs to produce blunt ends (Fig. 2.8b). However, for ligation purposes (described in Sect. 2.5), more amounts of ligase enzyme as well as DNA are required for ligating the blunt-ended DNA molecules efficiently [8]. One may also use additional tools such as adaptors (explained in Sect. 2.5.3) for efficient ligation of blunt end DNA.

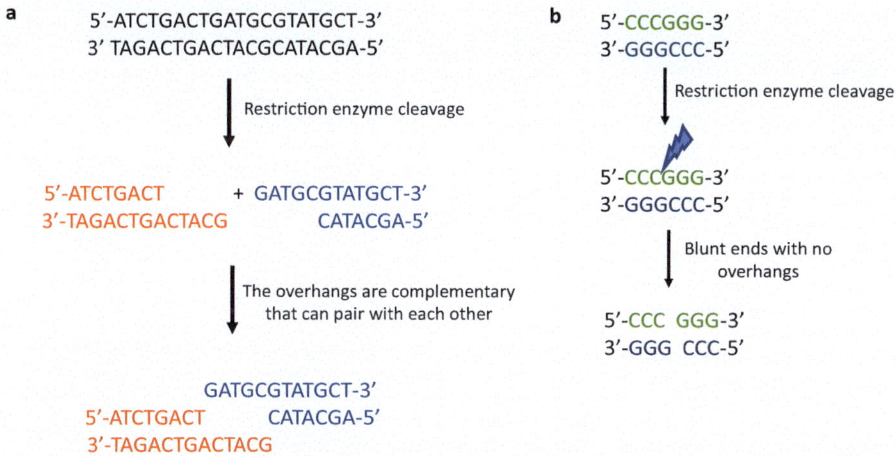

Fig. 2.8 Types of restriction enzyme cuts. (**a**) Generation of sticky or cohesive. Digestion of the DNA with a sticky end-generating restriction enzyme results in the formation of complementary staggered ends that have the capacity to pair up with each other. (**b**) Generation of blunt ends. Digestion of the DNA with a blunt end-generating restriction enzyme results in straight-cut cleavage and terminates both the strands in a base pair. There are no unpaired bases at the 5′ and 3′ prime ends

2.4.2 Steps and Tips for Restriction Digestion

General instructions:

1. The DNA for restriction digestion must be pure and devoid of impurities like EDTA, ethanol, and phenol, which are usually used to purify DNA.
2. Restriction enzymes should always be stored in a freezer. During the laboratory work, they can be kept in a benchtop cooler only for a limited amount of time.
3. To ensure optimal activity, restriction enzymes are used with appropriate buffers that are provided by the manufacturers. Some restriction enzymes, in addition to the buffer, require bovine serum albumin (BSA) for their optimal activity. BSA is usually supplied by the manufacturers at 100× concentration, which is then diluted to 10× in autoclaved distilled water before use.
4. The incubation temperature for most restriction enzymes is 37 °C. However, carefully read the reagent datasheet before incubating the reactions. Set the incubator or water bath at 37 °C or the recommended temperature for the restriction digestion reaction.
5. Incubation time varies depending on the amount of the enzyme used and the source of the DNA template. Usually 45 min to 3 h incubation is sufficient to digest any viral or bacterial DNA under the optimal conditions of incubation; however, eukaryotic DNA requires an overnight incubation.
6. Double digestion is a common procedure in restriction digestion, during which a piece of DNA is digested by two enzymes at the same time (Fig. 2.9). By using

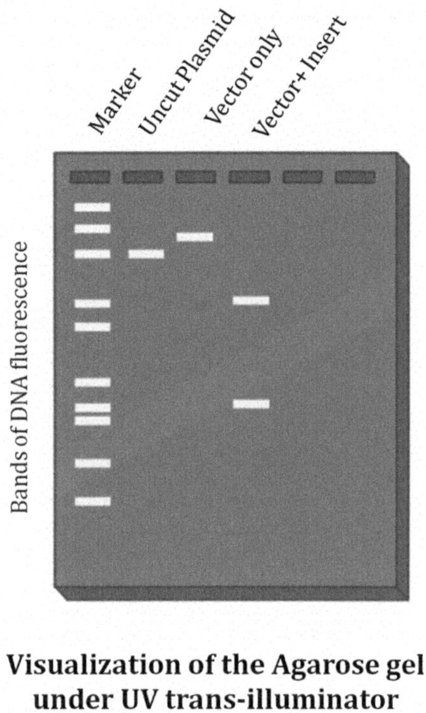

Fig. 2.9 Analyzing digested product-size on a gel. The generalized expected results after restriction enzyme digestion of the recombinant DNA product are depicted above. An appropriate molecular weight standard is used as a reference to determine the correct size of the vector and insert. A single digestion should result into linearization of the vector (linear DNA travels slower than supercoiled plasmid in the uncut lane), while a successful double digestion should result in the release of a lower molecular weight insert

one enzyme, the vector gets linearized and a single band is observed. However, digestion with two restriction enzymes (in a sequential manner) releases the insert and two bands corresponding to vector backbone as well as insert are observed. In double digestion, it is essential to choose a buffer that ensures optimal activity for both the enzymes used. Furthermore, if BSA is required for either of the enzymes, it must be added to the double digestion reaction. The advantage of using BSA is that it will not inhibit the activity of the other enzyme that does not require it. The information regarding suitable buffer for setting the double digestion reaction can be obtained from the website of the manufacturer. In a case where no single buffer is found for a double digestion reaction, the digestion must be done sequentially. First, the reaction is digested with one enzyme + buffer combination, followed by a second digestion step with the second enzyme + buffer combination.

Protocol: Setting restriction enzyme digestion

1. Thaw all reagents on ice.
2. Prepare the reaction mixture of about 50 μL in a microfuge tube.
3. Add reagents in following order: molecular grade nuclease-free water, buffer, BSA (if mentioned), DNA template, and restriction enzyme.
4. Gently mix by tapping the tube. Briefly centrifuge to settle the contents of the tube.

5. If required, prepare positive control reaction with DNA template of known restriction site corresponding to the respective restriction enzyme of your choice.
6. Typical incubation time and temperature is 37 °C for 1 h, though the time and temperature may vary depending on the restriction enzyme used.
   Note: Incubation time and temperature will vary depending on the enzyme as well as the concentration of the DNA template taken.
7. Restriction enzymes are then inactivated by incubation at high temperature (65–70 °C for 10–20 min).
8. Analyze the results of your restriction digestion using agarose gel electrophoresis (Fig. 2.9).
9. Typical restriction digestion reaction conditions are:

10× buffer	2 μl (1×)
DNA template	10 μl (2–4 μg)
Restriction enzyme 1 unit	1 μl
Autoclaved distilled water	7 μl
Final reaction volume	20 μl

Note: 10× denotes the concentration of the stock solution of any reagent; it is generally 10 times the concentration of the reagent that is supposed to be used in a particular reaction

2.5 Ligation

2.5.1 Introduction

Ligation of DNA is an important and final step in the construction of a recombinant plasmid. It involves joining of the DNA fragments (insert) to a compatible vector backbone that is digested with proper restriction enzymes. Both the insert and the vector need to have complementary overhanging base pairs or sticky ends (generated with the use of restriction enzymes during digestion) for the ligation reaction to take place (Fig. 2.10). Usually, digestion using two different restriction enzymes (one at the 5′ end and the other at the 3′ end) is preferred before ligating an insert into a vector. The pair of restriction enzymes used for the digestion should be the same for the vector as well as insert digestion so as to generate complementary overhangs. This allows the insert to be joined in the correct orientation to the vector and it also prevents the vector from self-ligating during the ligation process.

Apart from its application in cloning, *Non-Cloning Ligation* reactions have also found some popularity in other techniques. This form of ligation is basically adapted in Library preparation for Next Generation Sequencing (NGS) wherein a ligation step is typically incorporated to add bar-coded adapters to fragmented DNA [11]. It is also used in many novel detection or diagnostic methodologies, where the ligation of DNA probes followed by PCR amplification (Ligase Chain Reaction – LCR) has been used to detect single nucleotide polymorphisms (SNPs) [11].

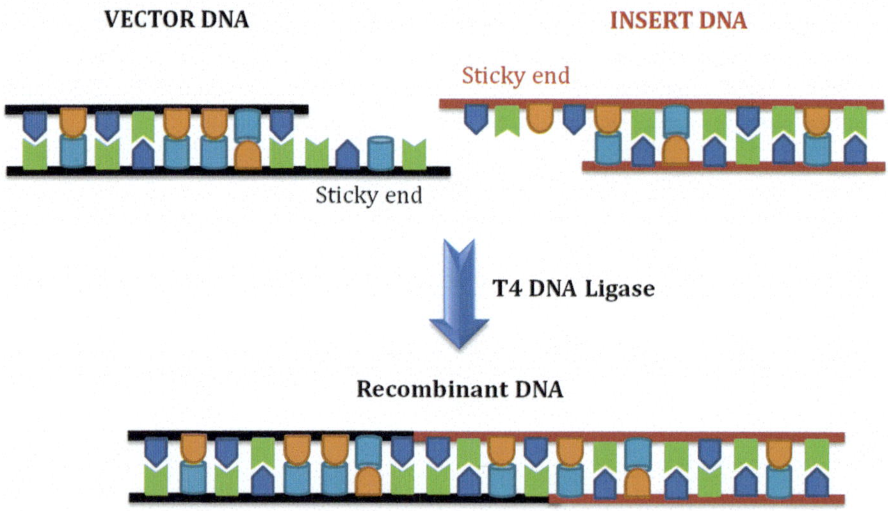

Fig. 2.10 Ligation reaction. The vector and insert are digested with the same pair of restriction enzymes prior to setting a ligation reaction. The complementary overhangs in both the DNA molecules, in the presence of the DNA ligase T4 enzyme, help in efficient ligation of the insert into the vector in correct gene orientation

In both the forms of ligations, cloning as well as non-cloning, the ligation reactions are primarily been catalyzed by enzymes called DNA ligases. However, this chapter will focus mainly on ligation pertaining to cloning genes of interest for producing recombinant proteins mainly in the bacterial host system for their further characterization in a laboratory setup.

2.5.2 DNA Ligases

For decades, DNA ligases have been studied for their role in joining the gaps that form DNA replication, recombination, and DNA repair. DNA ligases catalyze the formation of a phosphodiester bond between the 3′ hydroxyl and 5′ phosphate of the adjacent nucleotides resulting in the concomitant hydrolysis of ATP to AMP and inorganic phosphate [16]. A ligation reaction proceeds in three stages, where initially there is a transfer of an adenylyl group (AMP) from ATP to the ε-amine group of a lysine residue in the ligase enzyme. This results in the formation of an enzyme-nucleotide intermediate, with the release of pyrophosphate from ATP. In the second step, the adenylyl group is transferred from the enzyme to the 5-′-phosphorylated end of the "donor" DNA strand, thus activating the enzyme. The third step involves a nucleophilic attack of the 3′ hydroxyl group of the acceptor DNA to the adenylated donor end of the other DNA strand resulting into the formation of the phosphodiester bond between the two strands with concomitant release of AMP (Fig. 2.11). However, DNA ligases can only form this covalent

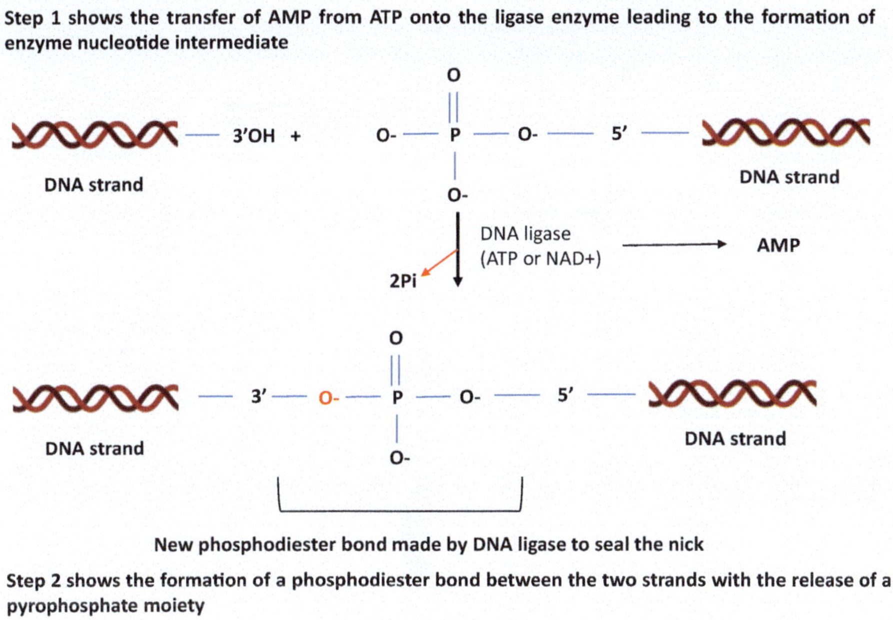

Fig. 2.11 Phosphodiester bond formation by DNA ligase. AMP is transferred from ATP to the ligase enzyme, resulting in the release of pyrophosphate from ATP. A nucleophilic attack of the 3′–OH group of the acceptor DNA strand to the adenylated donor strand results into the formation of the phosphodiester bond

linkage in a duplex molecule (i.e., when joining a nick in dsDNA or joining of an RNA to either a DNA or another RNA in a duplex form), but will not join single-stranded nucleic acids. For decades, molecular biologists have been exploiting DNA ligases for their efficiency in ligating DNA. T4 DNA Ligase, derived from bacteriophage T4, is the most commonly used DNA ligase and is found to be 400-fold more active than the bacterially derived *E. coli* DNA ligase. Hence, it is the enzyme of choice for most of the molecular cloning experiments [16].

2.5.3 Ligation Using Linkers and Adaptors

Although *E. coli* DNA ligase is an extremely popular enzyme for pasting a foreign gene into a vector, its application is somewhat limited by its inability to join a blunt-ended DNA cuts. To circumvent this problem, very large concentration of recombinant DNA molecules was used earlier. The presence of a highly concentrated DNA insert would increase the probability of its interaction with the ligase enzyme and hence ligation. This phenomenon, also known as "molecular crowding" [14, 17], however did not provide any promising solution to the problem that researchers encountered for ligating blunt-ended DNA molecules.

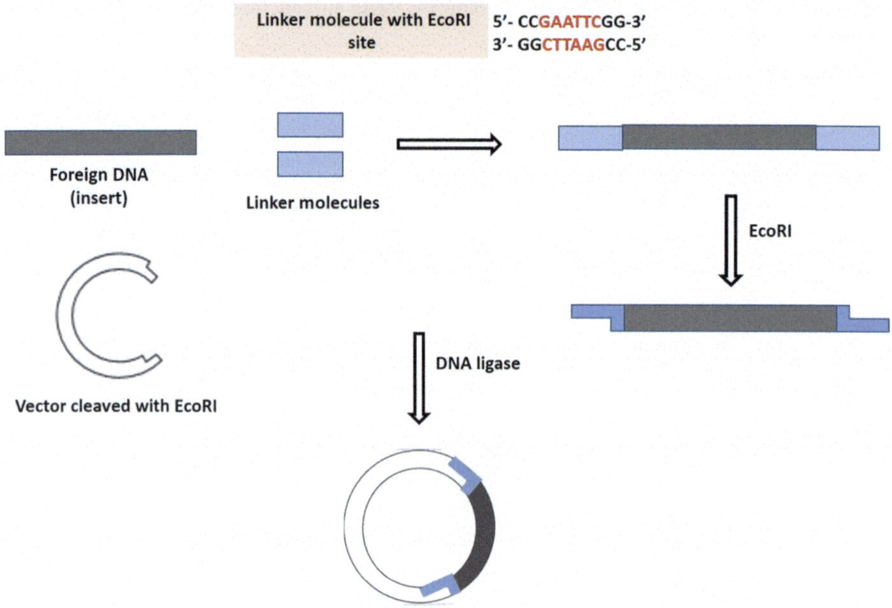

Fig. 2.12 Schematic representation showing attachment of linkers to DNA. A decameric linker molecule containing a site for restriction enzyme EcoRI is attached to a blunt-ended DNA insert. Upon individually digesting the vector and insert with EcoRI, it produces cohesive ends in both. The compatible overhanging ends of insert and vector facilitate the ligation required to produce the recombinant clone

Eventually, with the advancement in the recombinant DNA technology, a better approach to this problem was formulated where a linker sequence is attached to the blunt-ended DNA molecule (Fig. 2.12). This linker has a recognition site for a restriction enzyme that would produce sticky ends when cleaved. Once sticky ends are produced, ligation becomes easier [14, 15].

Another popular method of blunt-ended ligation is the use of adaptor molecules. Unlike linkers, adaptors are pre-formed cohesive-ended DNA fragments that are attached to the ends of blunt-ended DNA molecules, thus easing the ligation reaction [14]. Adaptor molecules with a free 5′ hydroxyl (-OH) group (Fig. 2.13) are used initially while ligating them to the DNA. Since the free 5′ phosphate end is a trigger for self-polymerization of DNA, it is replaced with -OH group. Once they are ligated to the DNA, 5'phosphate group is added to the adaptor ends in order to facilitate the next step of ligation reaction. The phosphate moiety is then added with the help of an enzyme polynucleotide kinase that uses phosphate group from ATP [18].

Homopolymer tailing is another approach that can be used for blunt end ligation. Polymeric tails of the same nucleotides are added to the population of DNA molecules. If there are two different populations of DNA molecules to work with, opposing homopolymer tails are added (for example, poly d(A) tailing on one set of molecules and poly (T) on the other), thus facilitating the annealing of DNA

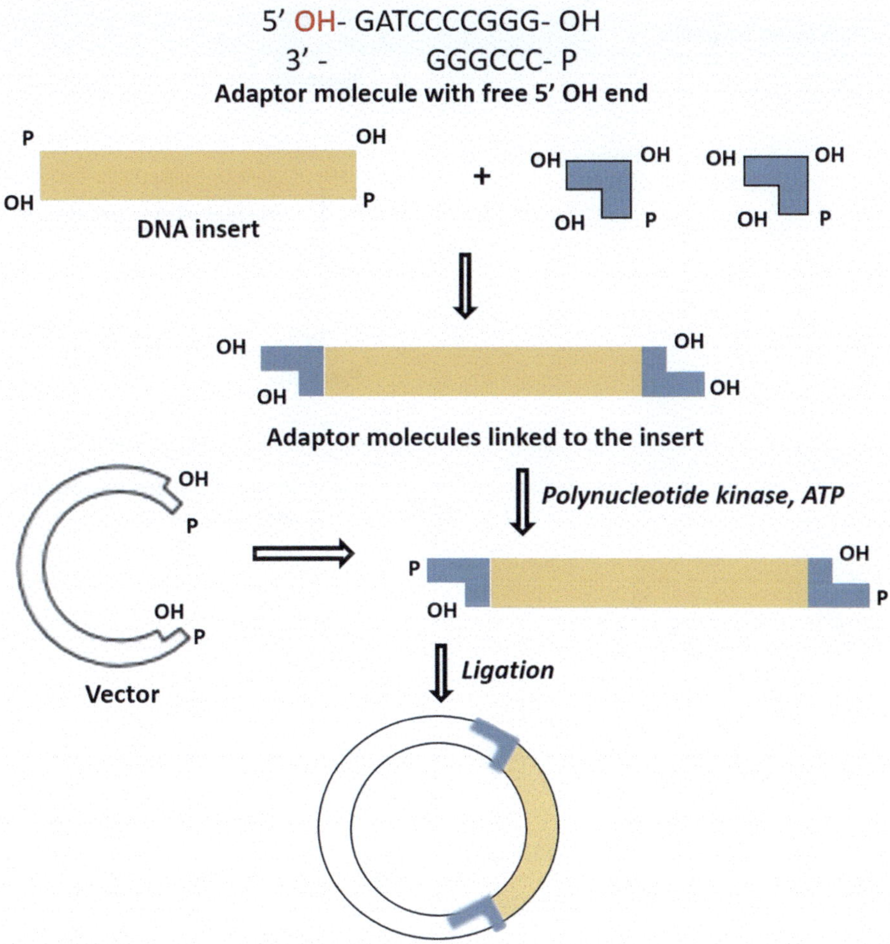

Fig. 2.13 Schematic representation showing attachment of adaptors to DNA molecule. The ends of the adaptor molecules contain a 5'-OH group instead of a 5′-phosphate group to avoid self-polymerization. Once the adaptors are linked to the insert, 5′-phosphate groups are added back with the help of polynucleotide kinase and ATP. For ligation reaction, the vector molecule is digested with the restriction enzyme that generates same compatible sticky ends as that in the adaptor molecule

molecules [14]. For synthesizing homopolymeric tails or extensions, the 3'-OH group of the DNA molecule is first exposed by cleavage with an exonuclease enzyme. This exposed DNA molecule then acts as a substrate for deoxynucleotidyl transferase (often purified from calf thymus), an enzyme that continuously adds specific nucleotide to the exposed 3'-OH end of DNA (Fig. 2.14).

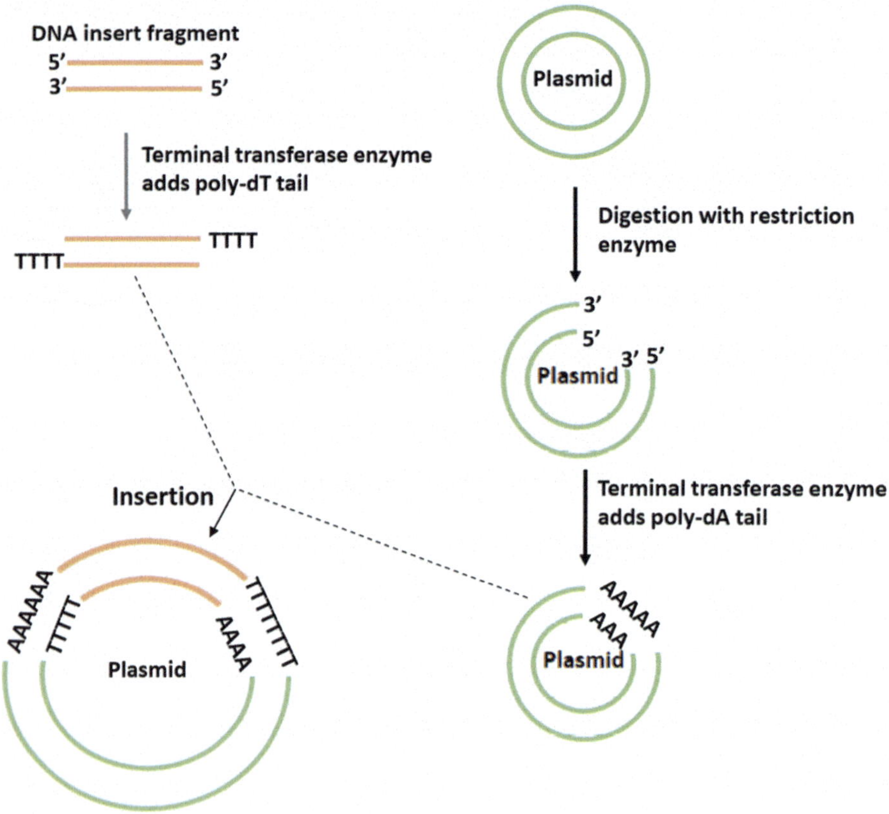

Fig. 2.14 Schematic representation of homopolymer tailing. When the gene insert is treated with exonuclease enzyme, it exposes the 3'-OH group of the insert. This region then is acted upon by deoxynucleotidyl transferase that adds specific nucleotides to generate homopolymer tails. Vector and insert consists of opposing homopolymer tails required for compatible blunt end ligation [14]

2.5.4 Standardizing the Ligation Reaction

The most important step in a ligation reaction is to optimize the amount of cut insert and vector to be used for the reaction. The vector to insert ratio used for a particular ligation reaction depends on the types of vectors used such as cDNA and genomic cloning vectors, as well as on the size and concentration of the vector and the insert used. For most standard cloning and ligation reactions (where the insert is smaller than the vector), a molar ratio of 1:3 of the vector to the digested insert DNA is usually recommended; however, one can also work with 1:1 and 1:2 molar ratio of vector to insert. In case of complicated cloning, where these ratios are not working, the amount of insert and vector can be optimized to improve the ligation efficiency [19]. For a standard ligation reaction of DNA fragments with blunt or sticky ends, about 100 ng of digested vector is recommended, and the following formula is used to calculate the amount of insert to be used:

$$\frac{\text{Amount of vector (ng)} \times \text{Size of insert (kb)}}{\text{Size of vector (kb)}} \times \text{Molar ratio of } \frac{\text{insert}}{\text{vector}}$$
$$= \text{Amount of Insert (ng)}$$

For example,

The amount of insert DNA of 1 kb size required for the ligation with a 4 kb digested vector (50 ng) in 1:3 vector to insert molar ratio will be as shown in Eq. ((2.2):

$$\frac{50\text{ng vector} \times 1\text{kb insert}}{4\text{kb vector}} \times \frac{3}{1} = 37.5\text{ng insert} \tag{2.2}$$

One can also use different ligation calculators such as NEBioCalculator [11] to calculate the molar ratios and estimate the amount of DNA to be used.

2.5.5 Steps Involved in Ligation

1. Assemble the following reaction (20 μl) in a sterile microfuge tube kept on ice.

T4 DNA Ligase Buffer (10×)	2 μl
Vector DNA (50 ng/μl)	2 μl
Insert DNA	Appropriate amount depending on the concentration and molar ratio
T4 DNA ligase[a] (20 NEB units/μl)	1 μl
Nuclease-free water	Add to a volume of 20 μl

[a]NOTE: T4 DNA Ligase is usually supplied in concentrated solutions (e.g., 400,000 units/ml, from New England Biolabs—NEB) by most manufacturers. Therefore, initially, it should be diluted in T4 DNA ligase dilution buffer and stored at a concentration of 20,000 NEB units/ml (60 NEB units corresponds to 1 Weiss unit) as aliquots at −20 °C. As described by Bernard Weiss, Charles Richardson, and his colleagues, one Weiss unit is defined as the amount of enzyme required to catalyze the ATP-PPi exchange in the ligation reaction [20]. It is important to note that, while setting up the reaction, the aliquot should be kept in a benchtop cooler to prevent damage due to rapid freeze/thaw and should only be added at the end in the reaction mixture

2. Gently mix the contents by pipetting the solution and microfuge briefly for a few seconds.
3. Incubate the reaction according to the following conditions or according to the manufacturer's instructions:

16 °C overnight or room temperature for 10 min	Cohesive (sticky) ends
16 °C overnight or room temperature for 2 h	Blunt ends or single base overhangs

Table 2.3 Different controls used in ligation reaction

Control	Ligase	Requirement
Uncut vector	–	To check the viability of competent cells and to verify the antibiotic resistance of the plasmid as well as the quality of antibiotic used in the medium
Cut vector	–	To compare the background (number of colonies) obtained due to uncut vector
Cut vector	+	To compare the background obtained due to vector re-circularization—mostly useful in case of phosphatase treated vector (used for blunt end ligation strategies)
Insert or nuclease-free water	+	To recognize the contamination of the intact plasmid used in ligation or in the transformation reagents

4. Heat-inactivate the reaction at 65 °C for 10 min, if required.
5. Proceed with transforming 1–5 μl of the reaction mixture into competent cells of choice.

General tips:

1. It is always preferable to have appropriate controls for each of the ligation reactions as tabulated in Table 2.3 below.
2. T4 DNA ligase buffer contains ATP to drive the ligation reaction. To avoid degradation of ATP due to multiple freeze/thaw cycles, dispense the buffer into smaller aliquots of 5–10 μl and use one aliquot at a time. The whole step is to be performed on ice.
3. Polyethylene glycol (PEG) is usually known to promote ligation of blunt-ended fragments through macromolecular crowding [21]. Addition of about 2 μl of 50% (w/v) PEG 4000 in a 20 μl ligation reaction can be considered for blunt-ended ligations. However, while cloning cDNAs, one has to be careful with the concentrations used, as PEG can lead to formation of undesirable concatemers as well as residual PEG can be inhibitory to lambda packaging reactions (in vitro reactions used in the construction of cDNA libraries and genomic cloning of methylated DNA into λ-phage or cosmid vectors).

2.6 Ligation Independent Cloning (LIC)

2.6.1 Background

Conventional cloning steps such as restriction enzyme digestion and the subsequent ligation can become tedious at times. Ligation Independent Cloning (LIC) is a form of a cloning method that helps preclude the usage of ligase enzyme and thus evades the need for performing the tricky ligation step as involved in the abovementioned traditional cloning steps. The ligation-independent cloning method was developed in the early 1990s, and since then it serves to be a quick, easy, and relatively cheap method for producing protein expression constructs [22].

The primary aim of this method is to generate long complementary overhangs at the ends of the template/insert DNA. These overhangs are required for establishing a stable and stronger association between the two fragments of interest without any external use of ligase enzyme. This method makes use of the T4 DNA polymerase enzyme for this purpose. The 3′ exonuclease activity of this polymerase occurs in the presence of only a particular dNTP. Because of this property, it can create overhangs of varying length (typically 10–12 bp) in a sequence-specific manner. However, at the site of the first occurrence of the nucleotide (same as the added dNTP), equilibrium between 3′ → 5′ exonuclease and 5′ → 3′ polymerase activity is reached. The polymerase then stalls at this particular position and now the 5′ → 3′ polymerase activity of the T4 DNA polymerase takes over its 3′ exonuclease activity (Fig. 2.15). Thus, long well-defined single-stranded DNA overhangs are produced at the ends of the plasmid as well as the gene of interest. Further, the annealing happens by simply incubating the complementary overhangs-containing vector and insert together. Due to the long length of these overhangs, the annealing reaction between the template DNA and PCR-generated insert becomes highly specific and the recombinant product is quite stable for subsequent transformation without any prior need for ligation. The assembled DNA construct, however, remains nicked at the junction site of the individual pieces. This issue gets resolved inside the transformed bacterial cells, wherein the bacterial ligases quite efficiently repair the nicked sites of the assembled product during replication cycle.

2.6.2 Protocol for LIC

The LIC cloning method involves the following major steps:

1. Preparation of Vector DNA.
   (a) For linearization of the empty vectors used for LIC, typically type II restriction enzymes (e.g., BsaI) are used. These enzymes cleave the vector at a specified distinct position with respect to its recognition sequence (...5-′-GGTCTC(N1)/(N5)-3′..) [23] as shown in Fig. 2.15.
   For linearization of the LIC vector by BsaI digestion, the following components are added:

Reagent	Amount
10× buffer (for restriction enzyme)	5 μl
LIC vector DNA	5 μg
BsaI (10 units/μl) (to be added at the end)	2.5 μl
Nuclease-free water	Add to a volume of 50 μl

Incubate the digestion mixture at 50 °C for 1 h. The linearized vector generated upon digestion will then be separated from the reaction mixture by agarose gel

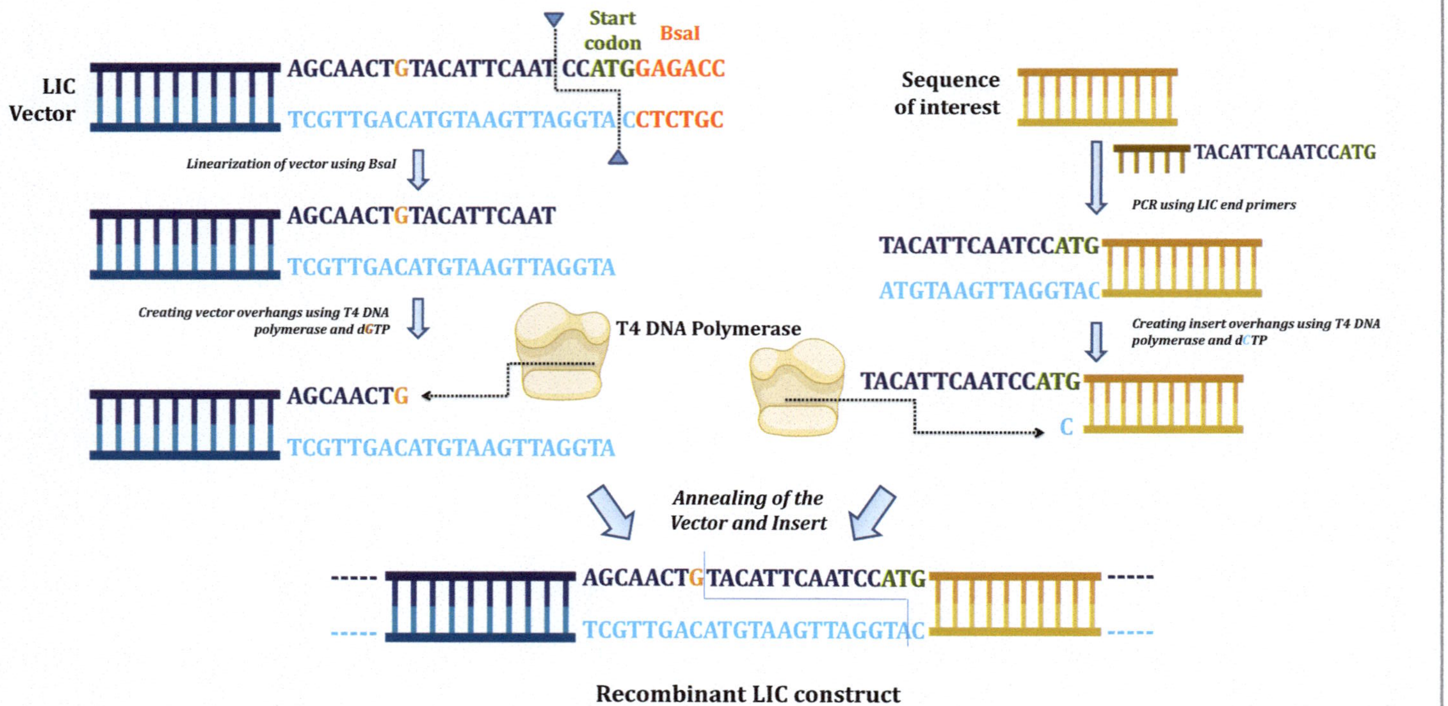

Fig. 2.15 Basic steps in LIC. The vector is linearized by using type II restriction endonucleases such as BsaI, which cleaves at a distinct site that is few nucleotides away from the recognition site. The vector and insert overhangs are generated using the T4 DNA polymerase that exhibits the chew-back (exonuclease) mechanism only till it encounters the first "G" (in case of vector) or "C" in the sequence (in case of insert). The generated overhangs help in the annealing reaction to finally obtain the recombinant DNA product

electrophoresis. The digestion of the vector will remove any part of the MCS or any other portion of the vector if two BsaI sites are present. When visualized after agarose gel electrophoresis, there will be one band representing the linearized vector and the other will be the segment having two BsaI site at the ends. This is followed by extraction of the linearized vector by carefully excising the vector band only and performing gel purification using a DNA extraction kit.

NOTE: It is preferred to elute the purified DNA product in nuclease-free water instead of TE (Tris-EDTA) buffer, to avoid any interference of high salts in the subsequent reactions

The concentration of the vector DNA can be determined using the absorbance at 260 nm. When measured using a spectrophotometer having 1 cm pathlength, the optical density for a 50 μg/mL solution of any dsDNA at 260 nm (OD260) equals 1.0 [24]. Thus, we can calculate the vector DNA concentration using the following equation:

$$\text{dsDNA concentration} = 50\ \mu\text{g/mL} \times \text{OD260} \times \text{dilution factor}$$

The purity of the nucleic acid is estimated by calculating the OD260/OD280 ratio. The OD260/OD280 ratio of pure DNA is around 1.8, while that for pure RNA the ratio is around 2.0. Lower ratios could be because of protein (in case of genomic/plasmid DNA extraction) or phenol contamination used during gel extraction.

(b) For creating overhangs at the end of the linearized vector, it is treated with T4 DNA polymerase, and a free nucleotide (e.g., dGTP) is added in the reaction (remember to exclude all the other nucleotides from the mixture of polymerase reaction). The enzyme chews back the sequence of the vector backbone until it encounters the first G nucleotide in the sequence. As the polymerase reaction is preferred over the exonuclease reaction in the presence of the dGTP, the polymerase will add back the guanosine residue and the exonuclease activity will stall. This is the state where the equilibrium between the two reactions is reached, as shown in Fig. 2.15.

For this type of T4 DNA polymerase reaction (40 μL mixture) in the LIC protocol, the following components are added [23].

Reagent	Final concentration	Volume (μL)
10× buffer (for polymerase)	1×	4
Gel extracted vector DNA	10–50 ng/μl	20–30
dGTP (100 mM)	2.5 mM	1
DTT (100 mM)	5 mM	2
BSA (10 μg/μl)	0.25 μg/μl	1
T4 DNA polymerase (to be added at the end)	0.075 units/μl	1
Nuclease-free water	Add to a volume of 40 μl	

Mix these components and incubate the reaction at 22 °C (or room temperature) for about 30 mins. After incubation, end the reaction by heating to 75 °C for 20 mins for inactivating the polymerase. Measure the final vector concentration through absorbance (the concentration obtained should be around 10–20 ng/μl) and store at −20 °C or lower until further use.

2. Preparation of the Insert DNA.
   (a) For amplification of the insert DNA, PCR is performed using suitable forward and reverse primers that are designed complementary to the 5′ and 3′ ends of the gene of interest, respectively. Before proceeding to the next step of overhang generation, it is essential to remove all the free nucleotides from the PCR amplified product, as they may interfere in the exonuclease activity of the T4 polymerase in the following step.
   (b) For generating overhangs in the insert DNA, T4 DNA polymerase is used. Unlike in case of the vector, here the polymerase reaction is performed in the presence of dCTPs. Thus, the T4 DNA polymerase exhibits the exonuclease activity only till it encounters the first C (cytosine) nucleotide in the sequence.
3. Annealing of the insert and vector.
   The complementary overhangs that are created in the vector (step 1) and insert (step 2) are long enough for very strong and specific, enzyme-free annealing of the two DNA fragments.

2.6.3 Advantages

1. LIC serves to be a sequence-specific, ligase-free cloning method that is simpler and time-saving.
2. It is cost-effective and works efficiently over a broad range of DNA concentrations, even when the individual DNA fragments are not present in equimolar concentrations or in a particular ratio depending on their molecular sizes.
3. It is highly sequence specific and there is no issue of self-ligating plasmid or ligation in wrong orientation, as observed in the conventional ligation protocols.
4. It does not require the usage of T4 DNA ligase but depends on the strong interaction between the long complementary overhangs of insert and plasmid, as well as on the specific bacterial DNA ligases for joining of the remaining nicks.

2.7 Choice of Host Cells

After all the labor-intensive steps of cloning have been carried out, one needs to decide a suitable host organism that would replicate this newly designed plasmid clone. The gram-negative, rod-shaped *Escherichia coli* bacteria have been the commonly used lab organism for a variety of experiments since ages. Majority of the common and commercially available lab strains of *E. coli* used today have descended from two individual isolates, the K-12 [25] and the B strains [26]. The K-12 strain led to the common lab strains MG1655 and its derivatives DH5α and DH10b (alternatively known as TOP10), while the host cells used for protein expression such as BL21 strain [27] and its derivatives are obtained from the *E. coli* B strain [28]. For cloning, a number of different commercial strains of *E. coli* are currently available that can be chosen based on their characteristics for selection of suitable clones. The commercial strains are marketed with specialized properties such as fast growth, routine cloning, high-throughput cloning, maximum DNA yield, cloning of unstable DNA, preparing unmethylated DNA, and much more. For making these features possible, many mutations/genetic changes are made to improve the plasmid yield and DNA quality, confer resistance to any antibiotic, and improve uptake of foreign DNA. Thus, each strain is described by its "genotype" that suggests the particular insertions and deletions that the strain carries and this helps in determining its usage for the desired cloning application. Table 2.4 provides the details of some of the popular strains derived from *E. coli* K-12 and their primary uses in the lab.

More details about the other host cells used for cloning and protein expression has been described in Chap. 4.

2.8 Transformation

Once the process of cloning is successful, one needs to propagate the recombinant DNA molecule into suitable host systems such as bacteria, so as to obtain ample amount of the cloned DNA required for further studies. Transformation is the method by which exogenous DNA is transferred into the host cell. The idea of inducing the bacterial cells to take up the external DNA molecule and replicate as its own has revolutionized various aspects of molecular genetics [36]. Transformation refers to the uptake of DNA into bacterial, yeast, or plant cells, whereas transfection is typically used in mammalian cells. Prior to performing any transformation method, the host cells are made competent (able to take up exogenous DNA) with the help of different methods [37]. The concept of competence and the different methods used to prepare competent cells are described in detail in Chap. 4. Classically, the procedure for introducing a DNA construct into a host cell by transformation is either by chemical method or electroporation or by particle bombardment. Chemical transformation involves treating cells with divalent cations such as calcium chloride ($CaCl_2$) or rubidium chloride (RbCl), which makes the bacterial cell wall more permeable to DNA. Heat shock is used to create temporary pores in the cell membrane, allowing exogenous DNA to be transferred into the cell. A mild electrical pulse is used in electroporation to make the bacterial cell temporarily permeable. Particle bombardment is generally employed for the transformation of plant cells where the DNA construct coated gold or tungsten particles are forced into the cell physically by using gene gun. Herein, we discuss the method of chemical transformation used for bacterial cells.

Table 2.4 Popular strains of *E. coli* used in cloning

Strain	Description	Resistance	Usage
MG1655	Wild type K-12 strain	–	First published sequence of an *E. coli* K-12 strain.
DH5alpha	Derived from *E. coli* K12	–	Most commonly used for general routine cloning and storage of common plasmids, used for blue/white screening [29]
HB101	Hybrid of *E. coli* K12 and *E. coli* B strains, with maximum part (98%) of K12	Streptomycin	Common lab strain for cloning and storage of pBR322 and many other plasmids [30]
DB3.1	HB101 derivative, strain with a specific mutation in the DNA gyrase	Streptomycin	Used for propagating plasmids expressing the ccdB gene (lethal gene that targets DNA gyrase) [31, 32]
MC1061	Parent of DH10B/TOP10 and other derived strains	Streptomycin	Common lab cloning and storage strain, leucine auxotroph, galactose, lactose, arabinose nonutilizing strain [33]
TOP10	MC1061 derivative	Streptomycin	General cloning and storage, blue/white screening, leucine Auxotroph [33]
DH10B	MC1061 derivative	Streptomycin	General cloning and storage, blue/white screening, leucine Auxotroph [33]
JM109	Partly restriction-deficient		Good strain for cloning repetitive DNA, can be used for plasmid maintenance and blue/white screening [34]
JM110	Allows for methylation sensitive restriction enzymes to cut the plasmid after preparation.	Streptomycin	For storing plasmids that should not be Dam (DNA adenine methylase mutation (GATC)) or Dcm (DNA cytosine methylase mutation (CCWGG)) methylated [35]
NEB Stable	Strain from NEB		Used for cloning into and storage of lentiviral and retroviral vectors; cloning of repeated sequences with the potential to recombine [35]
XL1 blue	Nalidixic acid resistant (antibiotic that inhibit a subunit of DNA gyrase and topoisomerase IV)	Tetracycline	Blue/white screening and routine cloning.
XL10 gold		Tetracycline and Chloramphenicol	Hte phenotype provides high competency required for cloning and propagation of large plasmids, ligated DNA, and libraries [34]

2.8.1 Protocol for Transformation

1. Remove cells from −80 °C freezer and thaw on ice.
   Add 1–5 μl containing 1 pg–100 ng of plasmid DNA to the cell mixture. The concentration of DNA used depends on the competency of the cells used; more competent the cells are, lesser is the DNA used. Carefully flick the tube 4–5 times to mix the cells and DNA. Do not vortex.
2. Place the mixture on ice for 30 min. Do not mix.
3. Heat shock at exactly 42 °C for precisely 60–90 s. Do not mix.
4. Post heat shock, immediately place the tube on ice for 3 min. Do not mix.
5. Add 700 μl of room temperature growth media into the transformation mixture. Allow the cells to grow by incubating the tube at 37 °C for 40–60 min.
6. Harvest the cells by centrifuging at about 5000 rpm for 5 min.
7. Resuspend the cell pellet in about 50–100 μl of the same media. Immediately spread plate onto a selection medium plate and incubate overnight (14–16 h) at 37 °C.

NOTE: The choice of antibiotics, other than ampicillin, may require some outgrowth before plating on selective media. Colonies develop more quickly at temperatures over 37 °C, although certain constructs at high temperatures may be unstable.

Details about the other methods of transformation and the troubleshooting involved have been provided in Chap. 4.

2.9 Colony Screening

After the steps of ligation and transformation, one needs to identify the colonies that have been successfully transformed with the recombinant DNA product. Antibiotic selection is one of the crude ways of identifying the plasmids that may carry the gene of interest. However, at times, self-ligated plasmids may also produce false positive results for the cloning, as the antibiotic resistance gene is present in the plasmid and not the gene of interest (insert). Hence, one needs to utilize more specific methods for screening the bacterial colonies transformed with the end product of the cloning reaction as described below:

2.9.1 Blue-White Colony Screening

2.9.1.1 Background

Blue-white colony screening method has been a classic way to detect the colonies that contain plasmid with an insert. It is an effective molecular biology tool that is widely used as a primary step in screening the final recombinant bacteria obtained from the cloning experiments. It is a negative selection system wherein bacterial lactose metabolism is used to indicate successful cloning.

This technique relies on the enzymatic activity of β-galactosidase, a tetrameric enzyme encoded by the lacZ α gene present in the well-characterized bacterial lacZ operon of *E. coli*. When lactose or its functional analog IPTG (isopropyl β-D-1-thiogalactopyranoside) is present in the cellular environment, it triggers the lacZ operon either by inducing the operon (lactose) or by inactivating the lac operon repressor (IPTG). Activation of the lac operon results in the generation of a fully functional β-galactosidase enzyme that metabolizes lactose into glucose and galactose. However, if the lacZ gene is been mutated or a part of it is deleted, β-galactosidase is not produced and the substrate remains intact [38].

Most plasmid vectors used for cloning purposes contain a short segment of lacZ α gene that codes for the first 146 amino acids of β-galactosidase, while the *E. coli* host strains used contains deletion mutation of the same segment, called lacZΔM15, which results into a nonfunctional β-galactosidase enzyme. Hence, when plasmid vector containing the lacZ α gene segment is taken up by such *E. coli* host strains, α-complementation occurs. Herein, the lack of lacZ α gene segment in the mutant bacterial cells is complemented by the α-peptide section present in the plasmid, resulting in the production of a functional enzyme. The plasmid vectors specifically used for cloning purposes also contain a multiple cloning site (MCS) present within the lacZ sequence (Fig. 2.16). Therefore, when an insert DNA is ligated into the plasmid vector, it disrupts the lacZ α gene segment, alpha complementation cannot occur upon transformation, and a functional β-galactosidase does not form. If the gene of interest is not inserted into the vector or is inserted at a different location other than MCS, the lacZ gene in the plasmid vector remains intact and this promotes the α-complementation process producing a functional enzyme.

For visualization of the β-galactosidase enzyme activity, a chromogenic dye-linked substrate called x-gal (5-bromo-4-chloro-3-indolyl-β-D-galacto-pyranoside) is added to the agar plate, along with the inducer IPTG. Production of β-galactosidase results into break down of x-gal into galactose and an insoluble blue pigment (5,5′-dibromo-4,4′-dichloro-indigo) [39]. Thus, as mentioned earlier, if the plasmid vector contains the insert, β-galactosidase is not produced and the resultant colonies are of whitish-cream color of standard *E. coli*. If cloning reaction is unsuccessful, the α-peptide remains intact and functional β-galactosidase enzyme is produced. X-gal in the medium is hydrolyzed by these non-recombinant cells to form 5-bromo-4-chloro-indoxyl, which spontaneously dimerizes to produce the blue pigment. Thus, the colonies formed by non-recombinant cells appear blue, which can be well distinguished from the recombinant ones that appear white.

2.9.1.2 Protocol for Blue-White Colony Screening

After performing the steps of ligation and transformation, the cells are plated onto media containing suitable amount of chromogenic substrate and IPTG. Different chromogenic substrates such as X-GlcA, X-Gal, and S-Gal are commercially available and the methods for addition of such products into the media differs; some are spread directly on LB (Luria-Bertani) agar plates along with IPTG, while others are incorporated into the medium before autoclaving and then agar plates are prepared [39, 40].

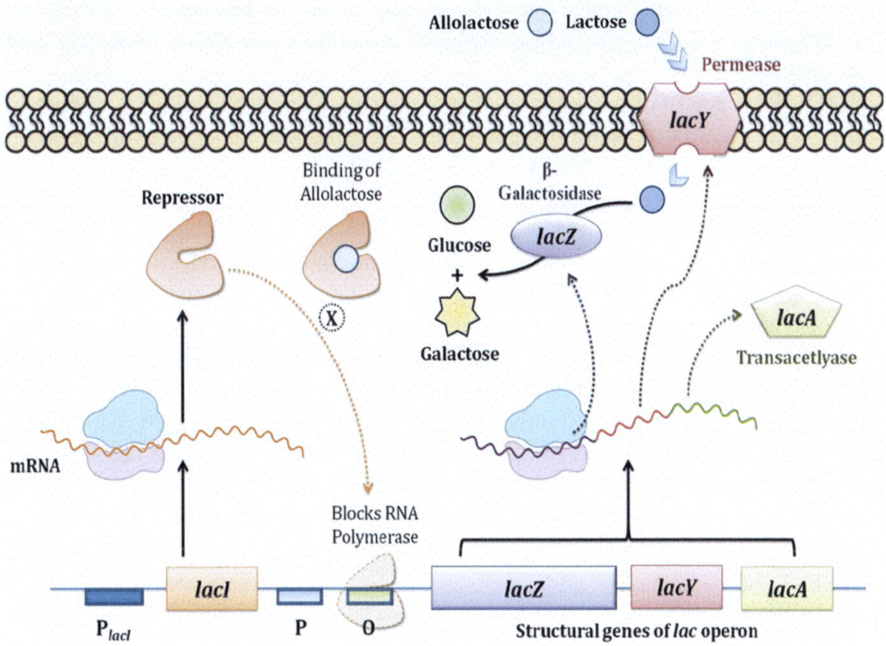

Fig. 2.16 The *lac* operon in *E. coli*. The three structural genes *lac*Z, *lac*Y, and *lac*A are under the control of a single promoter (P) and are together transcribed as a continuous mRNA. The repressor protein is produced constitutively by the *lac*I gene (through the upstream promoter P_{lacI}) and its function is to bind to the operator region (O). In the presence of inducers such as lactose (or its functional analog, Allolactose), repressor binding is prevented and the structural genes are transcribed. The gene product of *lac*Z is β-galactosidase—enzyme that cleaves lactose into glucose and galactose. *lac*Y encodes permease—a membrane channel protein that allows the transport of lactose into the cell at an increased rate, and lacA encodes transacetylase that acetylates galactosides other than lactose and prevents their cleavage by β-galactosidase

NOTE: The chromogenic substrates are light and temperature sensitive. They need to be prepared in the form of stock solutions and added to media only after autoclaving or according to the manufactures' protocol. If spread on top of LB plates, it should be evenly distributed and sufficient drying time should be provided before use.

In both cases, appropriate concentration of selected antibiotic is added to the medium and the following steps are followed:

1. Spread approximately 10–100 μl of the transformed *E. coli* cells onto the LB agar plates (containing IPTG and chromogenic substrate) using a flame-sterilized glass spreader.
   NOTE: Besides the recombinant product, transform empty plasmid vector (without insert) and spread plate. This plate serves as a good control indicating the quality of IPTG and chromogenic substrate.
2. Incubate the plates overnight at 37 °C for 24–48 h depending on the type of cells used.

3. After incubation, blue and white colonies appear on the agar surface. Select the recombinant cells in the white colonies to culture for DNA isolation and sequencing.
   NOTE: Presence of only white colonies is also not a reliable result. It is important to provide enough incubation for any intact β-galactosidase to be expressed and process the substrate into blue pigment (16–20 h).

2.9.1.3 Limitations

1. This technique is only a visual screening method and not a selection technique. Hence, it should be used in combination with other selection methods.
2. Due to the incorporation of some mutations in the lacZ gene of the vector while cell maintenance, the gene may sometimes be nonfunctional. Thus, the resulting colony may appear white, but will not be recombinant.
3. Blue-white colony screening only indicates the presence of an insert, which may not necessarily be the insert of interest. Disruption of the α-peptide DNA by any cloning artifact will also lead to false positive white colonies.
4. False negative cases are rare. However, if a small fragment is inserted in-frame, read-through can occur and lead to a functional β-galactosidase enzyme, giving rise to a blue colony.

2.9.2 Other Screening Methods

2.9.2.1 Positive Selection System

This method follows similar principle as that of the blue-white screening. Here, the positive selection vectors encode a lethal gene, such as any restriction enzyme that digests the host genomic DNA. Inserting the cloning fragment inserted into the center of this gene (present in the MCS) disrupts the lethality, and thus only the recombinant clones survive. Antibiotic selection can also be used in combination with this method to ensure that positive colonies do contain the plasmid with the lethal gene.

2.9.2.2 Screening by Plasmid Miniprep and RE (Restriction Enzyme) Digests

This method involves isolating the recombinant plasmid DNA from the clones and checking the presence as well as orientation of the insert by restriction enzyme digestion. The colonies obtained after transformation of the ligated product are inoculated in LB media supplemented with suitable antibiotics, and grown overnight. Plasmid DNA isolation is carried out from these miniprep cultures using the protocol, as described in Fig. 2.17. For high-copy plasmids, one can obtain 4–10 μg plasmid DNA per purification (1–5 ml). For low-copy plasmids, one needs to grow more amount of culture (10 mL) and can obtain 1–3 μg plasmid DNA per purification [19].

After isolating the plasmid DNA from the expected recombinant clones, the purified plasmid DNA is digested using restriction enzymes. Before using this

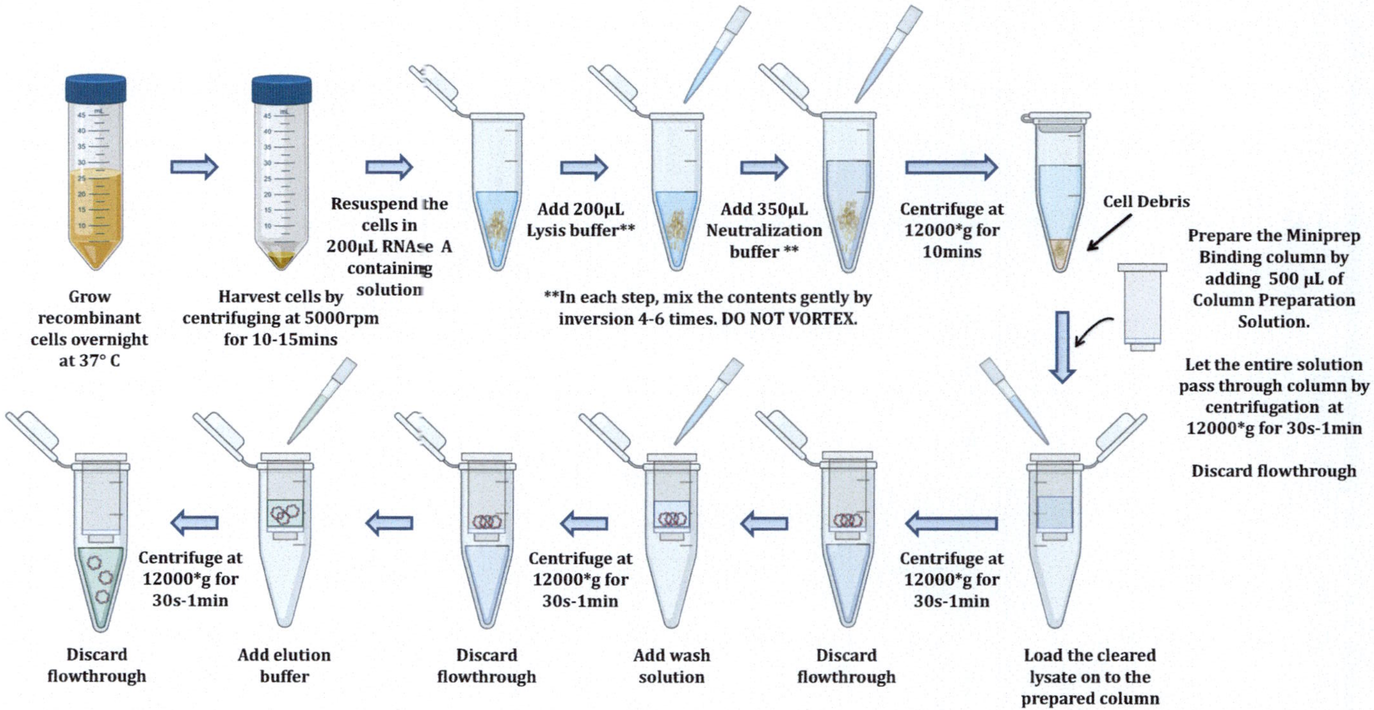

Fig. 2.17 Basic steps in Plasmid isolation. Miniprep cultures are grown overnight and the recombinant clones are harvested. Plasmid DNA extraction is then performed using a modified alkaline-SDS lysis method, followed by adsorption of the plasmid DNA on to silica matrix columns in the presence of high salts. Contaminants are then removed by a spin-wash step. The bound DNA is finally eluted in nuclease-free water or TE (Tris-EDTA) buffer

method for screening, one needs to perform restriction site mapping to identify restriction enzymes that can be particularly used to release the insert from the recombinant plasmid. Once the DNA is purified, about 0.5–1 μg of plasmid can be used for restriction enzyme digestion. The digested product can then be run on agarose gel to verify whether the vector backbone and insert are of the expected sizes.

2.9.2.3 Colony PCR

Colony screening with PCR is the most rapid and cost-effective screening test that helps determine the presence of insert DNA. It involves lysing the bacteria and amplifying a portion of the plasmid with specific primers. The most important step of this method is designing primers and determining their combinations to be used for PCR. There are three approaches for primer design depending on the requirement: (1) insert-specific primers, (2) backbone-specific primers, and (3) orientation-specific primers (Fig. 2.18) [41].

Fig. 2.18 Types of primers used in colony PCR. Designing specific primers is the most important step in colony PCR. Orange-colored box indicates the insert and the arrows indicate the primers. Depending on the need of the experiment, the specificity of the primers can be designed and the results may vary. Different combination of these primers can be used to detect the presence and orientation of the insert in a recombinant clone

In most cases, PCR can be performed either by using vector-specific primer or insert-specific primer or both. If one is doubtful about maintaining the orientation of the insert during cloning steps (especially for blunt end cloning), they can use orientation-specific primers for colony screening.

Once the primers are obtained, a standard PCR reaction (primers, dNTPs, and polymerase) is set up using a portion of the overnight grown culture of the transformed cells. Briefly the protocol comprises the following steps:

1. Lyse the cells to release the plasmid DNA by briefly boiling the sample and using the supernatant, or by directly adding the sample to the PCR master mix. The initial heating step of the PCR reaction helps in the lysis.
2. For amplification of the desired plasmid region, a standard Taq polymerase is sufficient.
3. Run the obtained PCR product with the specific controls on a 1% agarose gel to analyze the exact product size and the success of cloning.

This method allows screening of several colonies at a time and eliminates the need to first purify the plasmid DNA required for using as a template for PCR.

2.9.2.4 Sanger Sequencing

The final step in most of the cloning strategies is to verify whether the sequence of the insert, insert orientation, and the sequences of the junctions between the plasmid and insert DNA are correct. This can be achieved by sequencing the plasmid DNA using Sanger sequencing (also called as chain termination or cycle sequencing) [42, 43]. In addition to the reagents used in a standard PCR, four fluorescently labelled dideoxynucleotide triphosphates (ddNTPs: ddATP, ddGTP, ddCTP, and ddTTP) are also added in low ratio. Random addition of these distinctively labelled ddNTPs terminates the synthesis reaction. Due to the absence of hydroxyl group in ddNTPs, polymerase fails to form the phosphodiester bond with the next nucleotide and the reaction terminates. The basic steps involved in Sanger's sequencing are briefly described in Fig. 2.19. This method acts as a confirmatory step that allows one to identify any mutations that have been inadvertently incorporated in the cloning product. The sequencing is performed manually, or more commonly, in an automated fashion using a sequencing machine.

The sequencing data obtained is in the form of a color-coded electropherogram or chromatogram, which shows the fluorescent peak of each nucleotide along the length of the template DNA [44]. This can then be converted to nucleotide sequence by the computer. One can then use alignment tools like ClustalW to check the correct sequence of the insert or for the presence of any mutation in the insert.

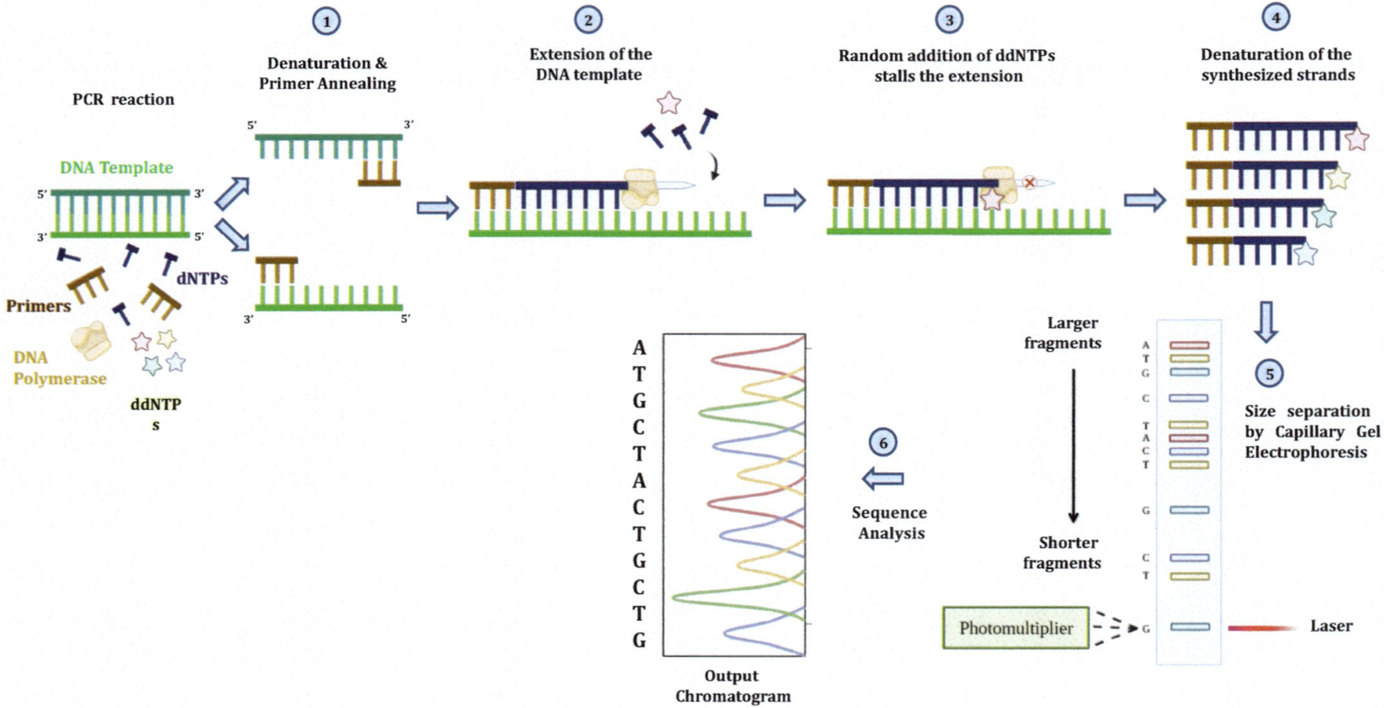

Fig. 2.19 Basic steps of automated Sanger sequencing. The dsDNA is denatured into two ssDNA and the respective primers bind at their 3′ ends. Extension of the new strand occurs until a termination nucleotide (ddNTP) is randomly incorporated. The resulting DNA fragments are again denatured into ssDNA and these are further separated by gel electrophoresis for determination of the sequence

2.10 Troubleshooting for Subcloning Experiments (Table 2.5)

Table 2.5 Alternative strategies to be used for solving cloning errors

Problem	Reason	Solution
Host cells do not contain plasmid after transformation	1. Less antibiotic used 2. Satellite colonies that are very small in size are chosen	1. Add adequate amount of antibiotic and use freshly made media and antibiotics 2. Selection of well-established bigger colonies is important
Host cells contain the wrong construct	1. PCR amplification was incorrect 2. Recognition site of restriction enzyme used for digestion site present in the gene 3. Mutations are present in the cloned DNA	1. Standardize the protocol for PCR; purify correct PCR product2. Check for the presence of restriction digestion site in the coding region of the gene by using certain software like Snapgene 3. Repeat sequencing of the clone; use a high-fidelity DNA polymerase in PCR reaction
No ligation product obtained	1. Insufficient ligation reaction	1. Temperature used for incubation of the reaction was not optimum 2. Ligase enzyme has been inactivated due to improper storage 3. Ratios of insert and vector need to be varied 4. Contaminants such as EDTA in DNA purification buffers need to be removed

2.11 Conclusions

In this chapter, we have seen how we can clone a gene of interest into a suitable vector and produce large copies of recombinant clones. One can generate gene libraries using either the genomic DNA or cDNA, utilize different restriction enzymes to cut the gene of interest and ligate it in a compatible vector using the ligase enzyme. These recombinant clones can be further used for the sequence analysis study, understanding the functional relevance of the gene by performing protein expression as well as for developing probes that are used for studying its expression within cells. The next two chapters (Chaps. 3 and 4) describe the nature of the vectors that can be used for different cloning purposes and the techniques employed for introducing the recombinant DNA molecules into suitable host cells in detail.

Acknowledgments The authors thank Advanced Centre for Treatment, Research and Education in Cancer (ACTREC) for providing necessary infrastructure and resources for successful completion of the chapter. The authors acknowledge Ms. Chanda Baisane, Bose Lab, ACTREC for formatting the manuscript.

Problems

Multiple Choice Questions

1. In lac operon, repressor molecules bind to __________
   (a) Enhancer
   (b) Promoter
   (c) Hormone Receptor
   (d) Operator
2. In Sanger sequencing, separation of the different sized templates is done through ___
   (a) Agarose gel electrophoresis
   (b) Capillary gel electrophoresis
   (c) SDS- PAGE
   (d) Electrophoretic mobility shift assay
3. Which of the following statements is false?
   (a) The competent cells with no recombinant plasmids are drug resistant
   (b) Plasmid replication is independent of cell division
   (c) The competent cells that take up recombinant plasmid molecules become drug resistant
   (d) The competent cells are originally drug resistant

Subjective Questions

1. A plasmid vector and gene under study were digested with the same restriction enzyme. The vector and the insert were ligated using DNA ligase. The ligation product was used to transform the host E. coli. The colonies were obtained on a suitable selection medium. What possible products are expected after isolating DNA from the colonies. What are the ways to determine these different products obtained?
2. Digestion of an insert with a restriction enzyme produces blunt ends. In order to ease the process and ensure that the insert ligates with the vector, what modifications can be made to the blunt-ended insert?

References

1. Dale JW, Schantz MV, Plant N. From genes to genomes: concepts and applications of DNA technology. 3rd ed. John Wiley & Sons: Chichester; 2011.
2. Russo E. Special report: the birth of biotechnology. Nature. 2003;421(6921):456–7.
3. Brown TA. Gene cloning and analysis an introduction. 6th ed. Chichester: John Wiley & Sons, Inc.; 2010.

4. Dale JW, von Schantz M, Plant N. From genes to genomes: concepts and applications of DNA technology. 3rd ed. Chichester: John Wiley & Sons, Inc.; 2011. p. 400.
5. Bajpai B. High capacity vectors. India: Springer; 2013.
6. Jeremy W, Dale MVS, Plant N. From genes to genomes: concepts and applications of DNA technology. Chichester: John Wiley & Sons, Inc.; 2011.
7. Sambrook J, Russell DW. Molecular cloning: a laboratory manual, vol. 2. 3rd ed. New York: Cold Spring Harbor Laboratory Press; 2001.
8. Coffin JM, Hughes SH, Varmus HE. Retroviruses. New York: Cold Spring Harbor Laboratory Press; 1997.
9. Bhagavan NV, Ha C-E. Essentials of medical biochemistry. 2nd ed. Netherland: Elsevier; 2015.
10. Eberwine J, et al. Complementary DNA synthesis in situ: methods and applications. Methods Enzymol. 1992;219:80–100.
11. New England Biolabs. https://international.neb.com/applications/cloning-and-synthetic-biology/dna-ligation.
12. A comprehensive list of all the known restriction endonucleases and their recognition sequences. REBASE: http://rebase.neb.com/rebase/.
13. Pray LA. Restriction enzymes. Nat Educ. 2008;1(1):38.
14. Primrose SB, Twyman R. Principles of gene manipulation and genomics. 7th ed. Oxford: Blackwell Scientific Publications; 2006. p. 672.
15. Lodish H, Berk A, Zipursky SL. Molecular cell biology. 4th ed. New York: W. H. Freeman; 2000.
16. Tabor S. DNA ligases. Curr Protoc Mol Biol. 2001;CH 3:Unit3.14.
17. Aumiller WM, et al. Chapter Five - Phase separation as a possible means of nuclear compartmentalization. Int Rev Cell Mol Biol. 2014;307:109–49.
18. Kleppe K, Lillehaug JR. Polynucleotide kinase. Adv Enzymol Relat Areas Mol Biol. 1979;48:245–75.
19. Promega. Subcloning notebook.
20. Weiss B, Thompson A, Richardson CC. Enzymatic breakage and joining of deoxyribonucleic acid. J Biol Chem. 1968;243(17):4543–55.
21. Zimmerman SB, Pheiffer BH. Macromolecular crowding allows blunt-end ligation by DNA ligases from rat liver or *Escherichia coli*. Proc Natl Acad Sci U S A. 1983;80(19):5852–6.
22. Haun RS, Serventi M, Moss J. Rapid, reliable ligation-independent cloning of PCR products using modified plasmid vectors. BioTechniques. 1992;13(4):515–8.
23. https://www.addgene.org/protocols/lic/.
24. Carlos F, Barbas III, et al. Quantitation of DNA and RNA. CSH Protoc. 2007;2007:pdb.ip47.
25. Bachmann BJ. Pedigrees of some mutant strains of *Escherichia coli* K-12. Bacteriol Rev. 1972;36(4):525–57.
26. Abedon ST. Selection for lysis inhibition in bacteriophage. J Theor Biol. 1990;146(4):501–11.
27. Daegelen P, et al. Tracing ancestors and relatives of *Escherichia coli* B, and the derivation of B strains REL606 and BL21(DE3). J Mol Biol. 2009;394(4):634–43.
28. Samuelson JC. Recent developments in difficult protein expression: a guide to E. coli strains, promoters, and relevant host mutations. Methods Mol Biol. 2011;705:195–209.
29. Grant SG, et al. Differential plasmid rescue from transgenic mouse DNAs into *Escherichia coli* methylation-restriction mutants. Proc Natl Acad Sci U S A. 1990;87(12):4645–9.
30. Carzaniga T, et al. Autogenous regulation of *Escherichia coli* polynucleotide phosphorylase expression revisited. J Bacteriol. 2009;191(6):1738–48.
31. Bernard P. New ccdB positive-selection cloning vectors with kanamycin or chloramphenicol selectable markers. Gene. 1995;162(1):159–60.
32. Maki S, et al. Partner switching mechanisms in inactivation and rejuvenation of *Escherichia coli* DNA gyrase by F plasmid proteins LetD (CcdB) and LetA (CcdA). J Mol Biol. 1996;256(3):473–82.
33. Durfee T, et al. The complete genome sequence of *Escherichia coli* DH10B: insights into the biology of a laboratory workhorse. J Bacteriol. 2008;190(7):2597–606.

34. Hoseini SS, Sauer MG. Molecular cloning using polymerase chain reaction, an educational guide for cellular engineering. J Biol Eng. 2015;9:2. https://doi.org/10.1186/1754-1611-9-2.
35. Ferenc BM. Plasmids 101: common Lab E. coli strains. Watertown, MA: Addgene Blog; 2014.
36. Cohen SN, Chan ACY, Hsu L. Nonchromosomal antibiotic resistance in bacteria: genetic transformation of *Escherichia coli* by R-factor DNA. Proc Natl Acad Sci U S A. 1972;69 (8):2110–4.
37. Hanahan D, Jessee J, Bloom FR. Plasmid transformation of Escherichia coli and other bacteria. Methods Enzymol. 1991;204:63–113.
38. Shuman HA, Silhavy TJ. The art and design of genetic screens: *Escherichia coli*. Nat Rev Genet. 2003;4(6):419–31.
39. Green MR, Sambrook J. Molecular cloning a laboratory manual. New York: Cold Spring Harbor Laboratory Press; 2012.
40. Padmanabhan S, Banerjee S, Mandi N. Screening of bacterial recombinants: strategies and preventing false positives. London: IntechOpen; 2011.
41. Kenkel BB. Plasmids 101: Colony PCR. https://blog.addgene.org/plasmids-101-colony-pcr.
42. Sanger F, Coulson AR. A rapid method for determining sequences in DNA by primed synthesis with DNA polymerase. J Mol Biol. 1975;94(3):441–8.
43. Sanger F, Nicklen S, Coulson AR. DNA sequencing with chain-terminating inhibitors. Proc Natl Acad Sci U S A. 1977;74(12):5463–7.
44. Dewey FE, et al. DNA sequencing: Clinical applications of new DNA sequencing technologies. Circulation. 2013;125(7):931–44.

Selection of Cloning and Expression Plasmid Vectors

3

Rucha Kulkarni, Roshnee Bose, and Kakoli Bose

Abstract

Purification of proteins using recombinant DNA technology is crucial for studying its structure, function, and interactions with other ligands/macromolecules as well as for therapeutic purposes. Thus, the need for purified proteins has been increasing day by day both in basic research labs and at the industrial level. Major developments in the field of recombinant DNA technology have been focused on improving the process of protein purification. However, the very first step in this process is cloning and expression of isolated protein coding genes through engineered plasmid molecules (self-replicating extrachromosomal circular DNA molecules) called vectors. Therefore, with the increasing demand for purified proteins, the need for more efficient and robust vectors was perceived. Today, a wide array of vectors for both cloning and expressing the gene of interest are being engineered that suit almost every requirement of the researcher. However, the challenge lies in choosing the correct vector for a specific requirement, and hence a thorough knowledge of all the available vectors used for different purposes becomes imperative. This chapter aims at guiding the researchers toward choosing appropriate vectors to cater to their various cloning requirements. It also provides troubleshooting tips and discusses latest advancements in this technology.

R. Kulkarni · K. Bose (✉)
Integrated Biophysics and Structural Biology Lab, Advanced Centre for Treatment, Research and Education in Cancer (ACTREC), Tata Memorial Centre, Mumbai, India

Homi Bhabha National Institute, BARC Training School Complex, Mumbai, India
e-mail: kbose@actrec.gov.in

R. Bose
Integrated Biophysics and Structural Biology Lab, Advanced Centre for Treatment, Research and Education in Cancer (ACTREC), Tata Memorial Centre, Mumbai, India

K. Bose (ed.), *Textbook on Cloning, Expression and Purification of Recombinant Proteins*, https://doi.org/10.1007/978-981-16-4987-5_3

Keywords

Plasmid · Expression vector · Cloning vector · Recombinant proteins · Gene · Therapeutic

3.1 Introduction

Gregor Mendel, the father of genetics, had put forth the concept of genes as carriers of phenotypic information and also described the different attributes of genetic inheritance [1]. Genes are the basic units of heredity and hence are responsible for passing on the traits from parents to offspring. They carry instructions for production of proteins that they encode. Molecular biology that deals with study of genes and their products has traveled a long way since the discovery of genetic material. Significant advances in this field over past several decades have enabled scientists to isolate a single gene, make copies of it, and express it in a different system all together for its characterization in detail.

The study of the human genome from its structural and functional perspectives provided a detailed insight into many of the human genetic disorders. In most of these diseases, either the protein product of a gene is faulty or the gene fails to express and produce the encoded protein. To repair such flaws in the genome, either the gene should be replaced with the original one prior to its expression or the final protein product of that gene should be introduced into the cells. This dire need, along with the demand of pure proteins for research and industrial use, has led to heterologous protein production inside another living system (host) such as bacteria. A host is any suitable non-pathogenic, fast growing living system that helps in replicating the cloned gene and expressing the protein in large amounts. Examples of different host systems including bacterium *Escherichia coli,* yeast, insect, fungi, and mammalian cells are used for production of recombinant proteins [2]. *E. coli*, however, has been the first choice to produce enzymes and other proteins for their in vitro characterization in research labs as well as for commercial use. Few best examples are large-scale production of human hormone insulin (treatment of diabetes), human growth hormone, and clotting factors [3–5]. The advantages of recombinant expression over obtaining therapeutic proteins from other animal sources are manifold; this technique ensures unlimited supply as well as safety and purity [6].

To study human proteins in vitro, a system that would carry the isolated protein-encoding gene to the host system, often known as carrier molecule or vehicle, is required. However, the caveat is that these carrier molecules should enable the host to make copies of the foreign gene using its own DNA replication machinery. Such carrier molecular entities that act as vehicles for the recombinant genes are termed as plasmids [6]. Plasmids are circular and self-replicating extrachromosomal DNA molecules, distinct from host cell's chromosomal DNA, that naturally exist in bacterial cells. Plasmids vary to a great extent in length from a few thousand base pairs to more than 100 kilobases. With the advancement of recombinant DNA technology, the innate properties of plasmids have been harnessed to produce an engineered version of these molecules to cater the needs of efficient cloning of foreign gene. Different varieties of these molecules with distinct properties are commercially available as plasmid vectors or plasmids. Therefore, the term plasmid or plasmid vector will be used henceforth in this chapter instead of vector to avoid confusion. In a nutshell, the plasmid vector and the host are manipulated genetically to replicate and optimally express the foreign gene for large-scale production of the recombinant protein.

Since recombinant proteins are produced in research laboratories for different purposes, plasmid vectors encoding their genes need to have specific properties to fulfill such research requirements. For example, proteins need to have certain tags (such as histidine tag (His_6), Glutathione S-transferase (GST) and Maltose binding protein (MBP)) at the extreme C- or N-terminus for ease of purification using affinity chromatography or for protein-protein interaction studies [7, 8]. Therefore, with various plausible combinations of plasmid components including replicons, selection markers, multiple cloning sites, and protein purification tags, the variety of available plasmid vectors has significantly increased. This certainly provides the researchers with the flexibility to opt for the plasmid vector that best suits their needs. Furthermore, plasmid vectors can be re-engineered to make them tailor-made for some specific experiments using molecular biology tools. This chapter elaborately discusses plasmid vectors used for cloning and expressing genes encoding proteins primarily in the bacterial host system, and categorizes them based on their types, important features, distinct behaviors, and applications in biological and industrial research. Furthermore, it provides relevant cloning protocols and cues for overcoming hurdles one faces during gene cloning.

3.2 Classification

Vectors can be classified under two categories:

- *Cloning vectors*: for cloning a gene to make its copies.
- *Expression vectors*: for expressing a gene to obtain its protein product.

There are certain important differences in these two types of vectors, which are quite evident from their names. While cloning vectors would only replicate and make copies of the insert, the expression vectors have certain elements engineered into them that help in expressing the insert gene to produce an optimum amount of desired proteins.

Thus, cloning vectors that are devoid of the necessary elements required for gene expression are capable of making only multiple copies of the target gene without giving the protein product [9]. Expression vectors, on the other hand, have necessary elements for transcription and subsequent gene translation [9].

3.3 Cloning: An Overview

The process that went into the creation of recombinant DNA molecules is termed as molecular cloning, the basics of which have been described in the previous chapter. Briefly, cloning is a process of generating identical copies, be it a single gene, a cell, or an organism [10]. This process, since its time of inception, has been extensively used and constantly standardized. Cloning emerged as a technique with the discovery of "restriction endonucleases" and "ligases"—the enzymes that are very specific in cutting and joining the DNA, respectively [11]. Discovery of these enzymes made the technique more sophisticated as well as yielded powerful tools for generating recombinant molecules. Thus, cloning using various vectors and their diverse features has become a common lab technique that helps devise strategies to answer critical biological research questions.

As a first step, it is necessary to isolate the specific protein-encoding gene from the source organism. Sometimes, if the available sample of desired DNA (gene) is limiting, performing a polymerase chain reaction (PCR) becomes necessary to amplify the gene so as to obtain sufficient amount of DNA [12]. The gene of interest is then cloned or inserted into plasmid molecules that provide necessary sequential elements for maintenance of the exogenous DNA as well as regulatory elements recognizable by the host for gene expression. Finally, the recombinant plasmid DNA is transformed into a suitable host system such as *E. coli*. The basic steps in cloning have been described and illustrated in the previous chapters of this book.

3.4 The Need to Choose a Plasmid Vector

For robust expression and purification, it is important to understand the intrinsic properties of the recombinant proteins of interest. The expression of foreign mammalian proteins in a microbial host further puts forth certain challenges before the scientists that need minute attention prior to cloning in a plasmid vector. For example, lack of post-translational modifications in bacterial systems often makes the expressed protein unfold, remain insoluble and unstable [13]. This has led to modification of purification protocols of these proteins with "*special needs*" that have been elaborated in the later chapters of this book. In addition, for ease with purification, certain affinity tags are genetically introduced at the extreme ends of the recombinant proteins through the plasmid vectors. Majority of the plasmids are suited for use in *E. coli* host since it is the most preferred host for protein production on a large scale. The bacterial strains have been modified such that they produce recombinant protein which accounts for almost 20% of their total cellular proteins. Thus, a researcher should have a thorough knowledge of the available plasmid systems for cloning and expressing proteins so as to prudently choose a system for best results.

The common important features of all plasmid vectors are described below as well as in Fig. 3.1.

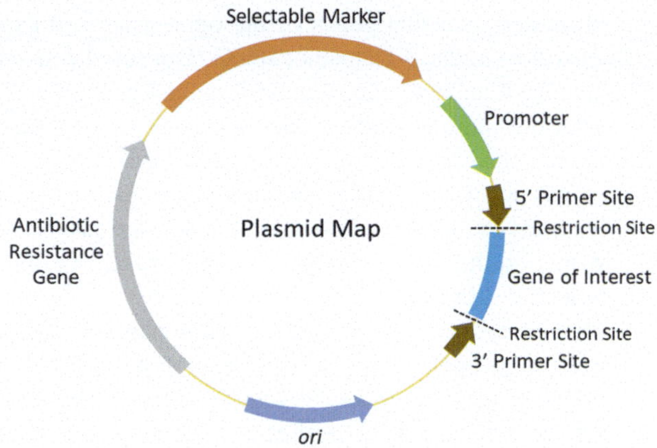

Fig. 3.1 Important features of a basic plasmid: the figure shows (a) *ori* element which is origin of replication sequence recognized by host replication enzymes; (b) promoter for gene expression; (c) selection marker which is mostly an antibiotic coding gene required for the selection of transformants; and (d) MCS which has many recognition sites for various restriction enzymes to clone any gene of interest

(a) *Ori*:
It stands for "origin of replication" that is required for replication and propagation of the insert inside the host. This is the site in a plasmid where replication machinery assembles and the process of replicating the entire plasmid initiates. It is a crucial component of the plasmid system without which the plasmid would get lost during cell division. Thus, *ori* sequence ensures that the plasmids replicate and are distributed or partitioned during cell division [14].

(b) Promoter:
Promoters play a pivotal role in gene expression. They are ~100–1000 base pairs in size and are located upstream of the gene of interest. The promoter sequence modulates binding of the RNA polymerase as well as other transcription factors, and hence choosing a proper promoter for a plasmid vector is crucial. It is important to ensure that the promoter is compatible with the plasmid vector as well as with the transcription machinery of the host system in which it is later introduced for expression. Bacterial promoters are less complex and distinct from their mammalian counterparts. For tight control of expression, synthetic promoters have been created that do not allow leaky expression of toxic proteins. Although constitutively active promoters are available, inducible promoters have gained tremendous popularity as they can be turned *on* or *off* (through heat, light, chemical, etc.) based on the individual experimental requirements [15]. A commonly used constitutive promoter is T7 (from T7 bacteriophage), which requires T7 RNA Polymerase for activation. A modified version of T7 is T7lac in which the RNA Polymerase is controlled by lac operon, thus making it an inducible promoter [15].

A routinely used negatively inducible promoter includes the pLac promoter. Although this promoter is constitutively active, it can be repressed by LacI repressor from *E. coli* bacterial system. The functioning of this promoter has been described extensively in the colony screening section of Chap. 2. Briefly, under normal conditions when lactose is absent from the medium, the operon is in the repressed state. This is due to the binding of the repressor protein onto the promoter such that genes of the operon are not expressed. In the presence of activators such as lactose or IPTG (Isopropyl β-D-1-thiogalactopyranoside—an analog of lactose), the repression on the promoter is eliminated. Binding of the activator induces conformational change in the repressor, and thus the operon is expressed. Lac inducible system has been widely exploited by many bacterial expression vectors [15]. One more example of a negatively inducible promoter includes the pBAD promoter, which is the promoter for arabinose operon and works on the same principle as that of pLac operon [16].

(c) Selection Marker:
This component is required for maintenance of the plasmid inside the host as it imparts a special characteristic phenotype such as resistance against an antibiotic, e.g., ampicillin, kanamycin, or chloramphenicol resistance genes. This aids the researchers in specifically selecting bacterial colonies harboring the recombinants that are capable of growing on an antibiotic resistant plate [9].

(d) Multiple Cloning Site (MCS):
MCS, also known as a polylinker region, is an important component of a plasmid that contains recognition sequences for several restriction endonucleases for cutting the plasmid and inserting the genes of interest. Multiple restriction enzyme sites present in a plasmid gives ample choice or flexibility to use different enzymes for cloning the insert. Moreover, these enzyme sites are unique to the MCS region, as they are not present elsewhere in the vector. This is done to ensure that the cut would be introduced only in the MCS leaving the rest of the plasmid intact that otherwise would have been fragmented into pieces [17].

3.5 Types of Plasmid Vectors

3.5.1 Plasmids Used for Cloning

A cloning plasmid self-replicates inside the host organism to produce identical copies, also known as clones. Since cloning vectors are only capable of making copies of the desired inserts, they do not harbor the regulatory elements needed for transcription and translation of the insert into its protein product [11]. Thus, the aim of cloning plasmids is maintenance and amplification of the inserted gene so as to obtain large number of pure copies for further manipulations and studies.

At its inception, cloning was done with naturally occurring plasmids such as ColE1 and pSC101 [11]. ColE1 is named after the gene colicin E1 that it encodes. pSC101 was one of the first bacterial vectors used for cloning by Boyer and Cohen in 1973 [18]. Although they were capable of carrying out cloning experiments, they failed to provide the flexibility to the researchers for cloning in a specific way.

To fulfill distinct cloning requirements, several engineered plasmid vectors came into existence. pBR322 is one such plasmid vector. It is one of the oldest and most widely used cloning vectors due to its robustness and ability to provide a wide spectrum of choices. A large number of other plasmids (such as pUC) have been tailor-made on pBR322 skeleton for specialized cloning needs [11, 19].

Multi-copy plasmids are the most suitable ones to be used for cloning, as they enable synthesis of large copies of insert (e.g., pUC19). Some of the desirable properties that a plasmid to be used for cloning should have are:

1. The molecular weight of the plasmid should be low for both efficient transformation of the host and DNA delivery.
2. Plasmids must be able to confer on its host a selectable phenotype or selectable advantage, which would distinguish the plasmid-bearing host from other organisms that do not contain the plasmid.
3. Plasmid vectors must have a site where large number of restriction enzyme recognition sites are located for ease of cloning the gene of interest. However, it is crucial to ensure that these sites are not located elsewhere on the plasmid. In short, it must contain a unique multiple cloning site or MCS region with wide choice of restriction enzymes.

3.5.1.1 Criteria for Choosing a Plasmid for Cloning

There are certain important properties of plasmids, as mentioned below, that need to be taken into consideration before choosing one for cloning:

Size of Insert

Every vector has its own capacity with respect to foreign DNA (insert) uptake, which is termed as the "insert size." If the insert size is beyond the capacity of the vector, possibility of defects arises in the efficiency of plasmid replication, especially in case of plasmids that have a high copy number. Certain plasmids that are capable of taking up very large insert size are generally used for constructing genomic libraries as discussed in the Chap. 2 [20].

Copy Number

One of the desirable properties of a cloning plasmid would be able to produce large number of plasmid copies. This yield of plasmid from its host depends on the number of copies the plasmid has inside the host system. This is termed as the "*plasmid copy number*." Copy number of a plasmid depends on control of replication initiation at *ori* sequence. For cloning vectors, a high copy number is always desirable to get a good plasmid yield [21]. However, sometimes, a very high copy number proves to be undesirable as it might unnecessarily burden the host replication machinery. Moreover, there are several low copy number plasmids that have been engineered for cloning some genes with distinct characteristics that might otherwise be toxic to the host system [22].

Cloning Sites (MCS region)

Cloning usually involves digesting the plasmid and the insert DNA with the same set of restriction endonucleases for generating compatible sticky ends followed by ligation of the generated compatible ends (as mentioned in Chap. 2). The plasmids that are generally used today consist of a multiple cloning site (MCS) or the polylinker region on their backbone in which all the restriction enzyme sites are concentrated. It is also ensured that these sequences are not found elsewhere on the plasmid backbone. For choosing a proper restriction site at the MCS for insertion of the foreign gene, it is important to confirm that the entire target gene sequence does not contain the recognition sequence of the restriction enzyme to be used. Any of the restriction enzymes that fulfill this criterion can hence be used for cloning [9].

3.5.1.2 Types of Cloning Plasmids

Although there are several types of cloning vectors, this chapter will describe the most popular ones and their variants.

(a) pBR Plasmids

pBR, named after their discoverers—Bolivar and Rodriguez, are a group of plasmids that are widely used for cloning [11, 19]. Since the plasmid has been completely sequenced, it allows confirmation of the presence of cloned insert by sequencing. Moreover, the exact length of every inserted fragment can be calculated. Some of the important features of these plasmids are provided below with an illustrative representation as shown in Fig. 3.2.

- pBR322 is the most common vector of the pBR series. The small size (4362 base pairs) of the plasmid allows easy entry inside the cells, thus increasing the chances of plasmid uptake by the host cells.

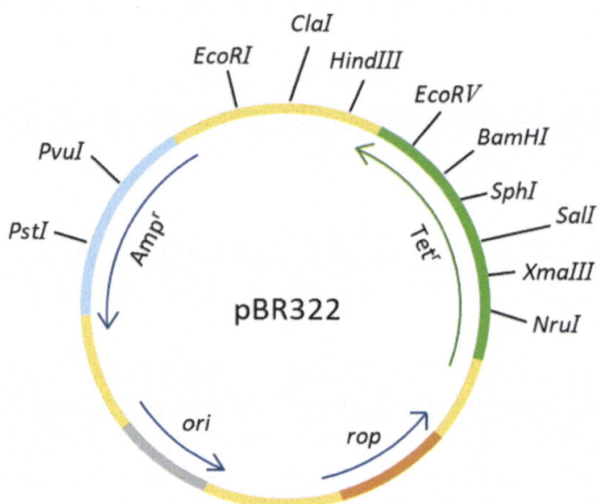

Fig. 3.2 Representative map for plasmid pBR322. Important features of pBR322 plasmid such as the *rop* gene for controlling the copy number of the plasmid, *ori,* which acts as an important recognition sequence for plasmid replication enzymes, tet^R gene that confers resistance against the antibiotic tetracycline, amp^R gene also called as *bla* (for β-lactamase) gene, which confers resistance against the antibiotic ampicillin. The MCS is distributed in the tet^R and amp^R regions and is represented by several restriction endonucleases

- The plasmid consists of *ori* from another plasmid pMB1, which is required for the replication of plasmid. This *ori* works only in *E. coli* system and not in any other organism.
- Presence of Rop protein (encoded by the *rop* gene within the plasmid) helps in the formation of a stable complex between RNA polymerase I and II. It is also responsible for controlling the plasmid copy number.
- The plasmid also contains genes for antibiotic resistance; the *bla* gene that encodes for β-lactamase provides resistance against β-lactam antibiotics such as ampicillin. The tet^R gene confers resistance against the antibiotic tetracycline.
- There are many sites for restriction endonucleases to act upon, within the antibiotic resistance genes. Whenever, any of these sites are used for cloning, there is insertional inactivation of the antibiotic resistance gene, aiding in the selection of recombinant plasmids [11]. Cloning elsewhere in the plasmid would be of lesser use as none of the antibiotic resistance gene would be inactivated; hence selection of recombinants would become difficult.
- The number of restriction enzyme sites in the pBR genome is around 40. Out of these, 11 sites are within the *tet*A (tetracycline resistant or tet^R) gene and six are contained within the amp^R gene. This gives flexibility in cloning an insert and choosing a suitable antibiotic resistance.
- There are three promoters P1, P2, and P3. Both P1 and P3 are for the β-lactamase gene. While P3 is the natural promoter, P1 has been engineered to create pBR322. P2 is located in the same region as P1 but on the opposite strand. It plays the role of transcription initiation in the direction of tet^R gene [23].
- pBR327 is one of the derivatives of pBR322 wherein there is deletion of nucleotides from 1427 to 2516 to reduce the size of this vector further. This deleted stretch was found to interfere with the insert expression in case of pBR322. The other features like amp^R and tet^R genes remain the same in both the vectors [24].

  Thus, pBR plasmids can take up an insert size of 4–5 kB which does not enable the cloning of very large inserts. The copy number of these plasmids is around 20–30 copies per cell. This copy number is lower as compared to many derivatives of pBR plasmids available today and that is the reason these derivatives are much more preferred. Some of the advantages of these plasmids are that since they are small in size, their transformation efficiency increases. Furthermore, there are also derivatives of these plasmids with increased copy number. Hence, a large amount of cloned DNA can be obtained. The two antibiotic resistance genes present with restriction enzyme sites for cloning increases the flexibility of cloning. Thus, pBR plasmids are more suitable for cloning or subcloning a moderate-sized insert but is not suitable for cloning large inserts, as the transformation efficiency and the copy number of plasmids are hampered if the cloned insert is beyond the capacity of the plasmid [25].

(b) pUC Plasmid Vectors

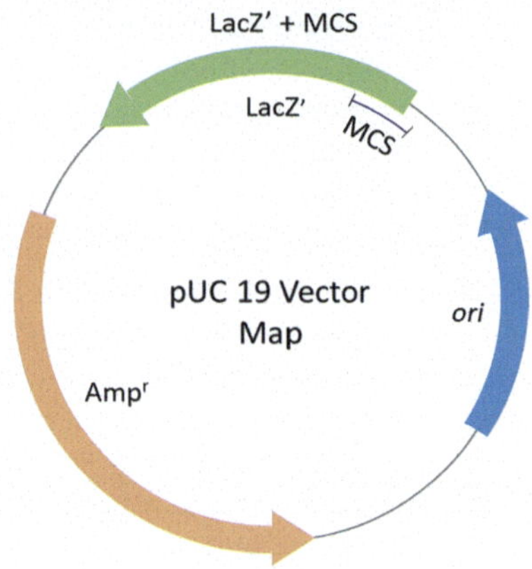

Fig. 3.3 Representative map for plasmid pUC—the plasmid consists of an *ori*, which is the origin of replication sequence recognized by the replication enzymes, amp^R that confers resistance against the antibiotic ampicillin and *lacZ* gene inside the MCS region for cloning gene of interest and selection of clones as described in the text above

These plasmids were prepared for the first time at the University of California from where they derive their name, pUC (Fig. 3.3) [26]. They are considered to be the derivatives of pBR322, which were engineered further to fulfill certain special cloning requirements as mentioned below:

- They are smaller in size (2686 bp) but have the capacity to carry a larger insert size up to 10 kb.
- They generally have a high copy number (500–700 copies per cell) as compared to pBR322 plasmids (20 copies per cell). Thus, they can produce large number of copies of the cloned fragment of DNA inside the host.
- They contain the selectable marker gene for ampicillin resistance and *ori* sequence both derived from pBR322.
- The vector contains the *lacZ* gene, which is derived from *E. coli.* The *lacZ* region encodes for the lac operon that produces the beta-galactosidase enzyme for lactose metabolism. A short stretch of *lacZ* gene (up to 146 amino acids) is present within the MCS region where the foreign gene is introduced. Post cloning of the foreign gene, when the plasmids are transformed inside a suitable host, functional beta-galactosidase enzyme cannot be produced due to the disruption of *lacZ* gene. On the other hand, failure in the cloning procedure would lead to synthesis of the functional enzyme [27]. This method helps in the selection of recombinant host cells through a method called the *blue white screening* where a chromogenic substrate X-gal is added to the agar plate whose hydrolysis by the functional enzyme gives blue coloration [9, 27]. This technique has been discussed elaborately in Chap. 2.

  Thus, a pUC series plasmid is used when a large insert size needs to be cloned but at the same time a high yield of plasmid is desired without compromising the transformation efficiency.

(c) Bacteriophage vectors

Another option of cloning is through bacteriophages (M13 and λ phage) [28, 29]. M13 is one of the filamentous bacteriophages that belongs to the family of Inoviruses [30]. They have a thread-like capsid architecture that packages a circular single-stranded DNA (ssDNA) genome. M13 has been engineered by deleting a part of its genome to make space for the inserts and thus restricting its use only with smaller size inserts. However, it provides a lot of flexibility in terms of position and number of inserts (within its size limit), it finds immense use as a scaffold in creating nanostructures [31].

In λ phage, since there is a limit to the length of DNA to be packaged (circa 53 kb), it has been engineered by deleting non-essential lysogeny gene so that larger target DNA can be accommodated [32]. However, it is important to note that, although ~40% of the wild-type λ phage genome is redundant for its survival, it cannot maintain its integrity if more than 25% of its wild-type genome is modified. Therefore, only foreign genes within certain size limit can be inserted [33].

λ phage vectors can be classified into two types:

(a) Insertion vectors
These are the basic types of lambda cloning vectors and considering the size constraints, can accommodate between 5 and 11 kb of foreign DNA inserts. However, to prevent non-specific cleavage, they are modified through mutagenesis to contain a unique restriction site where the target gene is cloned. However, the major limitation of this vector is its small insert size capacity.

(b) Replacement vectors
To circumvent the issue of small insert size, a modified version of the λ vector was developed where 40% of the vector genome is made up of artificially introduced "*stuffer sequence*" leaving rest 60% of the wild-type plasmid DNA intact. This method enables easy insertion of foreign DNA fragment by replacement of the stuffer element. Therefore, larger gene inserts (~10–20 kb) can be introduced through this method [11].

However, it is important to note that there is also a minimum size limit for inserts (~5 kb), since too tiny inserts do not get properly inserted and the cloning often remains unsuccessful. This unique characteristic has been harnessed for selecting recombinant plasmids from the ones that are empty (without inserts) [9, 33]; the latter might not effectively propagate.

(c) Cosmid vectors
Cosmids are types of hybrid plasmid vectors that contain a section of bacteriophage λ DNA with cohesive end site (*cos*), which has elements to incorporate DNA into λ phage vector. Cosmids are usually chosen to clone large gene inserts (~28–45 Kb) [33]. These vectors are then packaged into viral particles that subsequently infect *E. coli* where the cosmid vector self-replicates using the ColE1 origin of replication. An illustration of the entire process is shown in Fig. 3.4. This process can very successfully incorporate large foreign DNA into host system, which is otherwise not possible through transformation [11].

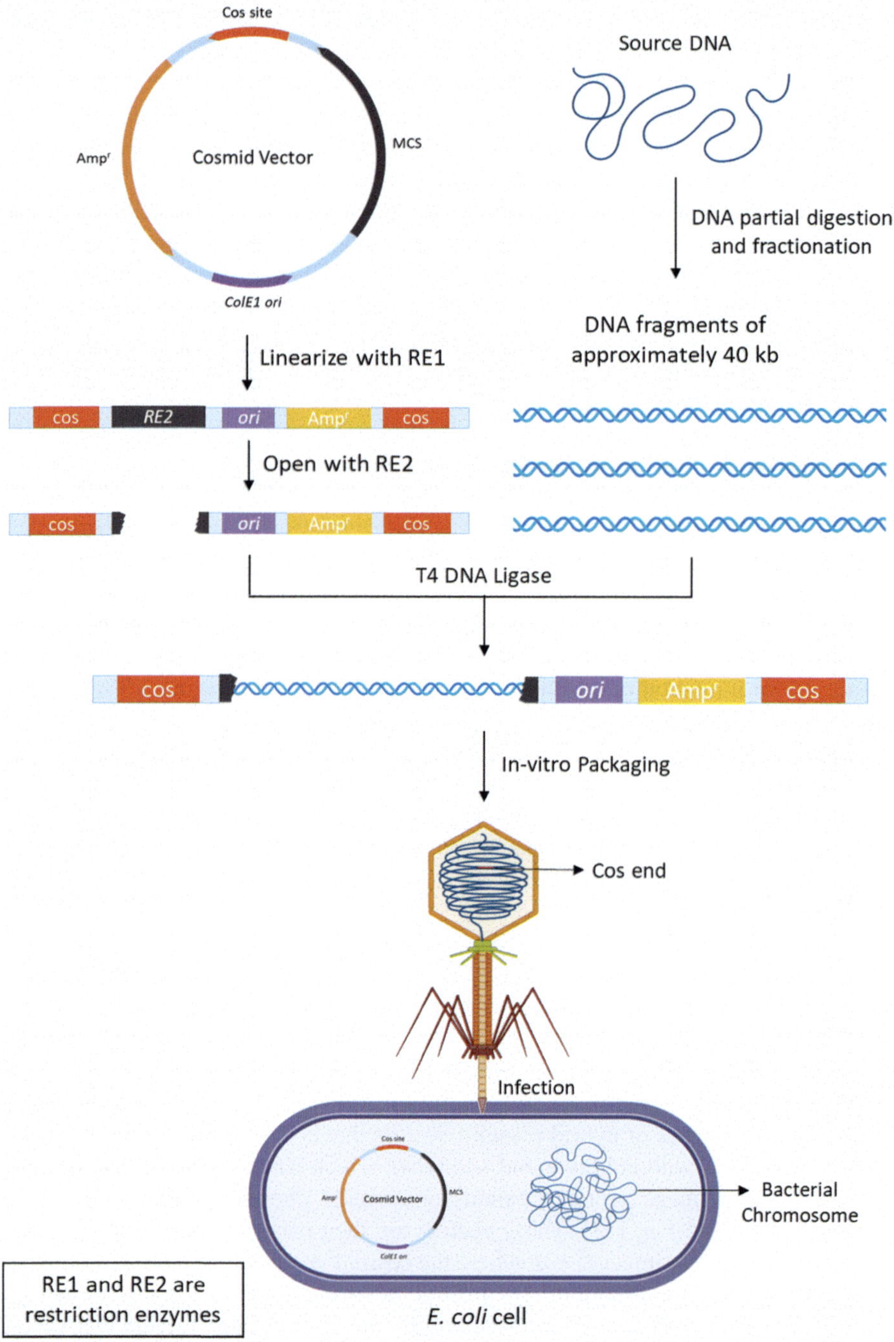

Fig. 3.4 Schematic representation of important features of a cosmid vector: The cosmid vector comprises ColE1origin for plasmid replication, polylinker or MCS region for cloning foreign DNA, Amp^R selectable marker for selection of recombinants, and Cos site for packaging the DNA into

(d) Bacterial artificial chromosome
When there is a requirement of a huge gene to be cloned and expressed, a plasmid specialized for this purpose becomes necessary. Bacterial artificial chromosomes (BACs) as shown in Fig. 3.5 are designed in a way to propagate ~1 MB-sized inserts in *E. coli*. However, the caveat is that the plasmid is maintained at a very low copy (1 copy per cell) within the host system so that it does not pose burden on the host replication machinery [33]. PAC is another similar carrier that is based on the P1 bacteriophage [34]. Although it can carry >100 kb gene insert, it has not gained popularity as BAC. These vectors are introduced within the host system through electroporation, i.e., high voltage pulse [20].
BACs have been used extensively for sequencing the human genome for the *Human Genome Project*. The F-plasmid, which is responsible for the process of conjugation between bacteria, has the ability to integrate itself into the bacterial chromosomal DNA. Thus, when a large fragment of foreign DNA is cloned into BAC, it gets incorporated into the bacterial chromosome. As the bacterial chromosome replicates, it subsequently leads to generation of copies of the insert DNA. The reason for BACs popularity is the stability of the insert without loss of any information. This is due to the fact that the fragment is maintained as a part of the genome rather than an extrachromosomal entity. BACs have been extensively used for cloning large fragments from the mouse and the human genome to prepare genomic libraries.
The first BAC vector that was constructed was pBAC108L. Derivatives of pBAC108L have been constructed, namely, pBeloBAC11 and pECBAC1. In these vectors, the cloning site has been replaced with *lacZ* gene with multiple cloning sites [35].

Table 3.1 summarizes the important features of the cloning vectors with respect to three important aspects—insert size, copy number, and applications.

3.5.2 Expression Plasmid Vectors

Post cloning, the foreign gene needs to be expressed for production of proteins and hence has to be cloned into an expression plasmid vector. Unlike cloning plasmids, these expression vectors contain appropriate regulatory sequences for transcription and translation of the gene of interest. Since the main aim is to express the gene product rather than obtaining large copies of the cloned gene, a common expression vector should comprise the following:

Fig. 3.4 (continued) phage head. The steps including introduction of foreign gene, packaging into viral particles, and infection of *E. coli* cells have been depicted as described in the text

1. Regulatory sequences to transcribe and translate the gene of interest.
2. Origin of replication.
3. Multiple Cloning Site (MCS).
4. Selectable marker for selection.
5. Tag for the separation and purification of expressed protein using affinity chromatography.

Thus, expression vectors are provided with the regulatory elements like the promoters for expressing the insert DNA. The promoters are usually under the control of an inducer, a molecule that when added to the culture system helps in expression of the insert DNA. In the absence of an inducer, the gene would fail to express. Such promoters that need an inducer for gene expressions are termed as "inducible promoters" and the ones that do not are called "constitutive promoters" [36]. The use of inducible promoters is recommended because constitutive expression of the insert DNA might interfere with bacterial growth as well as the protein product might damage the host cell due to toxic effects [37]. Common example of an

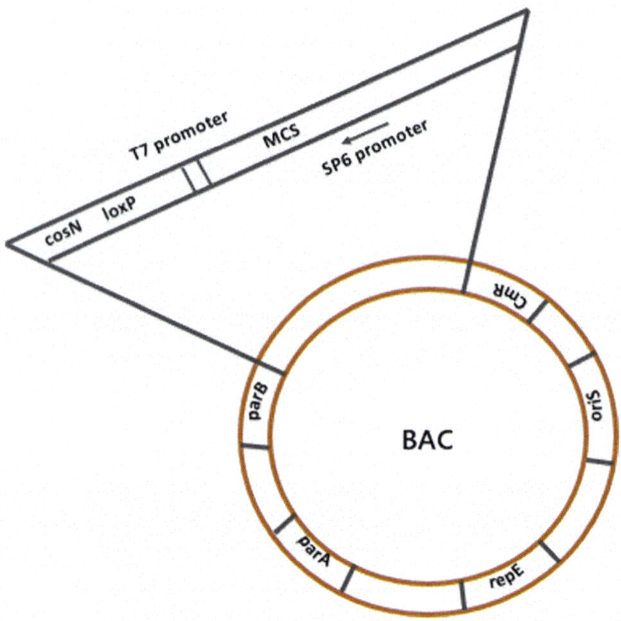

Fig. 3.5 Schematic representation of BAC vector derived from F plasmid. The unidirectional replication of the F plasmid is maintained by the repE and oriS elements. parA and parB elements help in maintaining the copy number of plasmid up to 1 or 2 copies per genome; the CmR gene codes for chloramphenicol resistance that acts as a selectable marker; the cosN and loxP sites help in recombination; MCS consists of the HindIII and BamH1 sites for cloning the gene of interest; the two promoters T7 and SP6 that flank the MCS also help in making RNA probes apart from gene expression [11]

Table 3.1 Important attributes of the cloning vectors

Plasmid	Insert size	Copy number	Applications
pBR322	4–5 kB	Low (~20)	Cloning, subcloning, highest transformation efficiency due to small size
pUC 19	~10 kB	High (~500–700)	Cloning, subcloning, moderate efficiency of transformation due to large insert size capacity and high copy number
λ replacement vectors	~10–20 kB	Vector with insert packaged into viral particles	Cloning large fragments of insert
Cosmids	~20–50 kB	Vector with insert packaged into viral particles	Enables the cloning of very large inserts
BAC	~1 MB	Very low (~1)	Used for human genome sequencing and for preparing genomic libraries

inducer is isopropyl β-D-1-thiogalactopyranoside (IPTG), the mechanism of which has been described above.

In addition to the above features, an expression vector also contains different tags. These tags are oriented in such a manner inside the vector that, when the protein of interest is produced, the tag is attached and expressed as a fusion extension—often a protein. Thus, the tag and the insert DNA are next to each other in the same orientation inside the vector. The tag can be attached either at the N- or C-terminus of the expressed protein. These tags are required for purification of the expressed protein through affinity chromatography using an appropriate resin that has an affinity toward the tag. Some of the commonly used tags include His_6, Maltose Binding protein (MBP), and Glutathione S-transferase (GST) tags [38, 39].

A few important bacterial expression plasmids and their applications are discussed below:

(a) pET Vectors

One of the most powerful systems developed for the purpose of cloning of genes of interest and expression of recombinant proteins in *E. coli* is the pET vector system [40]. This system of vectors is basically a derivative of the pBR322 series of plasmids. It has been engineered to make the best use of T7 bacteriophage and its features. The target gene that is cloned in this system of vectors is under the transcriptional and translational control of bacteriophage T7 RNA polymerase expression signals. The greatest advantage of using T7 RNA polymerase is its extremely high specificity for T7 promoter. Furthermore, the polymerase possesses a high level of activity and efficiency of translation through the translation initiation signals. Since the T7 promoter is not recognized by the host RNA Polymerase, the basal expression of the insert gene can be avoided. The expression is further controlled by conveniently adding an inducer (IPTG) when the bacterial culture reaches its log phase of growth [40].

Two different variants of the pET system are available—transcriptional and translational vectors. The genes that carry their own ribosome binding site and the ATG start codon are often cloned in the *transcriptional vectors*. These types of vectors are usually used for the cloning and expression of prokaryotic genes. The vectors that carry a ribosome binding site from phage T7 are termed as *translational vectors*. Generally, these vectors are used for cloning and expression of eukaryotic genes [41].

(b) pBAD system.
This is another example of a controlled protein expression system that is based on the *ara*BAD operon that is responsible for controlling the arabinose metabolism in *E. coli*. This operon is induced on addition of L-arabinose. Thus, in the pBAD vector, the gene of interest is cloned downstream of the araBAD promoter so that its expression can be induced by the addition of L-arabinose in the culture medium [16].

Some of the advantages of using pBAD vectors are:

1. *Tight control on expression*: The basal protein expression in the absence of the inducer is extremely low in this expression system as compared to that of pET plasmids. Therefore, it might be useful for expression of proteins that are toxic to *E. coli* systems.
2. *Strongly inducible promoter*: The expression of the insert increases manifold as soon as the inducer L-arabinose is added to the culture medium.
3. *Inexpensive protein production*: Since the inducer for these plasmids that is L-arabinose is inexpensive, large-scale protein production is possible in a very cost-effective way.

3.5.3 Affinity Tags for Protein Purification in Expression Plasmids

The pET expression vectors provide an option for tagging the protein with tags such as His_6, MBP, or GST. Tags are of importance, as they enable labeling of the protein of interest so that it can be separated from rest of the cellular proteins through affinity purification. To separate the tag from the protein after expression and purification, there are enzyme cleavage sites present between the protein gene cloned and the tag sequence. Enzymes such as TEV (Tobacco Etch Virus) protease are often used for optional tag separation [42]. The amino acid recognition sequence for the TEV protease is EDLYFQS, whose codon is introduced between the tag and the protein sequence. Apart from TEV, there are other protease recognition sites, viz., Factor Xa, Thrombin, and PreScission [43].

Certain vectors in the pET system can be modified such that they have two different tags on both sides of the cloned insert, e.g., His_6 tag and MBP tag. This is of importance when the larger tag such as MBP has to be cleaved off from the protein after first round of purification. Thus, pure protein can be obtained in the second round of purification using the His_6 tag, from a mixture that consists of cleaved MBP and his-tagged protein. A detailed manual can be obtained from [44].

There are two types of tags that can be engineered into plasmids.

1. Peptide tags
   - Since these tags are small in size (His_6 or S-tag), they are less likely to interfere with the biological activity of the cloned protein and hence need not be separated from the expressed protein.
   - Plasmids with such peptide tags are often preferred for cloning and expression when the protein is to be crystallized for determining its three-dimensional structure. It is always desirable to have the target protein to be crystallized in its pure form without any other interfering protein from tag.
   - Some of the pET series of vectors (e.g., pET-14b-19b) encode His_6 tag that expresses either at the N- or C-terminal end of the recombinant protein.
   - Expression of the encoded genes is usually under the control of T7 (pET-14b) or inducible T7lac promoters (pET-15b-19b).
   - Some of the pET vectors have more than one tag. For example, pET30b (+) encodes S tag and His_6 tag; pET41b(+) encodes S tag and His_6 tag as well as GST tag.
   - There is usually a protease cleavage site between tag and the protein of interest for optional tag removal.
   - The target protein is later purified using immobilized metal affinity chromatography (IMAC) technique discussed in Chap. 6 of this book.
2. Fusion protein tags
   - These tags are bigger in size and are often independent proteins, e.g., Maltose Binding Protein (MBP) and Glutathione S-Transferase (GST).
   - Plasmids with these tags would often have an enzymatic recognition site between the insert and protein. These tags need to be separated as they may interfere with the downstream processing of the protein.
   - The plasmids with these tags are mostly used when the protein is known to be insoluble or unstable and thus forms inclusion bodies [45]. Purification of proteins from inclusion bodies is discussed in detail in Chap. 10. Such unstable proteins are often stabilized by tags such as MBP and GST, the exact mechanism of which is largely unknown. However, these tags are thought to have intrinsic chaperone activity that might aid in folding and stabilization of the insert protein [45].
   - The pMAL and pGEX series of plasmids have MBP and GST tags, respectively. Amylose and GST resins are available for purifying the inserts cloned in pMAL and pGEX series of vectors [38, 39].

3.5.3.1 pGEX Plasmids with GST Tag for Protein Expression

Important features of pGEX plasmids are provided below:

- The protein gene cloned into this plasmid system is tagged with GST and expressed as a fusion protein.

- The expression of the fusion protein is under the control of the tac promoter (Ptac), which is an IPTG-inducible promoter that maintains a tighter control on its recombinant protein expression.
  GST-tagged proteins are also found to stabilize and solubilize certain unstable and insoluble proteins and thus can be an option for cloning and later purifying such difficult proteins
- Commercially there are 13 pGEX vectors available, the details will be available in the Sigma-Aldrich pGEX vector manual [46]. These plasmid vectors subtly differ from each other in terms of choices of restriction as well as protease cleavage sites. Nine of these plasmid vectors have recognition sites for different proteases (e.g., Thrombin, PreScission, Factor Xa) between the protein gene and the GST tag [43]. This site enables separation of the GST-tag from the protein after purification.

A representative pGEX vector and its fused gene product are shown in Fig.3.6.

3.5.3.2 pMAL Plasmids with MBP Tag for Protein Expression

Important features of pMAL plasmids are provided below:

- The site for cloning the insert is located downstream of the gene encoding the MBP tag (*malE* gene). Thus, induction of protein expression, which is majorly done using IPTG, produces the fusion protein (tag + protein of interest).
- The yield of the protein obtained by using pMAL vectors is very high and this is due to the use of a very strong promoter Ptac**,** which is an inducible promoter and the inducer used is IPTG [47]. The obtained protein is then purified using amylose resin that has affinity toward MBP tag.
- Fusion of protein gene with MBP is known to stabilize the protein and is also found to solubilize certain insoluble proteins. Thus, cloning some of the *difficult-to-purify* proteins in these plasmids can be of help in certain cases [48].
- Similar to pGEX plasmid vectors, several variants of pMAL vectors are commercially available with a variety of different restriction sites and protease recognition sites such as thrombin and Factor Xa. A detailed description of the types of the plasmid vectors with their minute details will be obtained at New England BioLabs pMal vectors manual [49] (Fig. 3.7).

Furthermore, proteins are cloned with tags sometimes to perform protein-protein interaction studies through pull-down assays. GST-, MBP-, and His_6 pull down assays are routinely used to identify interacting partners of proteins in the biochemical pathways [50].

3.5.3.3 Duet Vectors

Duet vectors are a type of expression vectors recently developed for the co-expression of two genes together. This is mostly desirable when the interaction between the two proteins needs to be studied and characterized. The host system used is *E. coli* and the origin of replication used is mostly from the ColE1 plasmids.

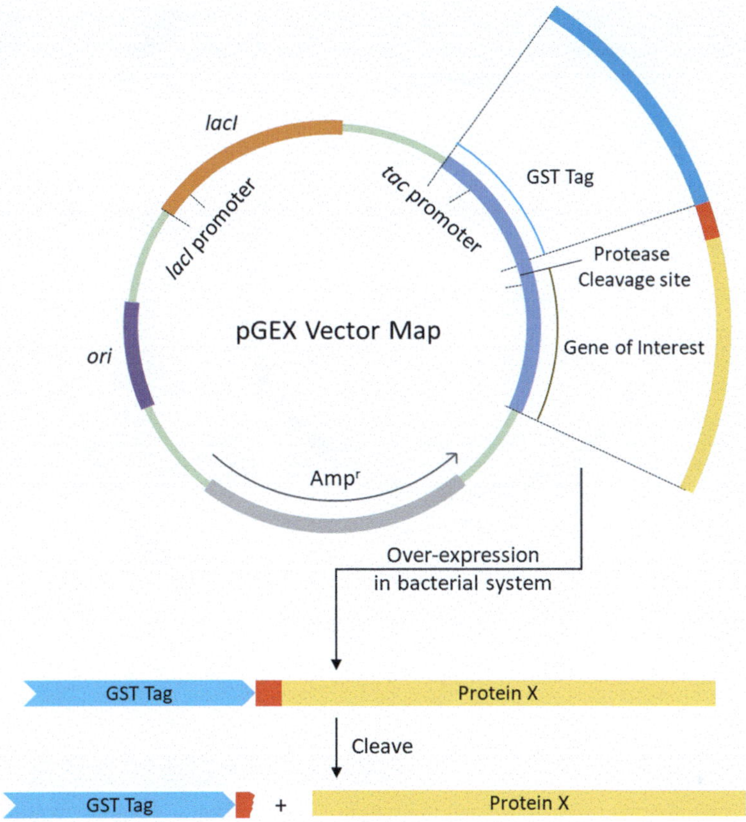

Fig. 3.6 Map of pGEX plasmid vector. The figure depicts the important features of a pGEX expression plasmid that produces GST-fused protein of interest. The labeled figure is self-explanatory

The copy number of Duet vector is up to 40 copies per host cell. The selectable marker gene is Ampicillin for selection of recombinants. These vectors contain two MCS with different tags (e.g., His tag and S tag). The commercially available S-tag system [51] comprises N-terminal 15 residues of the S-peptide and immobilized S-protein on agarose beads for purification of the protein of interest. Like most of the tags, S-tag vectors also encode a protease cleavage site for removal of the tag if desired by the end-user. T7 promoter controls the expression of the cloned genes. Some Duet vectors contain two different inducible promoters to specifically express one protein at a time [52].

3.5.3.4 Cell-Free Expression Systems

A very efficient and fast method of recombinant protein expression has been made possible by the development of cell-free protein production system [53]. In this in vitro procedure, cell lysate of *E. coli* comprising the essential cellular components

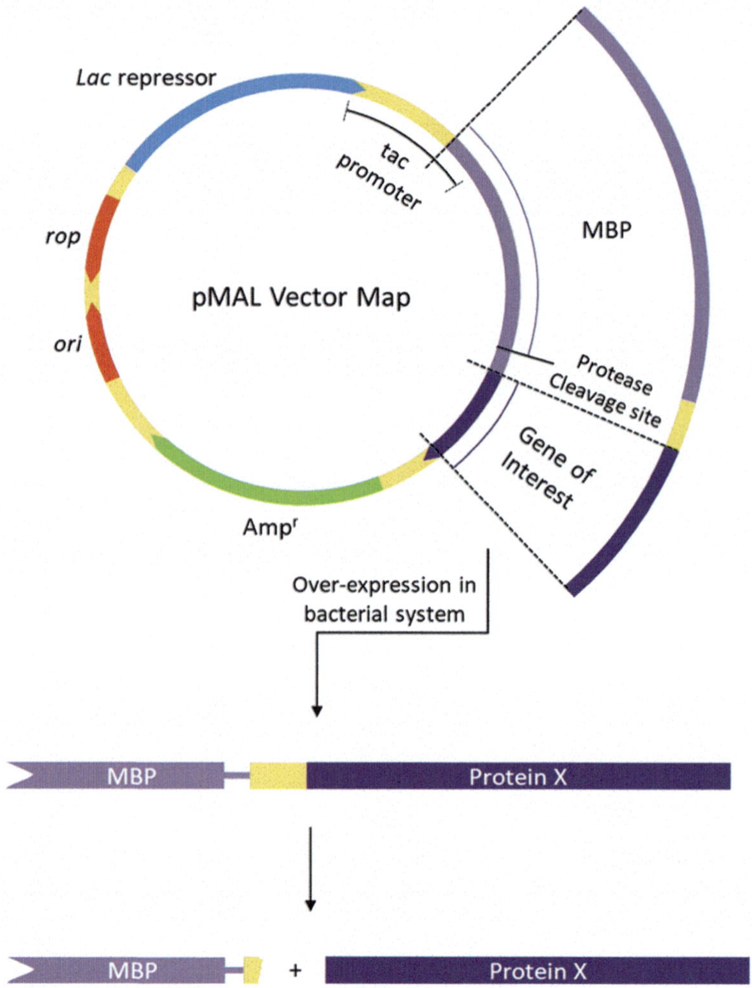

Fig. 3.7 Map of pMAL plasmid vector. The figure depicts the important features of a pMAL expression plasmid that produces MBP-fused protein of interest. The labeled figure is self-explanatory

for transcription and translation are necessary for production of the target protein. This method bypasses the lengthy cell culture step, thus significantly speeding up recombinant protein expression and production. Regular plasmid vectors as well as specially designed ones are used in this system. The major limitation of this system is that it is quite expensive.

3.5.3.5 Other Expression Systems

Apart from the abovementioned expression plasmids, there are other specialized mammalian vectors for primarily expressing mammalian proteins such as adenoviral

Problem	Reason	Solution
1. No protein obtained after using the pET series of expression vectors	The protein might be unstable and insoluble	The insert must be cloned in pMAL or pGEX series of vectors as the MBP or GST tags help in stabilizing the expressed protein
2. Low amount of plasmid obtained after cloning the insert in a cloning plasmid	The copy number of the plasmid is low	Use a high copy number plasmid (e.g., pUC series of plasmids)
3. Failure to clone the insert into the vector	Insert size is too large for the plasmid	Use a plasmid with a high capacity for insert (e.g., cosmid vectors)
4. Tag not separated from the protein	a. Absence of TEV enzyme recognition site between the tag and the protein sequence or the recognition site is faulty; confirm through DNA sequencing b. The tag might stick to the protein	a. Introduce a TEV enzyme recognition site between the insert and the tag and confirm through sequencing b. Use a different protein tag that would help stabilize the cloned protein and simultaneously would easily get separated from the protein. (Use GST tag instead of MBP which might easily separate post TEV cleavage)

systems, retroviral vectors, the pSV and pCMV series of vectors as well as the baculovirus [54]. All of these systems generally adopt promoters for cytomegalovirus and SV40 for expression of the gene of interest [54]. They all have their advantages and disadvantages and are usually used in mammalian host systems or cell lines, which is beyond the scope of this book.

3.5.4 Conclusion

With the exponential surge in rDNA technology research, introduction of newer and better plasmid vectors for optimal cloning of foreign genes and subsequent expression of target proteins has also gained importance. The availability of this wide array of vector-host combination has allowed researchers to hand-pick plasmids that best suit their specific needs and purpose. However, wider spectrum of choices also comes with the necessity to have precise knowledge of these plasmids, their pros and cons as well as their effective combination with the most suitable variant of the host system. This chapter, therefore, provides a detailed review on the types of plasmids to be used in the bacterial host system along with routine troubleshooting cues.

Troubleshooting:

Acknowledgments The authors thank Advanced Centre for Treatment, Research and Education in Cancer (ACTREC) for providing necessary infrastructure and resources for successful completion of the chapter. The authors acknowledge Ms. Chanda Baisane and Mrs Snehal Pandav Mudrale, Bose Lab, ACTREC for critical inputs and formatting of the manuscript.

Problems

Multiple Choice Questions

1. The site on an expression plasmid where transcription factors bind is known as:
   (a) *Ori*
   (b) Promoter
   (c) Polylinker
   (d) *rop*
2. The most suitable vector for constructing genomic libraries would be:
   (a) Cosmids
   (b) λ replacement vectors
   (c) BAC
   (d) pET vectors
3. The protein that binds to the pLac promoter to prevent the expression of lacZ gene is known as:
   (a) Lactose
   (b) IPTG
   (c) X- gal
   (d) Repressor

Subjective Questions

1. A gene encoding protein X was cloned into a plasmid containing 6×-His tag. After induction of gene expression, the protein was found to be expressed. When the protein was purified using an appropriate resin, it was found to be unstable and insoluble. Suggest a way in which such a protein can be stabilized and purified.
2. A gene 50–60 kB in size needs to be cloned in an appropriate vector for further study. Which cloning vector would be suitable for cloning such a moderately large fragment without loss of the gene and at the same time ensuring efficient host transformation?

References

1. Schacherer J. Beyond the simplicity of Mendelian inheritance. C R Biol. 2016;339(7–8):284–8. (1768-3238 (Electronic))
2. Brondyk WH. Selecting an appropriate method for expressing a recombinant protein. Methods Enzymol. 2009;463:131–47. (1557-7988 (Electronic))
3. Johnson IS. Human insulin from recombinant DNA technology. Science. 1983;219 (4585):632–7. (0036-8075 (Print))
4. Von Fange T, et al. Clinical inquiries: can recombinant growth hormone effectively treat idiopathic short stature? J Fam Pract. 2008;57(9):611–2. (1533-7294 (Electronic))
5. Manco-Johnson MJ. Advances in the care and treatment of children with hemophilia. Semin Thromb Hemost. 2003;29(6):585–94. (1878-1926 (Electronic))
6. Koçoğlu T, Yalçinkaya T. Recombinant DNA technology. Mikrobiyol Bul. 1992;26 (2):177–88. (0374-9096 (Print))
7. Mishra V. Affinity tags for protein purification. Curr Protein Pept Sci. 2020;21(8):821–30. (1875-5550 (Electronic))
8. Kimple ME, Brill AL, Pasker RL. Overview of affinity tags for protein purification. Curr Protoc Protein Sci. 2013;73:9.9.1–9.9.23. (1934-3663 (Electronic))
9. Brown TA. Gene cloning and DNA analysis: an introduction. 6th ed. Hoboken, NJ: Wiley-Blackwell; 2010.
10. Lodish H, Berk A, Zipursky SL. Molecular cell biology. 4th ed. New York: W.H. Freeman; 2000.
11. Primrose SB, Twyman R. Principles of gene manipulation and genomics. 7th ed. Hoboken, NJ: Wiley-Blackwell; 2006.
12. EMBL. Cloning Per Strategy. https://www.embl.de/pepcore/pepcore_services/cloning/pcr_strategy/index.html.
13. Tokmakov AA, et al. Multiple post-translational modifications affect heterologous protein synthesis. J Biol Chem. 2012;287(32):27106–16.
14. del Solar G, et al. Replication and control of circular bacterial plasmids. Microbiol Molecul Biol Rev MMBR. 1998;62(2):434–64.
15. Amann E, Brosius J, Ptashne M. Vectors bearing a hybrid trp-lac promoter useful for regulated expression of cloned genes in *Escherichia coli*. Gene. 1983;25(2–3):167–78. (0378-1119 (Print))
16. Guzman LM, et al. Tight regulation, modulation, and high-level expression by vectors containing the arabinose PBAD promoter. J Bacteriol. 1995;177(14):4121–30. (0021-9193 (Print))
17. Berg JM, Tymoczko JL, Stryer L. Biochemistry. 7th ed. New York: WH Freeman; 2012.
18. Thieman WJ, Palladino MA. Introduction to biotechnology. San Francisco: Pearson/Benjamin Cummings; 2004.
19. Balbás P, Soberón X, et al. Plasmid vector pBR322 and its special-purpose derivatives--a review. Gene. 1986;50(1–3):3–40. (0378-1119 (Print))
20. Bajpai B. High capacity vectors. Adv Biotechnol. 2013;86048301:1–10.
21. Wood WN, et al. Enhancing yields of low and single copy number plasmid DNAs from *Escherichia coli* cells. J Microbiol Methods. 2017;133:46–51.
22. Dumon-Seignovert L, Cariot G, Vuillard L. The toxicity of recombinant proteins in *Escherichia coli*: a comparison of overexpression in BL21(DE3), C41(DE3), and C43(DE3). Protein Expr Purif. 2004;37(1):203–6. (1046-5928 (Print))
23. Balbas P, Soberon X, Bolivar F, Rodriguez RL. The plasmid, pBR322. 1988. PMID: 3061523; https://doi.org/10.1016/b978-0-409-90042-2.50007-6.
24. Potapova IA, Repin VE, Shchelkunov SN. Maintenance stability of the pBR322 and pBR327 vector plasmids in Escherichia coli cells during multiple passages. 1985. PMID: 3886482.
25. Chan V, et al. The effect of increasing plasmid size on transformation efficiency in *Escherichia coli*. J Exp Microbiol Immunol. 2002;2. (0378-1119 (Print))

26. Vieira J, Messing J. The pUC plasmids, an M13mp7-derived system for insertion mutagenesis and sequencing with synthetic universal primers. Gene. 1982;19(3):259–68. (0378-1119 (Print))
27. Maas S. Efficient and rapid procedure for blue-white screening of recombinant bacterial clones. Biotechniques. 1999;27(6):1126–8. (0736-6205 (Print))
28. Chauthaiwale VM, Therwath A, Deshpande VV. Bacteriophage lambda as a cloning vector. Microbiol Rev. 1992;56(4):577–91.
29. Messing J. Cloning in M13 phage or how to use biology at its best. Gene. 1991;100:3–12. (0378-1119 (Print))
30. van Regenmortel M, Mahy B. Encyclopedia of virology. New York: Academic Press Inc.; 2008. p. 3192.
31. Ngo-Duc T-T, et al. M13 bacteriophage spheroids as scaffolds for directed synthesis of spiky gold nanostructures. Nanoscale. 2018;10(27):13055–63.
32. Glick BR, Pasternak JJ, Patten CL. Molecular biotechnology: principles and applications of recombinant DNA. Washington, DC: ASM Press; 2010.
33. Preston A. Methods in molecular biology. In: *E. coli* plasmid vectors. Totowa, NJ: Humana Press.
34. Sternberg N. Bacteriophage P1 cloning system for the isolation, amplification, and recovery of DNA fragments as large as 100 kilobase pairs. Proc Natl Acad Sci U S A. 1990;87(1):103–7. (0027-8424 (Print))
35. Coren JS. Retrofitting the BAC cloning vector pBeloBAC11 by the insertion of a mutant loxP site. BMC Res Notes. 2017;10(1):344.
36. Anilionyte O, et al. Short, auto-inducible promoters for well-controlled protein expression in *Escherichia coli*. Appl Microbiol Biotechnol. 2018;102(16):7007–15. (1432-0614 (Electronic))
37. Rosano GL, Ceccarelli EA. Recombinant protein expression in *Escherichia coli*: advances and challenges. Front Microbiol. 2014;5:172.
38. Duong-Ly KC, Gabelli SB. Affinity purification of a recombinant protein expressed as a fusion with the maltose-binding protein (MBP) tag. Methods Enzymol. 2015;559:17–26.
39. Harper S, Speicher DW. Purification of proteins fused to glutathione S-transferase. Methods Molecul Biol (Clifton, NJ). 2011;681:259–80.
40. Mierendorf RC, et al. Expression and purification of recombinant proteins using the pET system. Methods Mol Med. 1998;13:257–92. (1543-1894 (Print))
41. Shilling PJ, et al. Improved designs for pET expression plasmids increase protein production. Commun Biol. 2020;3:214.
42. Raran-Kurussi S, et al. Removal of affinity tags with TEV protease. Methods Mol Biol. 2017;1586:221–30. (1940-6029 (Electronic))
43. Waugh DS. An overview of enzymatic reagents for the removal of affinity tags. Protein Expr Purif. 2011;80(2):283–93.
44. Mierendorf RC, Morris BB, Hammer B, Novy RE. Expression and purification of recombinant proteins using the pET system. 1998. PMID: 21390849; https://doi.org/10.1385/0-89603-485-2:257.
45. Costa S, et al. Fusion tags for protein solubility, purification and immunogenicity in Escherichia coli: the novel Fh8 system. Front Microbiol. 2014;5:63.
46. Aatsinki JT, Rajaniemi HJ. An alternative use of basic pGEX vectors for producing both N- and C-terminal fusion proteins for production and affinity purification of antibodies. 2005. PMID: 15766870; https://doi.org/10.1016/j.pep.2004.11.012.
47. Branco LM, et al. Bacterial-based systems for expression and purification of recombinant Lassa virus proteins of immunological relevance. Virol J. 2008;5:74.
48. Lee SB, et al. Production of bioactive chicken follistatin315 in *Escherichia coli*. Appl Microbiol Biotechnol. 2014;98(24):10041–51. (1432-0614)
49. Labs NEB. pMAL™ protein fusion and purification system. Ipswich, MA: New England Biolabs Inc.

50. Yuan W, et al. Effect of different tags on pulldown assays implemented by LMO_2 fusion protein. Sheng Wu Gong Cheng Xue Bao. 2008;24(5):887–91. (1000-3061 (Print))
51. Raines RT, et al. The S. Tag fusion system for protein purification. Methods Enzymol. 2000;326:362–76. (0076-6879 (Print))
52. Gauttam R, Mukhopadhyay A, Singer SW. Construction of a novel dual-inducible duet-expression system for gene (over)expression in Pseudomonas putida. Plasmid. 2020;110:102514.
53. Gregorio NE, Levine MZ, Oza JP. A user's guide to cell-free protein synthesis. Methods Protoc. 2019;2(1):24.
54. Makrides SC. Vectors for gene expression in mammalian cells. New Comprehen Biochem. 2003;38:9–26.

Transformation and Protein Expression

4

Shubham Deshmukh, Rucha Kulkarni, and Kakoli Bose

Abstract

Transformation is an important step in recombinant DNA technology that allows transfer of genetic material inside cells. Often, this transfer accompanies a change in trait of the cells depending upon the genetic material used. Together, the transfer of genetic material and change in trait of cells is defined as transformation. Since its discovery, this technique is credited with countless profound findings in biology and holds an important position in a biologist's toolbox for manipulating DNA and cells. In this chapter, fundamental information essential for in vitro transformation, together with a series of principles and protocols that are routinely used in transforming bacterial cells, are discussed. Dedicated sections have been provided to preparation of competent cells, in vitro cellular transformation methods, and posttranslational protein modification in bacterial expression systems. Further, acknowledging the popularity of bacteria as "protein-production factories," special sections have been devoted to using different types of bacteria and optimizing gene expression in them. Additionally, discussion on approaches to troubleshoot difficulties and standardize experimental protocols provided comprehensiveness to the chapter. Several fundamental problems related to this topic have been discussed at the end of the chapter for the readers to further enhance their understanding on the topic by actively involving themselves in an experience that resembles the routine experimental protocols pertaining to recombinant protein expression.

S. Deshmukh · R. Kulkarni · K. Bose (✉)
Integrated Biophysics and Structural Biology Lab, Advanced Centre for Treatment, Research and Education in Cancer (ACTREC), Tata Memorial Centre, Mumbai, India

Homi Bhabha National Institute, BARC Training School Complex, Mumbai, India
e-mail: kbose@actrec.gov.in

K. Bose (ed.), *Textbook on Cloning, Expression and Purification of Recombinant Proteins*, https://doi.org/10.1007/978-981-16-4987-5_4

Keywords

Transformation · Posttranslational modification · Competent cells · Protein expression · Bacterial expression systems

4.1 Introduction

A groundbreaking discovery that revolutionized the field of molecular biology is the ability to modify the genetic material of cells. Genetic alteration requires that the recipient cells allow entry to foreign genetic material, its successful incorporation, and stable expression so as to bring changes in cellular behavior. Let us imagine that a geneticist wants to investigate the functions of a gene that he thinks might be responsible for a particular behavior of a cell or a tissue. In particular, the geneticist is inquisitive about the role of this gene in a human disease. He also wants to understand the design (nucleotide sequence), regulation, and mutations that contribute to the distinct functions of the gene! To address the abovementioned questions, the geneticist should not only find a way to obtain this gene in sufficient amount in the genome, but also be able to artificially introduce it into the cells to express it so as to study and manipulate its functions. Interestingly, the technique of in vitro transformation solves this problem by allowing transfer of naked fragments of foreign genetic material inside target cells by artificially permeabilizing the cell membrane. In most cases, the genetic material is a plasmid harboring complementary DNA (cDNA) or a gene that is inserted in target cells using chemical, physical, or biological methods. As a result of the transformation process, the genotype of the recipient cells is modified.

Historically, the discovery of natural genetic transformation was one of the key events in biology that stems from the work on *Streptococcus pneumoniae* in 1928 by Frederick Griffith. This work laid the foundation for the identification of DNA as the genetic material in most living organisms [1]. As the field advanced, transformation was demonstrated to be a frequently occurring natural process and was soon accepted to be a common mechanism for generating and maintaining genetic diversity in bacteria. With advancement of research in this area, this naturally adept molecular biology tool was soon introduced into biomedical and industrial applications. The scope of this technique for manipulation of genotype was further increased by inspecting eukaryotic cells for transformation. Since the genetic code is fairly conserved across the three domains of life, protein coding genes from eukaryotes can be expressed in prokaryotes and vice versa. The oldest published report on artificial transformation of animal cells comes from the injection of DNA from the tissue of one bird into an individual of the same species by Benoit et al. in 1957 [2]. It should be noted that the term "transformation" in the context of mammalian systems is generally used to describe the transition of a cell from normal to cancerous phenotype [3]. Hence, in mammalian systems the term "transfection" is coined for the addition of exogenous DNA to the cell. However, this chapter will confine itself toward understanding of "transformation" into the bacterial hosts for recombinant protein purification in a laboratory setup.

Recombinant DNA technology coupled with transformation/transfection provides us the opportunity to express proteins in bacterial, archaeal, yeast, mammalian, and plant cells for a multitude of purposes that include studying the functional implications of proteins in specific diseases, and large-scale production of enzymes for therapeutic and industrial use. In view of the fact that several different organisms have been adopted as hosts for expression of genes, the choice of the expression system depends on multiple factors. Prime considerations while selecting an appropriate host include the quantity of the protein required, size of the protein, and any disulfide bonds or posttranslational modifications of the protein. For example, the use of *E. coli* for protein production is a simple and economical method of producing bulk quantities of proteins that do not require posttranslational modification (PTM) for functioning; however, proteins that require PTM to function should be expressed in a eukaryotic host system. In this chapter, we discuss the different methods of transformation and the types of expression systems that can be used for protein production, however focusing majorly on the bacterial host systems.

4.1.1 Competence and Competent Cell Preparation

Before a cell could undergo transformation, it has to be "competent." Competence in terms of genetic transformation is defined as the inducible trait of cells for both importing and processing foreign genetic material [4]. The second component in the definition of competence is essential to generate new genotype because the internalized foreign DNA faces three fates inside a recipient cell, which decide the success of transformation. Firstly, the DNA might be rejected and degraded by nucleases; secondly, it might be inserted into the chromosome; and lastly, it might co-exist with chromosome. Competence in most naturally transformable organisms is genetically regulated by dedicated proteins that cause the uptake and processing of DNA. These proteins known as *competence-specific proteins* are a collection of membrane-embedded DNA-binding proteins, various nucleases, DNA importer enzymes, methylases, and recombinases [5, 6]. The molecular machinery involved in regulating competence might perhaps not be constitutively active inside bacteria but instead can respond to specific extracellular or intracellular signals in order to develop competence. Such a form of competence is termed as spontaneous transformation. On the other hand, some bacterial species could lack transformation-specific DNA-uptake and processing genes and thus might not be naturally transformable. However, both bacterial and eukaryotic cells can be artificially forced to enter a transient state of competence along with introduction of the desired DNA in the growth medium to enable in vitro genetic transformation [4]. Competence is thus a transient opportunity for gene introduction (Fig. 4.1). Extracellular DNA that is impure, often damaged, and in trace concentrations is not a rare sighting in a cell's natural environment, where it has naturally evolved. The probability of transformation in natural conditions is thus reduced by natural barriers such as adsorption of

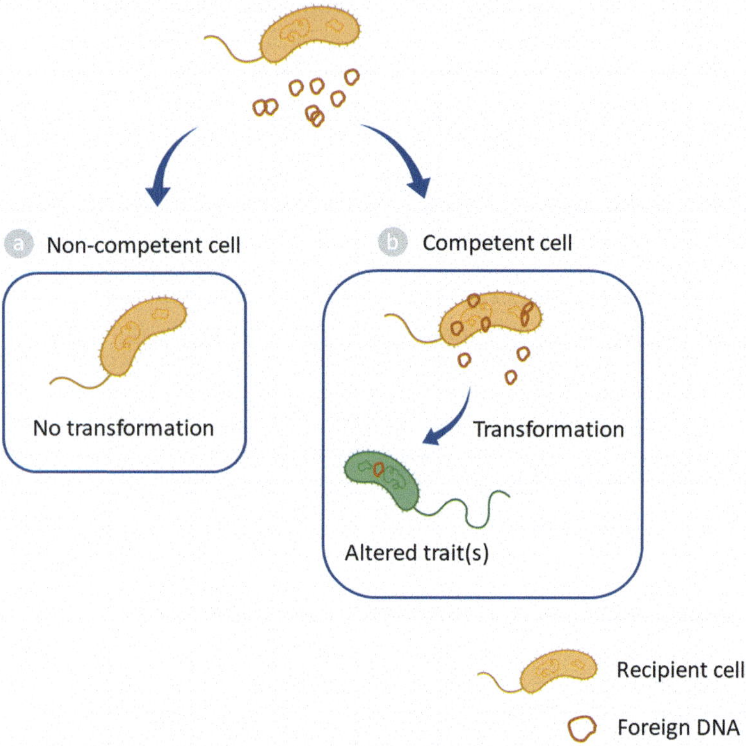

Fig. 4.1 Cellular competence and its role in transformation. (**a**) A cell such as a bacterium might not be naturally able to import and process foreign DNA. Such a cell is termed non-competent. (**b**) DNA could be imported and processed by a competent cell, leading to change in its trait(s). Competence could be artificially generated

DNA by particles in soil and suboptimal temperature, pH, osmolarity, and shear force [7, 8].

In contrast, optimization of the abovementioned parameters in laboratories provides much higher transformation efficiencies. Having appreciated the key importance of cellular competence in transformation, let us now explore how cells could be made competent artificially.

4.1.2 Competent Cell Preparation

In cloning step, the most commonly used bacterial species for transformation is *E. coli*. The history of artificial bacterial transformation began with the thought that *E. coli*, a commonly used laboratory organism, was resistant to transformation. However, in 1970, Mandel and Higa demonstrated that after treatment with calcium chloride ($CaCl_2$), *E. coli* might be induced to take up DNA from bacteriophage λ without the use of a helper phage [4]. Two years later, another group of scientists showed that $CaCl_2$ treatment is also effective for transformation using plasmid DNA [5]. The original method was further improved by Douglas Hanahan.

Since then, numerous methods involving chemicals have been used to make cells competent such as by using other monovalent and divalent cations, polyethylene glycol (PEG) or dimethyl sulfoxide (DMSO) [9]. Other than these chemical methods, electroporation has also been extensively developed and used to induce competency [10–12]. However, the requirement of special equipment has limited the popularity of this method. Nonetheless, the use of cations has proven to be the most effective chemical treatment to bring about bacterial transformation. Among various cations, use of Mn^{2+}, Ca^{2+}, Ba^{2+}, Sr^{2+}, and Mg^{2+} has proven to be effective [13].

Although the factors that regulate competence differ among various genera and remain as the key for the process, a simple procedure to acquire cells of reasonable competency using the chemical method is described below.

4.1.3 Chemical Method

Principle: High polarity and an overall negative charge of DNA pose a barrier for its transfer across the cell membrane. Hence, to facilitate the entry of DNA inside bacteria, its charge must be neutralized. Additionally, the cell membrane of bacteria must be transiently modified to neutralize charges as well as create pores using cations followed by a brief pulse of heat. This method is known as the heat-shock method of transformation [9]. Therefore, prior to exposure of cells to DNA and varying temperatures, they must be made competent using chemicals such as calcium chloride. Other key component in this protocol is Glycerol. The process of calcium chloride-based competence generation encourages rapidly growing bacterial cells to uptake DNA from the surrounding environment. The exact mechanism of how this process works is still largely unknown, but there are hypotheses on the different aspects of the procedure. The main role of calcium ions in the cell suspension is hypothesized to be a cation bridge that reduces repulsion between the phosphate backbone of foreign DNA and phosphorylated lipid A in lipopolysaccharide (LPS) of bacterial cell, owing to the negative charges on both. While the calcium ions neutralize the charge, they can also serve to cause folding of DNA molecule into a ball-like structure that enters cells easily. Glycerol brings DNA close to the surface of the cell due to molecular crowding, and also protects cells when stored at freezing temperatures. During the subsequent heat-shock transformation procedure, the heat pulse at 42 °C is believed to cause temporary pores in lipid membrane through which foreign DNA enters a cell. 10^5 to 10^7 colonies of transformed *E. coli* per μg of DNA generated by this method is more than enough for routine work such as mass-producing plasmid DNA or protein production [14]. However, for special purpose of cDNA library creation that includes environmental DNA sample, a higher number of transformed colonies are desirable.

Based on this method, we shall now discuss the protocol for generating competency in bacteria using *E. coli* as an example.

Protocol

This protocol for competence generation in bacteria is based on the original protocol published by Hanhan et al. [14]

Materials

- 2 mL of *E. coli* starter culture grown overnight
- 100 mL of sterile fresh Lysogeny Broth (LB) medium, pre-warmed to 37 °C
- 500 mL of sterile ice-cold Transformation Buffer 1 (TfB1)
- 100 mL of sterile ice-cold Transformation Buffer 2 (TfB2)

Equipment

1. 37 °C shaking incubator
2. Spectrophotometer/Colorimeter at 600 nm wavelength light
3. Centrifuge

Methods

1. $CaCl_2$ Buffer preparation
   (a) Transformation buffer 1 (TfB1)
   - Potassium Acetate 30 mM
   - Potassium Chloride 100 mM
   - Calcium Chloride 10 mM
   - Glycerol 15% (v/v)
   - Manganous Chloride 50 mM

   Make up the volume to 500 mL using deionized water (dH_2O) and autoclave.

   (b) Transformation buffer 2 (TfB2)
   - MOPS 10 mM
   - Potassium Chloride 10 mM
   - Calcium Chloride 75 mM
   - Glycerol 15% (v/v)

   Make up the volume to 100 mL using dH_2O and autoclave.
2. Growing cultures overnight
   Inoculate 2 mL of LB with a single colony of *E. coli* and incubate at 37 °C and 200 rpm in a shaker incubator for 12–16 h.
3. Subculturing overnight culture
   - Add 2 mL of overnight grown culture to 100 mL of fresh LB with no antibiotics.
   - Shake incubate at 37 °C and 200 rpm for 3–4 h until the optical density (OD) at 600 nm wavelength light reaches 0.5–0.7.
4. $CaCl_2$ wash (generating competency)
   - Stop bacterial growth by gently swirling the flask in an ice-water bath for 10–15 min.
   - Transfer the culture to appropriate container and centrifuge at 3000 × *g* for 15 min at 4 °C.
   - Discard the supernatant.

- Resuspend the cell pellet with gentle pipetting in 20 mL of ice-cold TfB1 and centrifuge at 3000 × *g* for 10 min at 4 °C.
- Discard the supernatant.
- Resuspend the cell pellet with gentle pipetting in 2 mL of ice-cold TfB2.
- Aliquot 100 μL cell suspension in each chilled sterile 0.5 mL microcentrifuge tube.

These competent cells can be used immediately for transformation or stored for future use. For future use, immediately snap freeze the tightly closed tubes by immersing them in liquid nitrogen followed by storing at −80 °C. The stored cells can remain competent for several months with minimal loss in transformation efficiency.

Important Notes

1. It is important to not let the Optical Density at 600 nm (OD_{600}) go higher than 0.6 for *E. coli* since maximum transformation efficiency is achieved when cells are in their log phase of growth. Therefore, the OD should be frequently monitored accurately using a spectrophotometer or a colorimeter, especially when it gets above 0.3, as bacterial cells grow exponentially. A 100 mL subculture medium inoculated with 2 mL starter culture usually takes 2.5 h to reach an OD of 0.6.
2. It is also crucial to keep the cells, buffers, and vessels at 4 °C for the rest of the procedure.
3. For storage, the prepared cells should be aliquoted in small volumes for single use since each freeze/thaw cycle reduces transformation efficiency. 10^5 to 10^7 colonies of cells transformed with 1 μg of plasmid DNA can be obtained using this simple and robust procedure to generate competent cells [15]. Such efficiency is usually enough for routine purposes such as extracting large amount of plasmid DNA, gene expression for protein production, and functional studies. However, much higher efficiencies are required when obtaining every possible clone is of utmost importance, for example, generating cDNA library from low concentrated DNA sample. Therefore, several modifications and optimizations of this basic procedure have been described in the literature [15, 16]. These improved procedures have generated competent cells with transformation efficiencies between 10^8 and 10^9 transformed colonies per μg of supercoiled plasmid DNA.

4.1.4 Preparing Electrocompetent Cells

Principle: The procedure for generating competent bacterial cells for transformation using the electroporation method is the easiest, fastest, most efficient, and highly reproducible compared to chemical methods. Just like the chemical method, bacteria are cultured till they reach the exponential phase, characterized by faster growth [17], chilled, centrifuged, washed extensively using dH_2O or a suitable buffer of extremely low ionic strength, and then suspended in 10% (v/v) glycerol. Cell viability is reduced due to arcing of electric current in the electroporation cuvette during DNA transfer by means of electroporation. Therefore, the ionic strength of electroporation buffer and DNA solution should be kept as low as possible to achieve high transformation efficiencies [18, 19]. Let us now discuss the protocol for generating competency in *E. coli* by using the method of electroporation.

Protocol

Materials

- 50 mL of *E. coli* culture grown overnight
- 500 mL of sterile fresh Lysogeny Broth (LB) medium, pre-warmed to 37 °C
- Sterile ice-cold dH_2O
- 250 mL sterile ice-cold 10% (v/v) Glycerol
- 10 mL of sterile fresh Glycerol Yeast extract Tryptone medium

Equipment

1. 37 °C shaking incubator
2. Spectrophotometer/Colorimeter at 600 nm wavelength light
3. Centrifuge

Methods

1. Preparing GYT medium ([20])
   - 0.25% (w/v) Tryptone
   - 0.125% (w/v) yeast extract
   - 10% (v/v) glycerol
2. Growing cultures overnight
   - Inoculate 50 mL of LB with a single colony of *E. coli* and incubate at 37 °C and 200 rpm in a shaker incubator for 12–16 h.
3. Subculturing overnight culture
   - Add 25 mL of overnight grown culture to 500 mL of fresh LB with no antibiotics.
   - Shake incubate at 37 °C and 200 rpm for 3–4 h until the OD at 600 nm wavelength reaches 0.5–0.7.
4. Generating competency.

   For maximum transformation efficiency, the temperature of bacteria should not rise above 4 °C at any step in the protocol given below:

- Stop bacterial growth by gently swirling the flask in an ice-water bath for 10–15 min. In preparation for the next step, place centrifuge bottles in an ice-water bath.
- Transfer the culture to the cold bottles and centrifuge at 3000 × *g* for 15 min at 4 °C.
- Discard the supernatant and resuspend the cell pellet with gentle pipetting in 50 mL of ice-cold dH_2O.
- Centrifuge at 3000 × *g* for 10 min at 4 °C.
- Discard the supernatant and resuspend the cell pellet with gentle pipetting in 250 mL of ice-cold 10% (v/v) Glycerol.

- Centrifuge at 3000 × *g* for 10 min at 4 °C. Supernatant should be discarded carefully since cells adhere loosely in 10% Glycerol.
- Discard the supernatant and resuspend the cell pellet with gentle pipetting in 10 mL of ice-cold 10% (v/v) Glycerol.
- Centrifuge at 3000 × *g* for 10 min at 4 °C.
- Discard the supernatant and resuspend the cell pellet with gentle swirling in 1 mL of ice-cold GYT medium.
- Take a small volume of the cell suspension, dilute it 1:100 with ice-cold GYT medium and measure its OD_{600}. Further, dilute the cell suspension appropriately to obtain a concentration of ~2.5 × 10^{10} cells/mL. Roughly, 1.0 OD_{600} = ~2.5 × 10^{10} cells/mL for most *E. coli* strains.
- To test whether arcing (electrical shorting that leads to burning of cells) occurs when electric current is applied to the cuvette, transfer a small volume of cell suspension to an ice-cold electroporation cuvette and apply a voltage of 13–15 kV/cm. If arcing occurs, remove excess salts from cell suspension by washing cells several times with ice-cold GYT.
- Aliquot 100 μL of ~2.5 × 10^{10} cells/mL concentration cell suspension in each chilled sterile 0.5 mL microcentrifuge tube.

These electrocompetent cells can be used immediately for transformation or stored for future use. For future use, immediately snap freeze the tightly closed tubes by immersing them in liquid nitrogen followed by storing at −80 °C. The stored cells can remain competent for several months with minimal loss in transformation efficiency.

Important Notes

1. It is crucial to keep the cells, buffers, and vessels at 4 °C for the rest of the procedure.
2. The dH_2O used should have low electric conductivity to avoid arcing during electroporation.
3. For storage, the prepared cells should be aliquoted in small volumes for single use since each freeze/thaw cycle reduces transformation efficiency.

4.1.5 Transformation Methods

Due to the constant requirement for introducing exogenous genetic material into bacteria, the scope of methods available for artificial transformation has broadened over time. The accessibility of such a broad scope has endowed several downstream applications of molecular cloning technology. This section attempts to describe the principles of various methods available for introducing DNA into bacterial cells and the factors that govern their efficiencies. A short comparison between the various methods has been made, which can serve as a starting point to determine the best method that could fulfill the specific requirements of a cloning application. Furthermore, in continuation of the previous section on competent cell preparation, the procedures for the two most commonly used methods of bacterial transformation such as the *Heat-shock* and *Electroporation* have been elaborated in the following sections.

The subsequent paragraphs describe various methods available for the transfer of exogenous DNA into a suitable host.

4.2 Heat-Shock Method

E. coli cells made competent by treating with a mixture of salts can be subjected to alternating high and low temperatures to facilitate the entry of DNA molecules through the outer and inner cell membranes. This method is based on the initial observation of Mandel and Higa, who demonstrated that log-phase bacterial cells treated with a solution of cations when briefly subjected to 37 °C or 42 °C could be easily transfected with λ bacteriophage DNA [15]. The exact mechanism of how the process of calcium chloride-based competence generation and heat-shock treatment encourages bacterial cells to uptake DNA is still largely unknown. However, there are a few hypotheses on the different features of the procedure. Mechanistically, in the heat-shock method, the role of calcium-rich environment and other divalent cations used for generating competence in bacterial cells is hypothesized to be a cation bridge that reduces electrostatic repulsion between the phosphate backbone of foreign DNA and phosphorylated lipid A in lipopolysaccharide (LPS) of bacterial cell. This neutralization of charge is necessary owing to the negative charges on both the DNA and LPS layer of the bacterial cell. While the calcium ions neutralize the charge, they can also serve to cause folding of DNA molecule into a ball-like structure. Condensation of DNA is necessary to cause its entry inside cell. This is consistent with the observation that super helical plasmid DNA is transferred more efficiently than linear DNA. Glycerol brings DNA close to the surface of the cell due to molecular crowding, and also protects cells when stored at freezing temperatures. During the subsequent heat-shock transformation procedure, the brief heat pulse at 37 °C or 42 °C is believed to cause temporary pores in lipid membrane through which foreign DNA enters a cell [21] (Fig. 4.2). The heat is also believed to enhance Brownian motion in cell suspension that further facilitates motion of DNA through cell membranes [21]. As per the hypothesized mechanism, owing to their thick cell wall, Gram-positive bacteria are transformed with low efficiency compared to Gram-negative bacteria using the heat-shock method. In this method, DNA transfer and therefore transformation efficiency is inversely proportional to the size and topological form of DNA [14]. 10^5 to 10^7 colonies of transformed *E. coli* per µg of DNA generated by this method is more than enough for routine cloning work such as mass-producing plasmid DNA or protein production. However, for special purpose of cDNA library creation, wherein each unique cDNA is important, a higher number of transformed colonies is desired. The procedure for heat-shock method of transformation is described in the succeeding paragraphs.

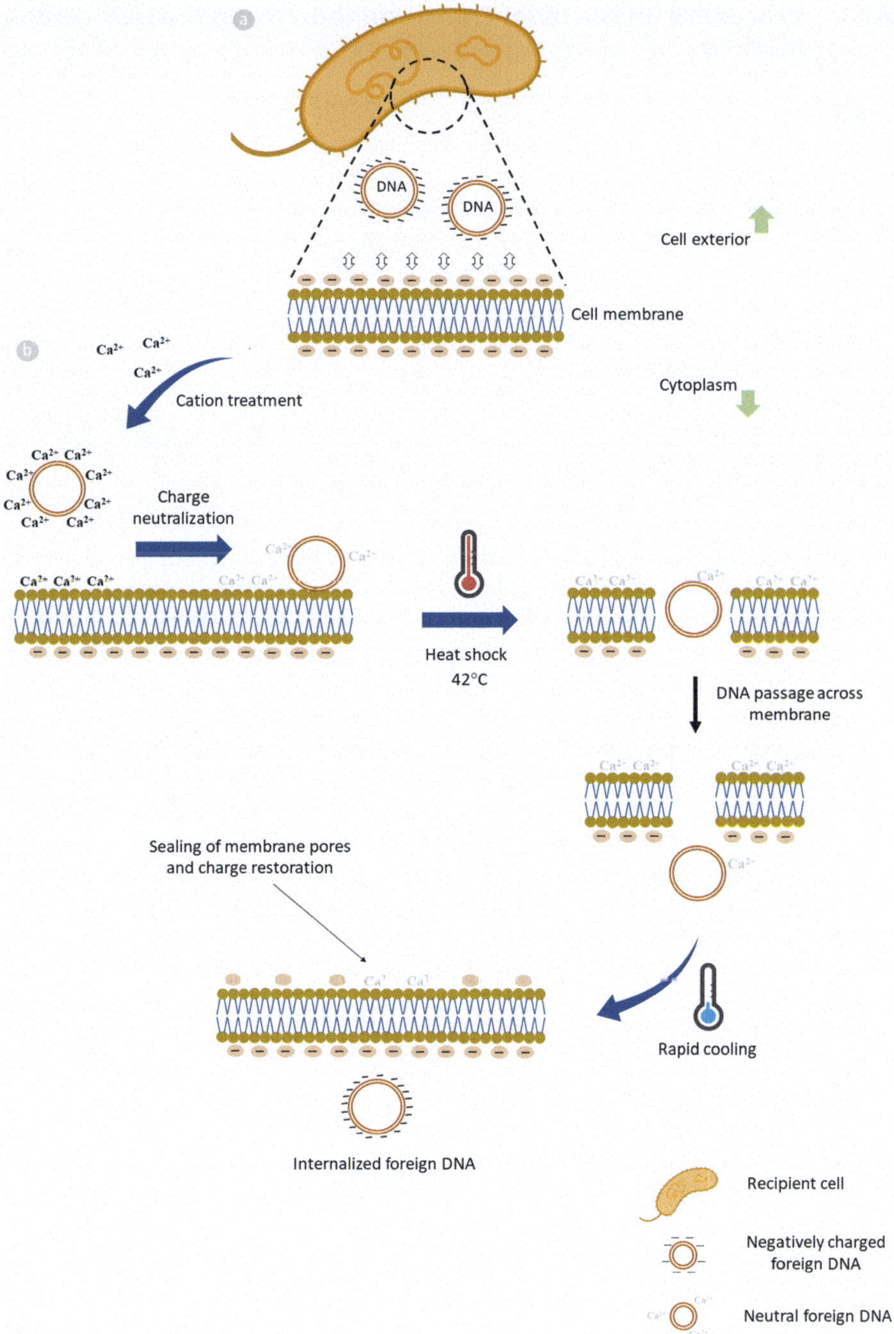

Fig. 4.2 Heat-shock method. (**a**) Charge repulsion between foreign DNA and cell membrane, together with low membrane porosity, pose a barrier for the entry of DNA inside recipient cells, as depicted in the magnified view. (**b**) While presence of cations such as Ca^{2+} reduces repulsion between the foreign DNA and cell membrane, a brief period of heating and cooling transiently

4.2.1 Procedure for Bacterial Transformation Using the Heat-Shock Method

Materials

- Freshly prepared or frozen competent bacterial cells
- Plasmid DNA solution
- Sterile fresh Lysogeny Broth (LB) medium, pre-warmed to 37 °C (~1 mL of this broth is needed per transformation aliquot)
- Lysogeny agar plates supplemented with appropriate antibiotic(s)

Equipment

1. 37 °C shaking incubator
2. Centrifuge
3. Water bath set to 42 °C
4. 37 °C static incubator

Method

- Using a micropipette add 50–100 ng of plasmid DNA to each 100 μL aliquot of competent cells. Gently mix the contents of the tube using the micropipette. When using frozen competent cells, allow cells to thaw gradually by keeping the tube in ice for 30 min. It is recommended to set both positive and negative controls for every transformation experiment. The positive control tube may contain competent cells that would receive a known amount of plasmid DNA of standard quality. On the other hand, the negative control tube may contain competent cells that shall not receive DNA at all.
- Place all the tubes in ice for 20–30 min to allow DNA-cell interaction. Meanwhile, a water bath with floating tube rack can be set to 42 °C for the next step.
- Transfer the tubes to the floating tube rack inside the preheated water bath. Place the tubes inside water bath for exactly 90 seconds without shaking. This heat-shock step is the major deciding factor of transformation efficiency.
- Immediately take out the tubes from the water bath and place in ice for 15 min. Meanwhile, a 1 mL aliquot of Lysogeny Broth (LB) medium can be warmed to 37 °C for the next step.
- Inside a bacteriological laminar air flow cabinet, add 700 μL of the pre-warmed LB medium to each tube.
- Place all the tubes in a shaker incubator and incubate for 60 min at 37 °C and 180 rpm shaking speed. Addition of antibiotics is not recommended to these small cultures since the cells need sufficient time to recover and express antibiotic resistance gene(s) present in the plasmid DNA.

Fig. 4.2 (continued) increases membrane porosity. The cumulative effect of cation treatment and varying temperatures enables entry of foreign DNA inside cells

- Transfer 100–200 μL of culture from each tube on separate Lysogeny agar plates supplemented with appropriate antibiotic(s), and spread uniformly using a sterile glass spreader. Alternatively, low density cultures can be centrifuged at 3000 × *g* for 5 min, resuspended in 100–200 μL of medium, and spread. When selecting cells that produce extracellular inactivators of antibiotics, cultures should be spread with low densities to prevent sensitive colonies from growing in the proximity of resistant cells.
- Place all the plates in inverted position inside a static incubator and incubate at 37 °C for 12–18 h.

4.2.2 Expected Observations

Test: Colonies of transformed bacteria may or may not appear on plates after appropriate incubation period.

Positive control: Several colonies of transformed bacterial cells should appear after appropriate incubation period.

Negative control: No colonies should appear on the plate. Appearance of any is an indication of the following possibilities:

1. The aliquot of competent cells was contaminated with antibiotic-resistant bacteria during the procedures of competence generation and/or transformation.
2. The plates were contaminated with antibiotic-resistant bacteria during their preparation and/or storage.
3. The antibiotic(s) added to the plate lost their potency either due to prolonged storage or because it was added to the medium when it was too hot.

4.3 Electroporation Method

The problems of low competency due to the genotype of bacterial cells or limited efficiency of the heat-shock method of transformation led to the usage of an alternative method of bacterial transformation. Since bacterial cells are electrically conductive, exposure to high intensity electric current physically distorts the cell membranes by causing their polarization. When a high-voltage electric current in the range 10–15 kV/cm is applied to bacteria, the physical distortion in the cell membranes causes formation of small temporary pores (hence, the name electroporation), which serve as entry routes for exogenous DNA (Fig. 4.3). First demonstrated by Neumann on mouse cancer cells [18], this method called Electroporation was later used to transfer DNA into other eukaryotic cells of fungi, plants, and yeasts and cells of both Gram-positive and Gram-negative bacteria with high efficiency [22]. This method, therefore, is independent of the type of host cell used in the transformation procedure. Since then, electroporation that is also known as electro-injection or electro-transfection has established itself as the fastest, easiest, most efficient, and highly reproducible method for introducing foreign DNA into various cell types [19].

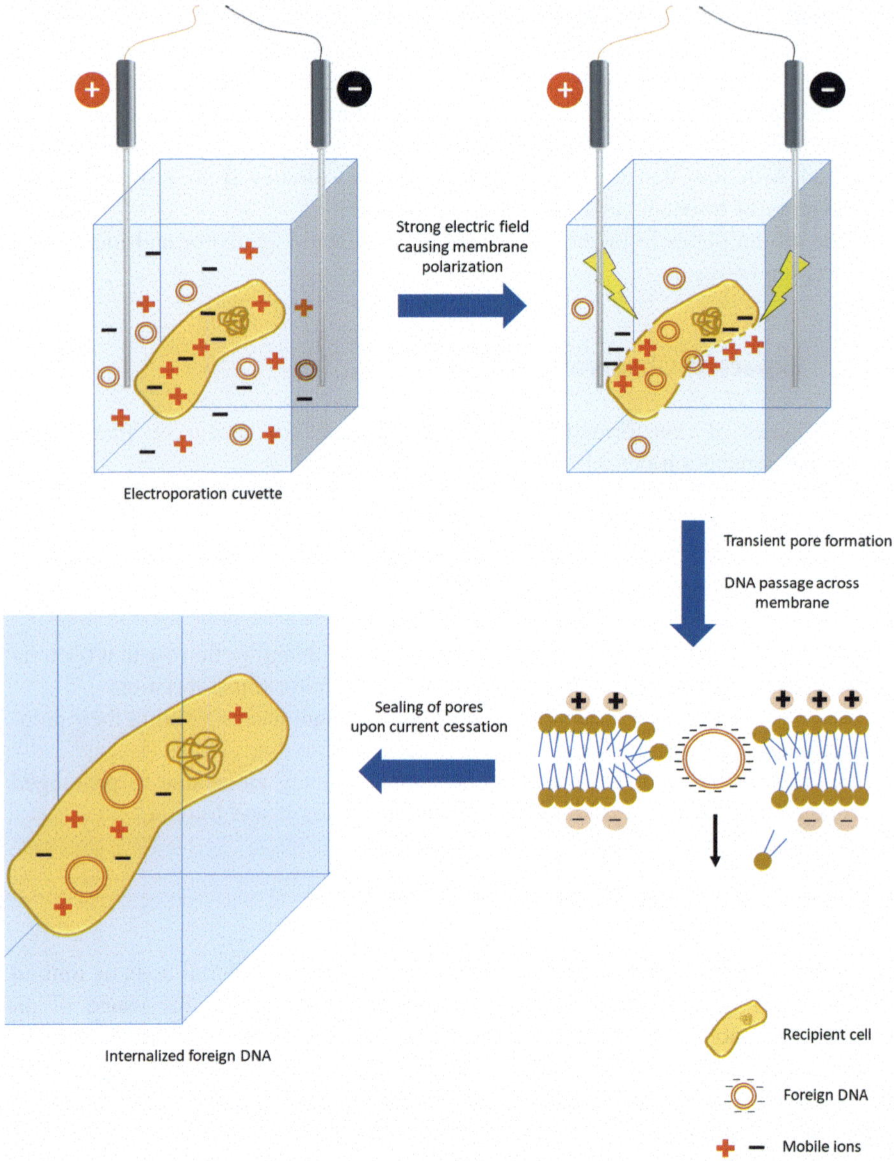

Fig. 4.3 Electroporation method. A suspension of cells and DNA is exposed to high intensity electric current inside an electroporation cuvette. Charge polarization by migration of ions caused as a result of electric field leads to formation of transient microscopic pores in the cell membrane, facilitating the entry of foreign DNA into cells. Subsequently, cessation of current causes the pores to seal, trapping the foreign DNA inside

The many advantages offered by this method over conventional ones are summarized below:

1. Plasmids ranging in size from 3 kb to 85 kb can be introduced in *E. coli* with transformation efficiencies in the order of 10^{10} and 10^7, respectively. This is extensive compared to the efficiencies of ~10^6 obtained by the routinely practiced heat-shock method.
2. Unlike the $CaCl_2$ heat-shock method of transformation, the efficiency of which is inversely related to the size and topology of DNA, the efficiency of electroporation method is related only to the concentration of input DNA. Therefore, a DNA as large as 150Kb has been transferred using electroporation [23].
3. Since the incubation step of host cells with DNA is eliminated, the procedure is rapid. Moreover, with no addition of chemicals such as PEG and DMSO, toxicity is near zero in this method.

4.3.1 Procedure for Bacterial Transformation Using the Electroporation Method

This procedure of Electroporation is applicable to DNA of size <15 kb and most strains of *E. coli* such as DH5α, DH10B, and XL-1 [23–29] to yield transformation efficiency that is suitable for routine applications of cloning.

Materials

- Freshly prepared or frozen electrocompetent bacterial cells
- Plasmid DNA solution
- Sterile ice-cold dH_2O
- Sterile and fresh Super Optimal broth with Catabolite repression (SOC) medium at room temperature
  (~1 mL of this broth is needed per transformation aliquot)
- Lysogeny agar plates supplemented with appropriate antibiotic(s)

Equipment

1. Sterile electroporation cuvettes
2. Electroporation apparatus
3. 37 °C shaking incubator
4. 37 °C static incubator

Method

- Clean electroporation cuvettes thoroughly using sterile ice-cold dH_2O.
- Pipette 50–100 μL of electrocompetent cells into a clean electroporation cuvette kept in ice. When using frozen competent cells, allow cells to thaw gradually by keeping the tube in ice for 30 min before transferring cells to the cuvette. It is recommended to set two control tubes for every transformation experiment by including an aliquot of competent cells that gets a known amount of super helical plasmid DNA of standard quality and another aliquot of cells that gets no DNA at all.

- Using a micropipette add 10 pg to 25 ng of plasmid DNA in appropriate volume (not exceeding 5 μL) to each aliquot of cells inside the ice-cold cuvettes. Gently mix the contents of the cuvettes using the micropipette and keep the cuvettes in ice for 1 min. The plasmid DNA preparation should be in dH_2O or TE buffer (pH 8.0).
- Set the electroporation apparatus to convey an electrical pulse of 2.5 kV, 200 Ω resistance and 25 μF capacitance.
- Wipe all the cuvettes from the outside to dry moisture. Place all the cuvettes in the apparatus and deliver an electrical pulse to the cells at the settings mentioned above.
- Immediately after the pulse, remove the electroporation cuvettes and add 1 mL of SOC medium to each cuvette.
- Aspirate the entire contents of the cuvettes and transfer to sterile microcentrifuge tubes inside a bacteriological laminar air flow cabinet.
- Place all the tubes in a shaker incubator and incubate for 60 min at 37 °C and 180 rpm shaking speed. Addition of antibiotics is not recommended to these small cultures since the cells need sufficient time to recover and express antibiotic resistance gene(s) present in the plasmid DNA.
- Given the higher transformation efficiency of this method, transfer small volumes (20–50 μL) from each tube on separate Lysogeny agar plates supplemented with appropriate antibiotic(s), and spread uniformly using a sterile glass spreader. Alternatively, a loop-full of culture can be streaked on plates.
- Place all the plates in inverted position inside a static incubator and incubate at 37 °C for 12–18 h.

For information on expected observations, please refer to the Heat-shock transformation method section.

Apart from the two most popular methods of transformation, few others have been developed recently that include electrospray [24, 25], sonoporation [26], and microinjection [26]. However, they are mostly used for transformation in mammalian and plant cell systems and hence beyond the scope of this chapter.

In summary, the description of a wide array of common options for transferring DNA into cells and tissues implies that each one of these was developed based on the identification of certain lacunae in the previous ones. While certain limitations have been overcome with the advent of newer methods, the new ones also have their own constraints. Therefore, it is difficult to make a "score board" of transformation methods with regard to their efficiencies alone. Therefore, the preference for one method over the other should depend primarily on the particular application.

Troubleshooting Guide

Since transformation procedures involve several steps, researchers might experience a number of problems that need to be addressed for downstream applications of cloning. This section on troubleshooting discusses the common problems that arise routinely in bacterial transformation experiments that employ the Heat-shock and Electroporation methods (Table 4.1).

Table 4.1 Troubleshooting of transformation procedure

Problem	Explanation	Recommendation
Few or no transformants	Poor transformation efficiency	1. Ensure care while preparing competent cells. Store cells at −80 °C without fluctuations in temperature. 2. Ensure care while performing the heat-shock step of transformation. 3. Ensure that cells chosen are compatible with the exogenous genetic material used for transformation. 4. Include a positive control using plasmid DNA of known compatibility.
	Poor quality of transforming plasmid DNA	1. If the plasmid DNA sample is product of reactions such as ligation, PCR, DNA probe attachment, etc., adequate amount of sample clarification can be done using mini filtration columns or gel electrophoresis followed by exclusion of DNA from the gel. For products of some reactions involving proteins such as enzymes, heat inactivation can increase transformation efficiency.
	Low concentration of transforming plasmid DNA	1. Use adequate amount of plasmid DNA suitable for the chosen competent cells. 2. Using excessive amount is shown to reduce transformation efficiency in some cell types.
	Inserted DNA or its gene product is toxic to host cell	1. The choice of genetic elements present on the plasmid DNA such as the inducible gene promoter, recombination sites, and the number of replication sites can be reviewed. A tightly regulated inducible promoter can be used to ensure minimal basal-level gene expression. 2. gene products toxic to the host can be avoided.
	Insufficient number of cells were plated	1. After the recovery step in transformation procedure, the culture can be centrifuged to increase cell concentration before plating.
	Wrong antibiotic or its high concentration in plate	1. Review the choice of antibiotic for the plasmid DNA's resistance marker. 2. Ensure that the plates have appropriate concentration of antibiotic(s) tolerable by resistant cells.
	Erroneous use of spreading tool	1. Ensure that the spreading tool used is cooled sufficiently before spreading cells.
Colonies do not contain desired plasmid DNA	Inserted DNA incompatible with host	1. Ensure that cells chosen are compatible with and do not cause exclusion of the exogenous genetic material used for transformation.
	Selected colony belongs to untransformed cells	1. Ensure that the plates do not have too low a concentration of antibiotic(s) to avoid

(continued)

Table 4.1 (continued)

Problem	Explanation	Recommendation
		growth of untransformed cells. 2. Avoid adding antibiotic(s) to agar while it is too hot. 3. Include a negative control. 4. Incubation time can be reduced to avoid growth of satellite colonies that benefit from antibiotic-inactivators secreted by true colonies. 5. Consider picking colonies from the center of the plate.
A lawn of cells appears	Large number of cells plated	1. Use appropriate volume of culture for plating or dilute if necessary.
	Long incubation period	1. Review if the chosen host is genetically modified for fast growth.

4.4 Recombinant Protein Expression in Different Bacterial Systems

With the rapid advancement of recombinant DNA technology, obtaining both prokaryotic and eukaryotic gene products from bacterial cells became plausible. This method simplified the method of protein synthesis for laboratory as well as industrial-scale applications since it eliminated the need of large amounts of animal and plant tissues or fluids. Production and purification of desired proteins in enormous quantity allows their biophysical and biochemical characterization and development of commercial materials. Mass production of recombinant proteins was made feasible mainly due to the constant research on the physiology and genetics of bacteria. Despite the remarkable expansion of the challenging protein-production field, the use of bacterial cells as protein-production mills has remained an indispensible tool. Since no universal protein production host has been found so far, the choice of the best host relies on several parameters such as the cellular source of a gene, type of gene expression vector, and protein production conditions such as temperature and cultivation media. Therefore, several types of bacteria were identified and developed to make them suitable for the varying needs of protein-production programs. The following sections of this chapter will elaborate on different species and strains of bacteria for recombinant protein synthesis.

Escherichia coli The spectacular potential of *Escherichia coli* as a gene expression host was first realized in the early 1970s after DNA of eukaryotic origin was propagated in it ([30]). Since then, *E. coli* has been used widely for protein production owing to its weak pathogenicity, easy genetic manipulation, simple and inexpensive culture media, low maintenance, and fast high-density growth, allowing proteins to be produced in less than one day. As a result, there are numerous molecular tools at our disposal for mass-production of proteins, such as several modified *E. coli* strains, broad inventory of gene expression vectors, optimized cultivation and transformation strategies. Different types of *E. coli* strains that have been developed and are commercially available have specific advantages and disadvantages, thus making them suitable for specific situations. For most applications the popular strains are described below.

BL21(DE3) This strain was first described by Studier in 1986 after various modifications of the "B" strain of *E. coli* [31]. Like the parental B strain, BL21 cells lack proteases Lon and OmpT that degrade foreign proteins. While Lon is a cytoplasmic protease, OmpT is an outer membrane protease that degrades extracellular proteins to salvage amino acids [32, 33]. These proteases pose a hurdle for producing foreign proteins in large quantities. For example, after cell lysis, the OmpT might degrade recombinant proteins in cell lysate, thereby reducing the total yield. Additionally, mutations in the Host specificity of DNA Subunit B (*hsd*SB) gene that disrupts DNA methylation and degradation prevents plasmid loss from transformed cells [34]. The DE3 designation means that this strain contains a λDE3 bacteriophage lysogen carrying a T7 phage RNA polymerase gene. The T7 RNA polymerase, which is many times faster than the bacterium's own RNA polymerase, is kept under the control of Lactose Operon repressor protein. Therefore, the phage RNA polymerase gene can be expressed by adding Isopropyl β-D-1-thiogalactopyranoside (IPTG) in growth medium. IPTG-induced production of T7 RNA polymerase gene in turn leads to expression of recombinant genes cloned downstream to the T7 promoter in a plasmid. However, the BL21(DE3) strain slightly expresses recombinant protein without the addition of IPTG. This phenomenon known as "leaky expression" becomes problematic for some proteins that are toxic to host cells often leading to protein misfolding. These misfolded proteins remain insoluble in the bacterial cells and termed as *Inclusion bodies* that will be discussed in a later chapter.

BL21(DE3)-pLysS Based on the original BL21(DE3) strain, the BL21(DE3)-pLysS strain contains a plasmid bearing the T7 Lysozyme gene. The T7 Lysozyme helps reduce leaky expression of recombinant proteins by inhibiting T7 RNA polymerase. This inhibition is overcome after adding appropriate amount of IPTG to bacterial growth medium.

Rosetta (DE3)pLysS Rosetta strains based on the BL21(DE3)-pLysS supply tRNAs that recognize codons, which are used more frequently in eukaryotes. Since tRNA population in *E. coli* reflects the codons that are frequently used by the bacterium's genes, translation of heterologous mRNA might be impeded due to lack of one or more tRNAs against eukaryotic codons. In Rosetta(DE3)pLysS cells, the genes encoding such special tRNAs are encoded in the same plasmid that carries the T7 lysozyme gene.

Although the BL21 and its derivatives are used commonly, the K-12 lineage of *E. coli* is also a popular choice for producing recombinant proteins.

4.5 Expression of "Difficult-to-Fold" Proteins in *E. coli*

One of the main strategies undertaken to reduce improper protein folding in bacteria is to slow down the rate of protein production, thereby allowing newly formed polypeptides enough time to fold properly. Growing cells at low temperatures reduces their growth rate that in turn keeps protein concentration low and decreases aggregation to facilitate proper folding. However, bacterial chaperones that assist in protein folding might perhaps not function efficiently at low temperatures. The Artic Express™ strain of *E. coli* that is modified to contain chaperonin Cpn60 and co-chaperonin Cpn10 from the psychrophilic (cold-loving bacterium with optimum growth at 15–20 °C) bacterium *Oleispira* spp. displays improved protein folding ability and *E. coli* growth at low temperatures [35].

Pseudomonas spp. While higher yield is important for a successful recombinant protein production program, the quality of the synthesized protein is equally important. Yielding adequate amount of properly folded, functional proteins that are free from host cellular contaminants, however, is challenging as it involves several complex and expensive gene expression and protein refolding strategies. To avoid the problem and allow easy purification of active proteins, scientists have explored the possibility of using a bacterial expression system with intrinsic ability to produce and secrete soluble recombinant proteins. Out of the multiple bacterial hosts available for recombinant protein production, *Pseudomonas* is specifically sought for production and secretion of proteins having complex folding requirements in high quantities due to its efficient protein secretion system [36]. By this way, the accumulation of inactive protein in cells that remains a major disadvantage is eliminated and recombinant protein can be produced without harvesting cells. Moreover, the non-pathogenic nature of this Gram-negative bacterium allows it to be used for producing pharmaceutically and agriculturally non-toxic proteins [37]. Further, the detailed knowledge of the genome of *Pseudomonas* indicates that the performance and scope of this expression system can be increased by genetic engineering of the bacterium to develop novel *Pseudomonas*-based protein production platforms.

Streptomyces spp. Although inappropriate folding of recombinant proteins and their arduous purification strategies can often be overcome with the use of the secretion system of *Pseudomonas,* other bacterial hosts with better protein secretion systems based on Gram-positive bacterium *Streptomyces* have also been developed. Among many other *Streptomyces* species, the easy acceptability of foreign DNA and weak protease activity of *S. lividans* have made it the most extensively used species for production and secretion of recombinant proteins [38]. With the knowledge of the genome of *Streptomyces*, a broad collection of vector systems have been constructed. Several of these are based on the plasmid pIJ101, such as pIJ702 and pIJ486, which are compatible with a wide variety of bacterial hosts. Such broad compatibility allows exchanging recombinant DNA between *Streptomyces* and other bacterial species. Moreover, unlike *E. coli* and *B. subtilis, Streptomyces* has shown exceptional proficiency in production of proteins such as Xyloglucanase from Actinobacteria, and Endoglucanase CelA from a thermophilic bacterium [39, 40]. This entails possibilities for many important and novel recombinant protein constructs that were either not studied or were de-prioritized due to strict dependence on conventional bacterial hosts, to be re-explored.

Rhodobacter spp. Proteins that are embedded in the plasma membrane perform crucial functions in both prokaryotic and eukaryotic cells. These proteins, called integral membrane proteins, comprise more than 50% of drug targets [41]. However, studies on these proteins are severely limited because their hydrophobic properties pose extreme difficulties in production and purification of functionally active forms in sufficient quantities. One of the emerging strategies to mass-produce membrane proteins makes use of natural coordination between synthesis of a membrane protein and lipid bilayer jackets in a bacterium's cytoplasm. These lipid bilayer jackets are used as platforms that harbor integral membrane proteins [42], thereby providing a suitable environment to hydrophobic proteins. However, common expression systems based on *E. coli*, for example, do not couple de novo membrane synthesis with protein production. Moreover, a high concentration of proteins can overwhelm the bacterium's secretory pathway leading to protein aggregation and/or cell death. Therefore, the physiological property of *Rhodobacter* species of photosynthetic bacteria to produce large amount of lipid jackets in cytoplasm is being exploited to develop strategies for production and purification of natively folded, functional membrane proteins. Naturally, the *Rhodobacter* species produce cytoplasmic lipid jackets to assemble transmembrane proteins of the photosynthetic apparatus.

4.6 Optimizing Gene Expression

One of the most fruitful applications of recombinant DNA technology is the capability to artificially produce large amounts of proteins in a host cell such as bacteria. Protein production is an indispensible component of the "protein design cycle" that is commonly known as "protein engineering." It involves an array of biochemical, biophysical, computational, and analytical techniques to study proteins and design their variants with desired characteristics for use both in fundamental research and in industry. Therefore, for the preparation of a protein in reasonable amounts and its analysis, the gene encoding the protein is expressed in a transformed host such as bacteria. It is to be noted that around 31 recombinant proteins were approved for use in therapies between year 2003 and 2006 [43] underscoring the importance of recombinant protein production.

However, the success of high-quality recombinant protein production depends on the proficiency with which the gene expression is carried out to complete the protein design cycle. Manipulation of the host's natural ability to recognize and express a foreign gene to an extraordinary degree is called "gene over-expression." When an over-expressed gene encodes for a protein, the host allows production of foreign proteins in amounts greater than those of the host's native proteins. This is an important advantage as it increases the chance of purification of a desired protein from a pool of host's proteins. While harvesting substantial amount of desired proteins is a prize, the proportion of useable protein depends upon how the host is manipulated for gene over-expression. Therefore, the goal of each protein production procedure is not only to identify conditions under which the host produces large amount of proteins, but also to identify conditions under which high-quality, natively folded, and functionally active proteins could be obtained.

The expression of foreign genes is primarily carried out in the host cell with the help of cloning vectors, as mentioned in the previous chapter. However, it is worth mentioning that the approach to use cell-free transcription and translation system that direct the synthesis of proteins without the need to grow and maintain host cells is also viable. In the following sections, we will learn how *E. coli* cells are manipulated to produce large amounts of proteins while ensuring that their structural integrity and biological activity are maintained.

4.7 Protein Production Protocol for Bacteria

Owing to the complexity of bacterial growth, a gene expression experiment can be improved by optimizing a great number of parameters. This is analogous to tweaking a factory's assembly line to produce foolproof products. With each change in parameters affecting the yield, folding, solubility, and activity of proteins, this task appears baffling. However, great efforts by several protein chemists and biophysicists have led to the following commonly accepted protocol that allows most proteins to be produced in *E. coli* [44].

The gene encoding desired protein (target gene) is cloned into a bacterial expression vector consisting of the T7 *lac*O promoter system of the Lactose Operon. The T7 promoter system provides strong transcription of adjacent target gene. Next, the expression vector is used to transform derivatives of the BL21(DE3) strain, such as Rosetta™(DE3)pLysS. The larger-volume culture, also called expression culture, is grown until the cells reach mid-log phase (OD_{600} of ~0.6). The temperature of the cultures is subsequently lowered to 30 °C and protein production is initiated using IPTG (Fig. 4.4). The lower expression temperature aids in production of properly folded and soluble protein. Finally, the cells are harvested by centrifugation and used for extracting and purifying proteins.

Let us now explore the factors that have an immense effect on the yield, folding, solubility, and activity of proteins produced in *E. coli*.

Rare Codons in Target Genes Often genes of human proteins fail to express in bacterial hosts due to the inability of the host to recognize certain codons. Such codons are labelled as "*rare codons*" owing to the fact that *E. coli* cells do not use

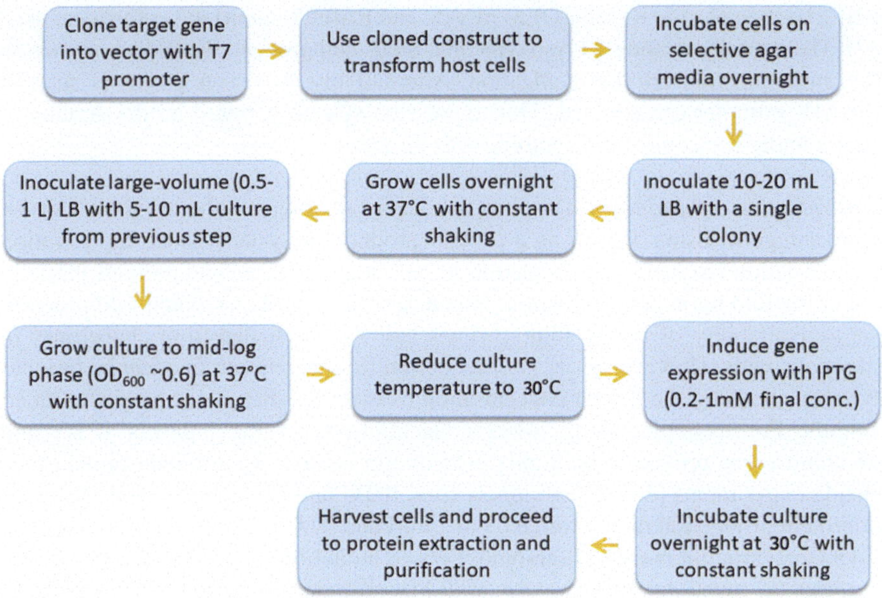

Fig. 4.4 Flowchart of a commonly followed protocol that allows most proteins to be produced in *E. coli* at a large scale

those frequently [45]. Rare codons include codons for Proline, Leucine, Isoleucine, and arginine. When an *E. coli* cannot recognize these codons due to lack of corresponding tRNAs, the process of translation stops midway and thus incomplete proteins are produced. Rarely, proteins might be produced with wrong sequence due to incorporation of wrong amino acids at the position of rare codons. Fortunately, we can now find whether a target gene has rare codons using a web tool (e.g., https://www.genscript.com/tools/rare-codon-analysis). If rare codons occur, the gene can be modified or a special *E. coli* host such as Rosetta(DE3)pLysS can be used. Such hosts co-express genes encoding the rare tRNAs with the unmodified target gene [46]. Both these approaches have overcome the problem successfully.

Leaky Target Gene Expression Production of target protein in most *E. coli* hosts is based on T7 RNA polymerase, which is several-fold faster than the bacterium's own RNA polymerase. The expression of T7 RNA polymerase gene in *E. coli* is often controlled via an inducible chromosomal copy of the gene under the control of the lacUV5 promoter. When the production of polymerase is not induced by an external inducer molecule such as IPTG, the target gene is not expressed. However, T7 promoter being a strong promoter, even minimal basal production of T7 RNA polymerase can cause "leaky" expression of the target gene. This is undesirable if the recombinant protein is prone to misfolding if produced at 37 °C when the growth of *E. coli* is fast. Secondly, leaky expression can lead to cell death if the protein is toxic to the host. Therefore, to overcome leaky expression, special *E. coli* hosts such as BL21(DE3) pLysS, Rosetta(DE3) pLysS, and Rosetta-gami(DE3) pLysS are used [47]. These strains encode T7 lysozyme that binds and prevents T7 RNA polymerase from initiating transcription elongation. Alternatively, a weaker promoter such as araBAD promoter can be used. This expression system is based on the Arabinose operon under the control of L-arabinose as an inducer ([44]).

Concentration of Inducers Despite its several advantages, there exists a major disadvantage to using *E. coli* as a protein production system. Since transcription and translation are rapid and coupled in *E. coli,* many mammalian proteins partially fold or misfold upon gene induction. In contrast, these proteins would fold properly due to availability of longer folding times and the assistance from chaperones in eukaryotic cells. However, in *E. coli,* reduction in transcription rate can be brought about by choosing the lowest concentration of inducers that yields properly folded proteins. For example, using lower concentrations of the inducers IPTG and L-arabinose has proven to be highly effective in routine recombinant protein production experiments [48]. The routinely used IPTG and L-arabinose concentrations for protein production range from 0.05 to 2 mM and 0.0002 to 2%, respectively. The wider concentration range of L-arabinose-based induction suggests that gene expression can be tuned finely using the araBAD promoter compared to the lacUV5 promoter of Lactose Operon.

Temperature In addition to reducing the concentration of inducers in expression culture, lowering the culture temperature is a routine approach for producing high-quality recombinant proteins. At lower temperature, cellular metabolism slows down leading to diminished rates of transcription and translation, hence reduced protein misfolding and aggregation in the host cell. Additionally, several proteases are less active at lower temperatures consequently minimizing the degradation of proteolytically sensitive recombinant proteins [49]. Due to the profound role of temperature in protein production, it is greatly advocated to grow the expression culture at 18 °C for certain types of proteins (as described above) in *E. coli.*

Using Molecular Chaperones Often, the proper folding of large eukaryotic proteins is assisted by special proteins called molecular chaperones. Even though molecular chaperones are naturally produced in *E. coli*, their contribution is severely limited either due to low concentration or low specificity toward foreign proteins during their over-expression. Thus, chaperones having broad specificity are co-expressed singly or in combination with the target protein in bacterial hosts. During folding of nascent polypeptide chains, exposure and binding of their hydrophobic surfaces to each other accelerate aggregation of partially folded proteins. Such aggregated complexes are tough to reverse and often pose burden to the host cells. Common chaperone systems, for example, the GroEL-GroES system, function by temporarily masking the hydrophobic surfaces of nascent target proteins from each other during the folding process.

4.8 Posttranslational Modifications in Bacterial Expression Systems

Often, successful episode of transcription and translation of a recombinant protein does not ensure that the product will be biologically active. This is because most proteins, especially those of eukaryotic origin, are naturally modified by the cell through posttranslational modifications (PTMs). Only after being modified, these proteins display function. Posttranslational modifications are chemical modifications made naturally to proteins after they are synthesized [50]. Via PTMs, a protein's activity is modulated by covalent modification of its backbone (cleavage of peptide bonds) or of its amino acid side chains. So far, we know more than 500 different types of PTMs in proteins. Commonly used chemical modifiers include phosphoryl, hydroxyl, acetyl, carboxyl, amide, methyl, adenylyl, palmitoyl, myristoyl, uridylyl, prenyl, sulfate, oligosaccharides, and adenosine diphosphate ribosyl groups ([50]). When an amino acid residue is modified, a novel property is introduced in the protein. Therefore, evolutionarily, PTMs have aided extension of the functions of proteins beyond those manifested by unique amino acid sequences. Hence, to unravel the functions of proteins, their PTMs must be mirrored when expressed recombinantly. Often, it is the inability of an expression host like bacteria to apply appropriate PTMs on a recombinant protein that causes it to be inactive, misfold, and become insoluble while being produced. While there are PTMs in prokaryotes like *E. coli*, for example, glycosylation, they are less common and slightly different in nature (Fig. 4.5). Consequently, there arises the need for either not choosing or modifying *E. coli* as an expression host for some proteins of eukaryotic origin.

Although it is interesting to learn about such a diverse range of PTMs in both eukaryotes and bacteria, to cover them all from the standpoint of recombinant protein production in bacteria is beyond the scope of this chapter. We shall therefore discuss one of the most common PTMs, i.e., glycosylation.

Glycosylation It is estimated that about two-thirds of proteins in eukaryotes are glycosylated [51]. Therefore, it could be assumed that most eukaryotic proteins of therapeutic importance might also require proper glycosylation for full functionality [52, 53]. This is why the majority of approved therapeutic proteins are expressed in mammalian cells. However, as mentioned earlier, mammalian systems have disadvantages such as high cost, low yield, product heterogeneity, slow growth, and complex manipulation of their characteristics. These disadvantages have a great effect on healthcare research since the high capital investment gives rise to products that are expensive and limited in stock during incidences like disease outbreaks. In view of this fact, numerous studies are currently focused on expressing affordable and high-quality complex human glycoproteins in simple systems such as *E. coli*. It is worth mentioning that the continuing efforts have been fruitful owing to the significant progress in regard to use of genetically engineered *E. coli* for production of antibodies [54].

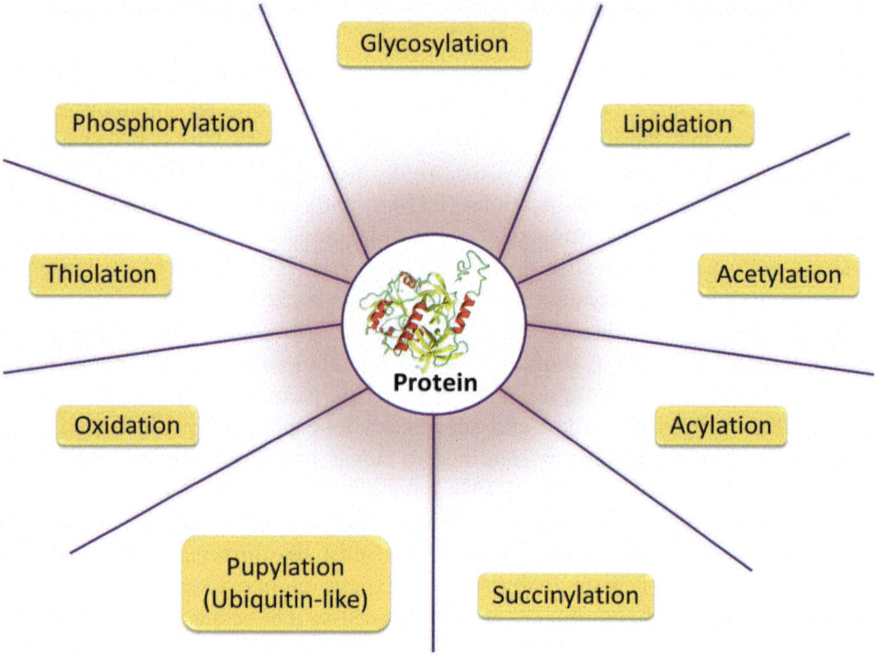

Fig. 4.5 Posttranslational Modifications (PTMs) of *E. coli* proteins expressed in *E. coli*

First thought to be a property of only eukaryotes, it is now well established that natural glycosylation of proteins also occurs in bacteria [55]. From the time of its discovery in *Campylobacter jejuni*, a bacterium that causes diarrhea in humans, the bacterial protein glycosylation machinery has been artificially transferred into *E. coli* [56]. This successful attempt gave rise to an area in biotechnology called "Bacterial Glycoengineering," and is on the rise since then.

Despite the patterns of protein glycosylation in bacteria differing slightly from the eukaryotic counterparts, transferring a broad glycosylation system in *E. coli* has allowed researchers to take advantage of this popular expression host for composing novel products such as vaccines and therapeutic enzymes besides using Glycoengineering for research [57].

4.9 Expression in Yeast Cells

While natural transformation is commonly found among various prokaryotes, there are very few reports which clearly show that eukaryotic microorganisms such as yeasts are naturally capable of taking up genetic material. According to one report, DNA uptake in yeast is active when ample sugar is metabolized in the absence of other nutrients, which suggests that such a condition is likely to occur under normal conditions in a cell [58]. Yeasts are single-celled eukaryotes that have gained appeal for studies of basic processes in molecular and cellular biology. This is due to the similarities between yeast and human enzymes making the study of human proteins in yeast systems more biologically relevant [59]. Therefore, there are increasing numbers of examples of human proteins that function properly when artificially expressed in yeast cells. This is mainly due to the fact that being eukaryotes, they provide chemical habitat for posttranslational modifications and secretion of proteins, resulting in a product that is similar or identical to the native protein [60]. Like bacteria, yeasts are simple to cultivate at industrial scales on inexpensive growth media, and there is an array of techniques already available for its genetic manipulation. Another important benefit of using yeast for recombinant protein expression is the safety of yeast-derived pharmaceutical preparations. Most yeast cell walls lack toxic pyrogens whereas mammalian cells might contain viral or oncogenic DNA and antigens. Moreover, the genome of the commonly used yeast *Saccharomyces* is rigorously characterized, which permits manipulation of specific regions of its chromosome to better understand eukaryotic biology [61, 62]. Taking into consideration the ease and practicality of the transformation technique of gene manipulation, the importance of yeasts as model organisms and industrial preparations has therefore been uplifted by application of the transformation technique to yeast biology. A number of other yeasts have often been used in preference to *Saccharomyces cerevisiae* for mammalian gene expression owing to the advantages in protein secretion efficiency, accurate posttranslational modifications, sensitive gene regulatory elements, and high yields. Therefore, a substantial section of the literature has been devoted to discussing other yeasts such as *Pichia pastoris, Hansenula polymorpha, Kluyveromyces lactis, Yarrowia lipolytica*, and *Schizosaccharomyces pombe* [63]. Despite having several yeasts in the toolbox of yeast recombinant technology, there are still problems that arise due to incompatibility between the yeast expression systems and the proteins that are being expressed. One such problem pertains to human proteins that are processed within cellular organelles (e.g., those taking part in the secretory pathway). Expression of those proteins in yeast cells might lead to incorrectly folded and/or glycosylated proteins. To overcome these problems, the secretory pathway of yeast *P. pastoris* has been genetically modified so as to mimic protein glycosylation in humans [64, 65].

Though important under certain specific conditions, bacterial system still outweighs the yeast expression system due to several factors that include homogeneity, ease of growth and purification, expression and finally the yield. Furthermore, proteins expressed in insect and mammalian systems are primarily used for cell biology studies where posttranslational modification gets priority over yield and homogeneity. Table 4.2 provides a simple comparison of different types of protein expression systems.

Table 4.2 Comparison of different protein expression systems

	Bacteria (*E. coli*)	Yeast	Mammalian cells	Insect cells
Advantages	• Easy manipulation • Low cost • Less cell doubling time (~30 min) • Robust, easy • High yield • Easy radiolabelling for structural studies	• Easy manipulation • Low cost	• Natural protein folding • Posttranslational modifications	• Posttranslational modifications
Disadvantages	• Most posttranslational modifications absent • Highly reducing cytoplasm thus production of disulfide-bonded proteins difficult • Production of membrane proteins difficult	• Minimum posttranslational modifications • Production of membrane proteins difficult	• Large cell doubling time (~24 h) • High cost, lower yield	• Large cell doubling time (~18 h) • High cost • Production of membrane proteins difficult

4.10 Conclusions

This chapter discusses two important components of recombinant DNA technology:

(a) Transformation methods.
(b) Bacterial Protein Expression systems.

It provides a detailed discussion on competent cell preparation, different methods of transformation including protocols and troubleshooting guides. Furthermore, it elaborates on different protein expression systems in bacteria along with protocols and their use for specific requirements. This chapter will therefore guide a researcher in choosing the appropriate transformation method and protein expression system to obtain an optimum amount of functional protein.

Acknowledgments The authors thank Advanced Centre for Treatment, Research and Education in Cancer (ACTREC) for providing necessary infrastructure and resources for successful completion of the chapter. The authors also acknowledge Dr. Mohan Shankar, Bose Lab, ACTREC, and Ms. Chanda Baisane, Bose Lab, ACTREC, for their critical inputs and formatting of the manuscript.

Problems

Multiple Choice Questions

1. Which factor among the following is important to ensure proper folding of the heterologous proteins:
   (a) Expression vector
   (b) RNA polymerase
   (c) Nutrient media
   (d) Temperature and inducer concentration
2. In the chemical method for competent cell preparation, $CaCl_2$ acts as:
   (a) Nutrient component
   (b) Cation bridge
   (c) Anion bridge
   (d) Buffering agent
3. The property of DNA that is a hindrance to its uptake by bacterial cells in the process of transformation is:
   (a) Double helical structure
   (b) Negative charge
   (c) Nitrogen bases
   (d) Use of chemicals during transformation

Subjective Questions

1. A gene of interest was successfully cloned in an expression vector; its sequence was confirmed and was transformed into E. coli DH5α cells for plasmid preparation. Sufficient amount of plasmid was isolated and the quality of the DNA was checked using agarose gel electrophoresis. The same construct was subsequently transformed into BL21 (DE3) cells at a concentration of 10 ng/μL. Although control plates showed colonies, no colonies were observed on the test ampicillin plates. What could possibly explain the observation?
2. A protein coding gene was successfully cloned in an expression vector that was confirmed by sequencing. This construct containing the gene was transformed into a protein expression host like BL21 (DE3) and colonies were obtained. After growing a large-scale culture and harvesting the cells, it was found that no protein was expressed. What could be the probable reason for such an observation?

References

1. Griffith F. The significance of pneumococcal types. Epidemiol Infect. 1928;27(2):113–59.
2. Martinovitch PN, et al. Teratological changes in the offspring of chicken embryos treated with Tyrode or with Tyrode plus DNA. Development. 1962;10(2):167–77.
3. Hanahan D, Weinberg RA. The hallmarks of cancer. Cell. 2000;100(1):57–70.
4. Chen I, Dubnau D. DNA uptake during bacterial transformation. Nat Rev Microbiol. 2004;2 (3):241–9.
5. Johnston C, et al. Bacterial transformation: distribution, shared mechanisms and divergent control. Nat Rev Microbiol. 2014;12(3):181–96.
6. Claverys J-P, Martin B, Polard P. The genetic transformation machinery: composition, localization, and mechanism. FEMS Microbiol Rev. 2009;33(3):643–56.
7. Lorenz MG, Wackernagel W. Bacterial gene transfer by natural genetic transformation in the environment. Microbiol Rev. 1994;58(3):563–602.
8. Nielsen KM, et al. Release and persistence of extracellular DNA in the environment. Environ Biosaf Res. 2007;6(1–2):37–53.
9. Chan W-T, et al. A comparison and optimization of methods and factors affecting the transformation of Escherichia coli. Biosci Rep. 2013;33:6.
10. Dower WJ, Miller JF, Ragsdale CW. High efficiency transformation of E. coli by high voltage electroporation. Nucleic Acids Res. 1988;16(13):6127–45.
11. Yoshida N, Sato M. Plasmid uptake by bacteria: a comparison of methods and efficiencies. Appl Microbiol Biotechnol. 2009;83(5):791–8.
12. Liu C, et al. Conducting nanosponge electroporation for affordable and high-efficiency disinfection of bacteria and viruses in water. Nano Lett. 2013;13(9):4288–93.
13. Aune TEV, Aachmann FL. Methodologies to increase the transformation efficiencies and the range of bacteria that can be transformed. Appl Microbiol Biotechnol. 2010;85(5):1301–13.
14. Hanahan D. Studies on transformation of Escherichia coli with plasmids. J Mol Biol. 1983;166 (4):557–80.
15. Mandel M, Higa A. Calcium-dependent bacteriophage DNA infection. J Mol Biol. 1970;53 (1):159–62.

16. Inoue H, Nojima H, Okayama H. High efficiency transformation of Escherichia coli with plasmids. Gene. 1990;96(1):23–8.
17. Monod J. The growth of bacterial cultures. Annu Rev Microbiol. 1949;3(1):371–94.
18. Neumann E, et al. Gene transfer into mouse lyoma cells by electroporation in high electric fields. EMBO J. 1982;1(7):841–5.
19. Prasanna GL, Panda T. Electroporation: basic principles, practical considerations and applications in molecular biology. Bioprocess Eng. 1997;16(5):261–4.
20. Tung WL, Chow K-C. A modified medium for efficient electrotransformation of *E. coli*. Trends Genet (Regular ed.). 1995;11(4):128–9.
21. Asif A, et al. Revisiting the mechanisms involved in calcium chloride induced bacterial transformation. Front Microbiol. 2017;8:2169.
22. Fiedler S, Wirth R. Transformation of bacteria with plasmid DNA by electroporation. Anal Biochem. 1988;170(1):38–44.
23. Eynard N, et al. Fast kinetics studies of Escherichia coli electrotransformation. Eur J Biochem. 1992;209(1):431–6.
24. Mehier-Humbert S, Guy RH. Physical methods for gene transfer: improving the kinetics of gene delivery into cells. Adv Drug Deliv Rev. 2005;57(5):733–53.
25. Lee Y-H, et al. Nanoparticles facilitate gene delivery to microorganisms via an electrospray process. J Microbiol Methods. 2011;84(2):228–33.
26. Miller MW. Gene transfection and drug delivery. Ultrasound Med Biol. 2000;26:S59–62.
27. Taniyama Y, et al. Development of safe and efficient novel nonviral gene transfer using ultrasound: enhancement of transfection efficiency of naked plasmid DNA in skeletal muscle. Gene Ther. 2002;9(6):372–80.
28. Han YW, et al. Identification and characterization of a novel adhesin unique to oral fusobacteria. J Bacteriol. 2005;187(15):5330–40.
29. Sheng Y, Mancino V, Birren B. Transformation of Escherichia coli with large DNA molecules by electroporation. Nucleic Acids Res. 1995;23(11):1990–6.
30. Morrow JF, et al. Replication and transcription of eukaryotic DNA in Esherichia coli. Proc Natl Acad Sci. 1974;71(5):1743–7.
31. Studier FW, Moffatt BA. Use of bacteriophage T7 RNA polymerase to direct selective high-level expression of cloned genes. J Mol Biol. 1986;189(1):113–30.
32. Lee I, Berdis AJ, Suzuki CK. Recent developments in the mechanistic enzymology of the ATP-dependent Lon protease from Escherichia coli: highlights from kinetic studies. Mol BioSyst. 2006;2(10):477–83.
33. McCarter JD, et al. Substrate specificity of the Escherichia coli outer membrane protease OmpT. J Bacteriol. 2004;186(17):5919–25.
34. Murray NE. Type I restriction systems: sophisticated molecular machines (a legacy of Bertani and Weigle). Microbiol Mol Biol Rev. 2000;64(2):412–34.
35. Ferrer M, et al. Chaperonins govern growth of Escherichia coli at low temperatures. Nat Biotechnol. 2003;21(11):1266–7.
36. Derouazi M, et al. High-yield production of secreted active proteins by the Pseudomonas aeruginosa type III secretion system. Appl Environ Microbiol. 2008;74(11):3601–4.
37. Landry TD, et al. Safety evaluation of an α-amylase enzyme preparation derived from the archaeal order Thermococcales as expressed in Pseudomonas fluorescens biovar I. Regul Toxicol Pharmacol. 2003;37(1):149–68.
38. Anné J, et al. Recombinant protein production and streptomycetes. J Biotechnol. 2012;158 (4):159–67.
39. Sianidis G, et al. Functional large-scale production of a novel Jonesia sp. xyloglucanase by heterologous secretion from Streptomyces lividans. J Biotechnol. 2006;121(4):498–507.
40. Bettache A, et al. Production of endoglucanase by a novel strain Streptomyces sp. TKJ2 and its optimization under submerged fermentation. Biotechnology (Rajkot). 2014;10:29–36.
41. Drew D, et al. Assembly and overexpression of membrane proteins in Escherichia coli. Biochim Biophys Acta (BBA)-Biomembranes. 2003;1610(1):3–10.

42. Laible PD, et al. Towards higher-throughput membrane protein production for structural genomics initiatives. J Struct Funct Genom. 2004;5(1):167–72.
43. Walsh G. Biopharmaceutical benchmarks 2006. Nat Biotechnol. 2006;24(7):769–76.
44. Schleif R. Regulation of the L-arabinose operon of Escherichia coli. Trends Genet. 2000;16 (12):559–65.
45. Sharp PM, Li W-H. The codon adaptation index-a measure of directional synonymous codon usage bias, and its potential applications. Nucleic Acids Res. 1987;15(3):1281–95.
46. Wakagi T, et al. Cloning of the gene for inorganic pyrophosphatase from a thermoacidophilic archaeon, Sulfolobus sp. strain 7, and overproduction of the enzyme by coexpression of tRNA for arginine rare codon. Biosci Biotechnol Biochem. 1998;62(12):2408–14.
47. Hayat SM, et al. Recombinant protein expression in Escherichia coli (E. coli): what we need to know. Curr Pharm Des. 2018;24(6):718–25.
48. Turner P, Holst O, Karlsson EN. Optimized expression of soluble cyclomaltodextrinase of thermophilic origin in Escherichia coli by using a soluble fusion-tag and by tuning of inducer concentration. Protein Expr Purif. 2005;39(1):54–60.
49. Pinsach J, et al. Influence of process temperature on recombinant enzyme activity in Escherichia coli fed-batch cultures. Enzym Microb Technol. 2008;43(7):507–12.
50. Knorre D, Kudryashova N, Godovikova T. Chemical and functional aspects of posttranslational modification of proteins. Acta Nat. 2009;1(3):3.
51. Apweiler R, Hermjakob H, Sharon N. On the frequency of protein glycosylation, as deduced from analysis of the SWISS-PROT database. Biochim Biophys Acta (BBA)-General Subjects. 1999;1473(1):4–8.
52. Solá R, Rodríguez-Martínez J, Griebenow K. Modulation of protein biophysical properties by chemical glycosylation: biochemical insights and biomedical implications. Cell Mol Life Sci. 2007;64(16):2133–52.
53. Sinclair AM, Elliott S. Glycoengineering: the effect of glycosylation on the properties of therapeutic proteins. J Pharm Sci. 2005;94(8):1626–35.
54. Lizak C, et al. N-linked glycosylation of antibody fragments in Escherichia coli. Bioconjug Chem. 2011;22(3):488–96.
55. Nothaft H, Szymanski CM. Protein glycosylation in bacteria: sweeter than ever. Nat Rev Microbiol. 2010;8(11):765–78.
56. Wacker M, et al. N-linked glycosylation in Campylobacter jejuni and its functional transfer into *E coli*. Science. 2002;298(5599):1790–3.
57. Baker JL, Çelik E, DeLisa MP. Expanding the glycoengineering toolbox: the rise of bacterial N-linked protein glycosylation. Trends Biotechnol. 2013;31(5):313–23.
58. Nevoigt E, Fassbender A, Stahl U. Cells of the yeast Saccharomyces cerevisiae are transformable by DNA under non-artificial conditions. Yeast. 2000;16(12):1107–10.
59. Rine J. The yeast Saccharomyces cerevisiae in molecular and cellular biology: a smaller but not lower eucaryote. Am Zool. 1989;29(2):605–16.
60. Sudbery PE. The expression of recombinant proteins in yeasts. Curr Opin Biotechnol. 1996;7 (5):517–24.
61. Cherry JM, et al. Saccharomyces genome database: the genomics resource of budding yeast. Nucleic Acids Res. 2012;40(D1):D700–5.
62. Mewes H, et al. Overview of the yeast genome. Nature. 1997;387(6632):7–8.
63. Baghban R, et al. Yeast expression systems: overview and recent advances. Mol Biotechnol. 2019;61(5):365–84.
64. Hamilton SR, et al. Production of complex human glycoproteins in yeast. Science. 2003;301 (5637):1244–6.
65. Choi B-K, et al. Use of combinatorial genetic libraries to humanize N-linked glycosylation in the yeast Pichia pastoris. Proc Natl Acad Sci. 2003;100(9):5022–7.

Introduction to Recombinant Protein Purification

5

Nitu Singh and Kakoli Bose

Abstract

Purified form of recombinant proteins is a prerequisite for undertaking in vitro biochemical and structural analyses of these macromolecules. Unfortunately, due to other proteins from the expression host, such as *E. coli,* the task of obtaining the desired protein from the heterologous system at highest purity and in sufficient quantity is arduous. With the increasing demand for recombinantly purified proteins both in basic and industrial research, over the last five decades, a plethora of research endeavors have been directed toward developing efficient protein purification techniques that would precisely amalgamate time and yield cost-effectively. However, it is extremely important to put careful forethought prior to developing a purification flow-scheme for a target protein to obtain the best possible output. This chapter outlines the general considerations to be undertaken while designing and streamlining these protocols with the help of recent advances in protein purification methodologies. It also provides an overview of the various chromatographic techniques that will be further elaborated in the succeeding chapters of this book.

Keywords

Purification · Protein stability · Salting · Inclusion bodies · Dialysis · Chromatography

N. Singh
Regional Centre for Biotechnology, NCR Biotech Science Cluster, Faridabad, Haryana, India

K. Bose (✉)
Integrated Biophysics and Structural Biology Lab, Advanced Centre for Treatment, Research and Education in Cancer (ACTREC), Tata Memorial Centre, Mumbai, India

Homi Bhabha National Institute, BARC Training School Complex, Mumbai, India
e-mail: kbose@actrec.gov.in

K. Bose (ed.), *Textbook on Cloning, Expression and Purification of Recombinant Proteins*, https://doi.org/10.1007/978-981-16-4987-5_5

5.1 Introduction

Protein purification gained major attention, when James B. Sumner in 1926 started purification and crystallization of urease from yeast [1]. Since then, biochemists, in particular, have put tremendous effort in developing convenient purification methods for obtaining recombinant proteins from the bacterial or eukaryotic expression systems using the differences in the physicochemical properties of the proteins [2, 3]. Over the past five decades, the methods for separation and purification of recombinant proteins have contributed immensely to the advancements in the fields of biosciences and bioengineering. This has been mainly possible due to the increased use of tags for separation of recombinant proteins, development of new chromatographic techniques, and use of computerized sophisticated instruments [4, 5]. The challenge in protein purification is self-evident, given the complex mixture of biomolecules present inside an expression host, commonly used for recombinant protein extraction. With the advent of new generations of chromatographic media and automated systems, gone are those days when an investigator used to spend several months in establishing a purification protocol. However, not all problems are resolved by using sophisticated column packing and laboratory equipment. Difficulties still persist in finding optimal conditions for sample pretreatment, choosing an appropriate buffer condition, or dealing with insolubility of the proteins. Therefore, prior to stepping into purification of a particular protein of interest, it is important to consider a few things such as the purpose of purifying the recombinant protein, its purity, and its required concentration as well as storage condition [6]. All these factors are critical in designing and executing a procedure for purifying a target protein to a sufficient degree in a cost-effective and timely manner. In this chapter, we discuss purification steps for isolating recombinant proteins from hosts, such as *E. coli*, emphasizing on the conditions for cell lysis, protein solubility and stability. It also includes an overview of protein separation techniques using different chromatographic platforms, with each individual technique discussed in more detail in the subsequent chapters of this book.

5.2 Databases and Tools to Determine Physicochemical Properties of Protein

Now that more than thousands of whole genomes have been sequenced, determining the physicochemical properties of the proteins from their amino acid sequences can help in designing an effective purification strategy for that protein. Some of the parameters that we can learn from the amino acid sequence of the protein using a sequence analysis software is described below.

5.2.1 Physicochemical Parameters Important in Initial Designing of the Purification Procedure

5.2.1.1 Molecular Weight of the Protein

The amino acid composition of the protein can be easily used to calculate the molecular weight of the polypeptide, such as for a cloned protein. This information is very helpful in the initial estimation of the level of protein expression on SDS-PAGE (in case the expressed protein is large enough to give a distinct band). However, it is important to note that the calculated molecular weight and the apparent weight based on the mobility on SDS-PAGE can sometimes deviate due to several other factors [7]. A more quantitative method of molecular mass estimation, such as mass spectrometry and analytical ultracentrifugation, can be used in such cases [8, 9]. In addition, the knowledge of the molecular weight is also important for the initial selection of the size-exclusion chromatography medium for separation of the proteins under native conditions [10]. In cases where the amino acid sequence is unavailable, one can combine the size exclusion chromatography (size-based separation technique) and the activity assays to estimate the molecular weight [10].

5.2.1.2 Isoelectric Point, pI

Theoretically, the isoelectric point of the protein can be estimated from the primary sequence or can be determined experimentally using isoelectric focusing [11]. The pI value is helpful toward selecting a suitable matrix for purification using ion-exchange chromatography (charge-based separation technique) [12]. Additionally, since a protein tends to have lowest solubility at its pI [13], one can also consider an isoelectric reversible precipitation step during purification considering that the target is not in a stable complex with other proteins [14]. This helps in isolating the recombinant protein form the cellular lysate of the host, such as *E. coli*, prior to its loading on a chromatographic column.

5.2.1.3 Molar Extinction Coefficient/Absorptivity Coefficient

Using the amino acid sequence of a protein, one can calculate its molar extinction coefficient. This is one of the important parameters for the estimation of the protein concentration. At a wavelength of 280 nm, tyrosine, tryptophan, and cystine (cysteine does not absorb much at wavelengths >260 nm, while cystine does) exhibit strong UV-light absorption. Cystine is formed when a pair of cysteine molecules join together by a disulfide bond. Using the Beer-Lambert equation, the extinction coefficient (ε) of the native protein is computed [14, 15]:

$$\begin{aligned}\varepsilon\ (\text{Prot}) = {} & (\text{Number of Tyr}) * \varepsilon(\text{Tyr}) + (\text{Number of Trp}) * \varepsilon(\text{Trp}) \\ & + (\text{Number of Cystine}) * \varepsilon(\text{Cystine})\end{aligned}$$

where ε(Tyr) = 1490, ε(Trp) = 5500, ε(Cystine) = 125; (for proteins in water measured at 280 nm)

The absorption of the UV-light is proportional to the aromatic amino acid content and total concentration of the protein. Using the estimated extinction coefficient (with fixed amino acid composition of a given protein), we can easily calculate the protein's concentration in solution from its absorbance. The online tool ProtParam (http://www.expasy.org/tools/protparam) gives two values based on the above equation [16]. The first value is based on the assumption that all cysteines in the sequence occur as half cystines (i.e., all pairs of cysteine residues form cystines), while the second value assumes that all pairs of cysteines are in reduced state. It is reported that this computation is considerably reliable for Trp containing proteins; however there may be around 10% error in proteins without Trp [17]. Nevertheless, this method is not valid if the protein sample contains nucleic acid contaminants or other components that show significant absorption at 280 nm such as iron–sulfur centers, or bound cofactors. Moreover, it is not accurate in case of complex protein solutions, such as cell lysates, which contain a mixture of proteins with unknown extinction coefficients.

5.2.1.4 Cysteine Content

Knowing the number of cysteines in the protein sequence will help decide whether the purification buffer should contain a reducing agent. The reducing agents such as beta mercaptoethanol or dithiothreitol in the buffer can prevent any unwanted intra- or inter-disulfide bond formation and aid in protein solubility as well as stability.

5.2.1.5 Stability

The knowledge of the protein stability with respect to pH, salt, temperature, proteases, or aggregation can be very helpful during purification and storage of the proteins [18]. However, most of these parameters are only experimentally derived. Using bioinformatics tools, such as ProtParam [16], one can estimate the in vivo half-life and instability index of the protein from its amino acid sequence. Half-life is the time taken for half of the amount of protein in a cell to disappear after its synthesis. The calculation of half-life in ProtParam relies on the N-end rule [19]. This rule that originated from the studies performed in model systems such as mammalian cells, yeast, and *E. coli* demonstrate importance of N-terminal residues in determining the stability of the protein [20].

The instability index (II) is estimated by calculating the frequency of occurrences of dipeptides in the protein of interest, which is then compared with a set of test proteins that are known to be stable or unstable [21]. The instability value less than 40 is predicted as stable, while a value above 40 indicates that the protein might be unstable. This information is useful during protein purification as a protein predicted to be unstable needs to be handled with special care such as use of low temperature during purification and addition of salts and/or protease inhibitor cocktails in the extraction buffers. Furthermore, a shorter purification and storage time could also be considered.

5.2.1.6 Hydrophobicity

It is possible to predict if the protein is hydrophilic or hydrophobic by analyzing the amino acids in the sequence. One such method was developed by Kyte and Doolittle that plots the hydropathy value over the length of the protein sequence [22]. Based on the hydrophobic or hydrophilic properties of the 20 amino acids, the hydropathy scale is calculated at each point in the sequence. This can help in determining the hydrophobic core of the protein and potential membrane spanning region, which will certainly be important in designing the purification scheme. The ProtParam tool uses this method to predict the GRAVY (grand average of hydrophobicity) value for a protein, which is calculated as the sum of hydropathy values of all the amino acids divided by the number of residues in the sequence [22].

5.2.2 Bioinformatics Resources

One of the most frequently used online tools for computing the physicochemical properties of proteins using the amino acid sequence is the ProtParam feature of ExPASy (Expert Protein Analysis Software) (http://www.expasy.org/tools/protparam) as mentioned above. ProtParam calculates parameters such as molecular weight, amino acid composition, molar extinction coefficient, atomic composition, instability index, aliphatic index, and GRAVY [16].

To use this tool, go to ExPASy ProtParam tools and either enter the Swiss-Prot/TrEMBL accession number (for example, Q9VFJ3- human mitochondrial serine protease HtrA2) or a sequence identifier (ID) (for example, HTRA2_DROME), or you can paste the raw amino acid sequence (in one-letter code) in the box and click "compute parameters." The result obtained can be saved or printed for further analysis.

Another commercial package is "Protean" from the Lasergene protein module of DNASTAR (http://www.dnastar.com). Similar to ProtParam, it also performs protein sequence analysis, wherein it computes parameters such as protein stability. In addition, it also allows for residue-specific change in the protein 3D-structure and helps in predicting whether these changes promote stabilizing or destabilizing effects. Using the Lasergene "Protein's advanced protein design software" one can make accurate protein stability predictions in minutes for the designed mutants.

5.3 Lysis and Protein Extraction

5.3.1 Source Material for the Protein

Although purification of proteins from their natural sources is relatively common, with the recent developments in the field of gene manipulation and recombinant expression, one usually targets overexpression systems such as bacteria, yeast, insect, or mammalian cells. Each individual host system has its own advantages and disadvantages. For instance, *E. coli* gives the highest yield, typically an average yield of 2.5 g/l of culture, while it lacks the posttranslational modifications, if any [23]. On the other hand, if the protein requires glycosylation, the yeast system provides more extensive glycosylation than insect and mammalian cells [24]. The choice depends on the properties of the protein as well as on its downstream application.

Proteins overexpressed in genetically modified organisms or cultured eukaryotic cells are majorly localized in the cytoplasm or can be targeted for secretion into the medium. In case of bacterial expression, such as in *E. coli*, they can also be targeted to the periplasmic space or selectively released in the growth medium by altering the growth conditions [25]. This can help in eliminating the need for cell lysis, thus providing a substantial level of purity at the initial steps. Therefore, choice of the host and overexpression strategy is one of the rate-limiting steps of protein purification.

5.3.1.1 Extraction Methods

In most cases, the extraction procedures depend on the source of the protein, which could be bacterial, yeast, or mammalian cells and either intracellular or extracellular. Extraction from an intracellular source often faces compromised recovery and purity. The main objective of the extraction should be to achieve the desired protein in a non-degraded or non-denatured form with minimal or no contaminants.

The extraction protocol is usually optimized by strategic variations in the parameters such as extraction medium, time, temperature, equipment for lysis, and energy input (agitation speed, pressure, etc.). It is to be noted that the choice of the method should be such that it is as gentle as possible because too vigorous or harsh conditions might denature the desired protein or release the endo-proteolytic enzymes and cause general acidification. Also, one should keep in mind that the extraction should be performed quickly at low temperatures (at 4 °C or on ice) in a suitable buffer to maintain the ionic strength and pH to stabilize the protein [3]. One of the major problems that is confronted during extraction is proteolysis or contamination with nucleic acids. However, to some extent this can be tackled by performing the extraction at low temperatures in the presence of protease inhibitors and with inclusion of nucleases in the extraction medium. Therefore, for an optimized design, it is essential to standardize preliminary experiments in a small scale for maximization of protein content and activity, which can later be scaled-up effectively.

The composition of the extraction medium should be such that protein remains stable and is effectively released from the cells with maximum recovery and purity. Following are the factors that should be taken into consideration while preparing the extraction medium/lysis buffer: buffer salt, pH, reducing agent, chaotropic agents, detergents, metal ions, proteolytic inhibitor, and DNase [3].

5.3.1.2 Extraction Medium/Lysis Buffer

Buffer Salt and pH

There are several factors that should be considered while selecting a buffer. Firstly, pH optimum of the protein is important in determining the best buffer condition. In

Table 5.1 Buffers and pKa values

Buffer	pKa (25 °C)	pH range for use
Phosphoric acid	2.12	1.1–3.1
Formic acid	3.75	2.8–4.8
Acetic acid	4.75	3.8–5.8
Sulfonic acid	6.91	5.9–7.9
Dihydrogen phosphate ion	7.21	6.2–8.2
Ammonium ion	9.25	8.2–10.2
Hydrogen phosphate ion	12.66	11.3–13.3

Table 5.2 Detergents used for protein solubilization [28]

Detergent	Ionic character	Critical micelle concentration (% w/v)
Triton-X 100	Non-ionic	0.02
Nonidet P-40	Non-ionic	0.012
Octylglucoside	Non-ionic	0.73
Tween 80	Non-ionic	0.002
Sodium deoxycholate	Anionic	0.21
CHAPS 3-[(3-Cholamidopropyl dimethyl amino propanesulfonic acid)	Zwitterionic	1.4

practice, the pH value is chosen such that the protein activity is maximum. Since the buffering capacity is maximal within one pH unit from the pKa value, the selected buffer preferably should be close to this value (Table 5.1, buffers with the pKa values). Secondly, good buffers should be relatively free of side effects. For example, TRIS and primary amine buffers can form Schiff base adducts with aldehydes and ketones and interfere with Bradford dye-binding assay [26]. Some inorganic buffers can remove metal ions by chelation, thus inhibiting metal-dependent protein activity. Lastly, in protein purification, the cost and compatibility of the buffer with different purification techniques are important parameters to be considered. Once an ideal buffer is chosen, one needs to select a suitable ionic strength of the buffer. Most of the proteins show maximum solubility and activity at low to moderate ionic concentrations, 0.05–0.1 M. It is to remember that proteins also act as buffers, and therefore it is important to carefully monitor the pH after addition of large amounts of proteins to a weakly buffered solution.

5.3.1.3 Detergents and Chaotropic Agents

In some cases, the desired protein could be associated with the membrane or might be aggregated due to its hydrophobic nature. In such cases, use of detergent or chaotropic agents helps in solubilization of the protein, thereby allowing its separation. The detergents are amphipathic molecules that during solubilization help the membrane proteins to partition into apolar lipid bilayers [27]. They also aid in masking the hydrophobic surface of the proteins, thus preventing protein aggregation. Some of the commonly used detergents are listed in Table 5.2. Many of these at a concentration below the critical micelle concentration (CMC), i.e., the concentration of a surfactant above which it starts to form micelle, do not denature or interfere with the protein's biological activity, with few exceptions such as SDS (sodium dodecyl sulfate) [29]. Mostly, detergents are added in the first step of purification, i.e., in the lysis buffer. This is because the presence of detergents in the later steps often complicates the purification process, especially in column chromatography [6]. In a few cases, there might be a requirement of the detergent throughout the purification process, leading to separation of purified protein-detergent complexes. Beside detergent, one can also use chaotropic agents, such as urea, guanidine hydrochloride, or polyethylene glycol (moderate hydrophobic organic compound), to solubilize the aggregates [30]. Chaotropes are agents that denature proteins by disordering the surrounding water molecules through disruption of the hydrogen bonding network among the water molecules. These chaotropic agents are commonly used in purification of the recombinant proteins from the inclusion bodies as discussed in the following sections.

5.3.1.4 Reducing Agents

Proteins with free cysteine residues (i.e., exposed thiol groups) can easily get oxidized to disulfides, sulfinic acid (—SOOH), or sulfonic acid (—SO_2OH) during the process of purification [31]. This is generally due to lower redox potential inside the cell as compared to the surrounding medium used for protein purification. The oxidation of the thiol groups could be problematic as it would affect the solubility and stability of the proteins. Reducing agents such as 1,4-dithioerythritol (DTE), dithiothreitol (DTT), beta mercaptoethanol (BME), or Tris (2-carboxyethyl) phosphine (TCEP) at concentrations 1–20 mM can be safely added to protect the thiol group, without reducing the internal disulfide bridges [32]. However, one should be careful while choosing the type of reducing agent. For instance, a protein sample should not be stored in BME for too long as it is highly volatile and susceptible to air oxidation. The oxidized form of BME can react with the reduced cysteines to form disulfide [31]. However, since BME is comparatively cheaper, it can be added during the purification process, but for storage, a more stable agent like TCEP should be used. Alternatively, the oxidation of the reducing agent can also be hindered by addition of chelating agents such as EDTA (ethylene diamine tetra acetic acid) in the concentration range of 10–25 mM [33]. However, one should not use it in combination with the divalent metal ion-dependent separation techniques such as IMAC (immobilized metal ion affinity chromatography) and AIEX (anion exchange chromatography) or with proteins that are dependent on divalent metal ions for their activity.

5.3.1.5 Stabilizing Additives

Several additives can be added to the extraction buffer to stabilize the target protein. The choice of the additive should be made considering its possible effect on the downstream chromatographic separation. Most often the additive is required only in the initial steps of lysate preparation and can be safely excluded in the subsequent steps of the purification. Commonly used additives are sucrose, glucose, glycerol, and other polyols. Glycerol in the concentration range of 5–50% is frequently used to prevent aggregation and stabilize the protein by increasing protein compactness and decreasing interactions of the hydrophobic surfaces [34]. Polyols such as mannitol, sucrose, propylene, and polyethylene glycol, typically at a concentration of 10%, can be added to stabilize the proteins [35].

5.3.1.6 Nucleases

Although not mandatory, addition of nucleases to the lysis buffer is beneficial in several ways. The nucleic acid released during the cell disruption process might cause aggregation of cell debris and hinder the chromatographic purification either by binding to the chromatographic medium or to the target protein. It might also lead to increase in the viscosity of the sample solution. Most effective engineered nuclease, Benzonase, that breaks down both RNA and DNA can be added in the lysis buffer at a concentration of 1–20 μg/ml before cell disruption [36].

5.3.1.7 Protease Inhibitors

A key threat to protein stability during purification is from the proteases of the source organism. The simplest way to protect against proteolytic degradation is by working quickly at low temperature, suitably on ice. In addition, a mixture of protease inhibitors should be added during purification, especially in the lysis buffer (Table 5.3 provides the list of protease inhibitors) [37]. In some cases, the proteases can be inactivated by adjusting the pH of the solution to a value at which the proteases are inactive, while maintaining the activity of the desired protein.

5.3.2 Clarification of the Extract

The crude extract is often turbid and contains insoluble residues and cellular debris, which might block a chromatography column. Therefore, in a purification protocol, which includes a chromatographic separation, it is important to perform pre-clearing of the crude extract. This could be achieved by centrifugation and/or filtration before applying the sample to the column

Table 5.3 Protease inhibitors used in protein purification

Protease inhibitor	Molecular weight (Da)	Target protease	Inhibitor type
AEBSF	239.5	Serine proteases	Irreversible
Aprotinin	6611.5	Serine proteases	Reversible
Bestatin	308.38	Amino peptidases	Reversible
EDTA	372.4	Metalloproteases	Irreversible
E-64	357.4	Cysteine proteases	Irreversible
Leupeptin	475.6	Serine and cysteine proteases	Reversible
Pepstatin A	685.9	Aspartic acid proteases	Reversible

Abbreviation: *AEBSF* 4-(2-aminoethyl) benzene sulfonyl fluoride, *E-64* l-*trans*-epoxy succinyl-l-leucylamido-(4-guanido) butane

5.3.2.1 Centrifugation

In a laboratory-scale setup, centrifugation is a common method for clarification of the crude extract. For larger volume of cell lysates, one should use the following setting: 40000 × *g* for 30 min at 4 °C. In case of small sample volumes, a benchtop centrifuge at the highest available g-force, such as 14,000 × *g* for 15–30 min, is sufficient. Important to note here is that the centrifuge should be pre-cooled along with the rotor before placing the samples in it [38].

5.3.2.2 Filtration

Alternative to centrifugation, clarification of the crude extract can also be achieved by filtration, which is relatively less time-consuming than the centrifugation method. The sample to be clarified is passed through a 0.45 μm pore size filter fitted to the syringe or filtration assembly. The 0.45 μm has been recognized as standard pore size membranes that can efficiently remove large bacteria or particulate matter. The common membrane filters with least nonspecific binding include cellulose acetate, polyethersulfone (PES), or polyvinylidene fluoride (PVDF). However, often the filters become saturated after being repeatedly used as they tend to have a certain capacity. Therefore, depending on the sample volume, the filter and syringe capacity should be carefully decided prior to setting up the experiment.

5.4 Checking Solubility and Designing Purification Strategies

5.4.1 Protein Solubility and Precipitation

Protein solubility is a key parameter in any protein purification. The solubility differs markedly from one protein to another, and is highly dependent on the solvent and physicochemical properties of the proteins. The parameters that largely influence protein solubility include solvent's pH, ionic strength, temperature, and the type of exposed amino acid side chains on the proteins' surface. Proteins with lower ratio of the solvent-exposed charged and hydrophobic amino acids tend to be less soluble [39]. Since it is difficult to accurately predict the solubility properties of a protein, one should carefully design pilot studies by varying different conditions and checking protein solubility on SDS-PAGE.

As discussed in Sect. 5.3.2, composition of the extraction medium is critical in protein solubility. For instance, addition of salt such as NaCl, pH (close to protein pI value), and/or inclusion of stabilizing additives should be considered to prevent aggregation and precipitation of the proteins. In cases, where the sample volumes are too large, one can also exploit the protein's tendency of differential precipitation in the presence of neutral salts (ammonium sulfate), polymers (polyethylene glycol), or organic solvents (e.g., ethanol or acetone). This process of purification is termed "salting out," which relies on the principle that at high concentrations of salt, the salt molecules compete with proteins for binding with water, thus leading to protein precipitation [40]. This happens because protein molecules preferentially interact with each other due to energetically favorable protein-protein interactions instead of protein-solvent interaction. Salting-out offers an alternate means to concentrate the proteins at a very early stage of purification, and with two–three-fold degrees of purity.

5.4.2 Salting-out

Salting-out, which is generally known as salt-induced precipitation or salt fractionation, is based on the interaction of protein with the salt (solute). Salts tend to dissociate in the aqueous solution (solvent), which forms the basis of the salting-out process [41]. In a condition where the salt concentration is increased, the water molecules start getting attracted by the salt ions, which in turn decreases the number of water molecules that can interact with the charged portion of the protein. As a result, protein molecules tend to associate with each other due to stronger protein-protein interaction than the solute-solvent interaction causing protein aggregation and subsequent precipitation. This process is known as salting-out (Fig. 5.1). Importantly, the salt concentration needed to precipitate a protein differs from protein to protein. One of the most common salts used for protein precipitation is ammonium sulfate.

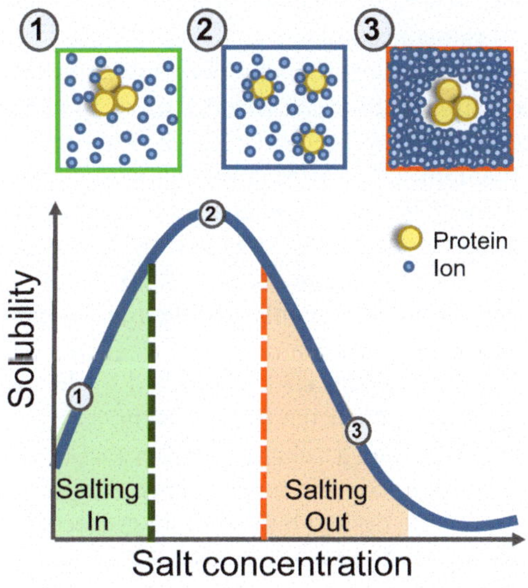

Fig. 5.1 Dependence of protein solubility on salt concentration. A schematic two-dimensional solubility curve of a protein as a function of salt concentration. The solubility curve divides the space in two areas—salting in (*green*) and salting out (*pink*). During the salting in process (phase-1), the salt molecules increase the solubility of the proteins by reducing the ionic interactions between the protein molecules. As the concentration of the salt increases, the excess ions start competing with proteins for the solvent. Gradually, post phase-2, the protein-protein interactions become energetically more favorable than protein-solvent interaction, and the proteins tend to precipitate and come out of the solution (phase-3). This effect is referred to as "salting-out" of the protein

5.4.3 Ammonium Sulfate Precipitation

Ammonium sulfate is the most common salt used for protein precipitation since it is relatively inexpensive, highly soluble in water, and very stabilizing to protein structure. The amount of salt needed for protein precipitation differs from one to another, and also varies with the temperature. Often it is desirable to perform precipitation at low temperatures to avoid protein denaturation. The concentration of ammonium sulfate required to precipitate a particular protein depends on the molecular weight of the protein, solvent pH, temperature, as well as number and position of polar groups [40]. Preferably screening for the percentage saturation of ammonium sulfate to precipitate the target protein or any contaminant should be first established in a small-scale setup. Following is a general protocol for the ammonium sulfate precipitation of a crude lysate [40].

1. Keep the clarified extract on ice and add pre-chilled 50 mM HEPES or Tris–HCl buffer, pH 8.0 to maintain the pH of the sample solution, since addition of ammonium sulfate acidifies the solution.
2. Slowly add fresh, desiccated powder of ammonium sulfate with gentle stirring. It is important to make sure that the salt is completely dissolved before adding more solid, and care should be taken to avoid foaming of the solution. Adding the salt very slowly ensures that local concentration around the addition site does not exceed the required salt concentration.
3. Carefully calculate the amount of ammonium sulfate to be added to attain the chosen percent saturation. Online calculators such as from EnCor Biotechnology Inc. (http://www.encorbio.com/protocols/AM-SO4.htm) can be used to calculate the amount of ammonium sulfate to be added to a specific volume of a solution to reach a particular percentage saturation at a specific temperature.
4. Allow the sample to precipitate for 30 min at 4 °C with continuous stirring.
5. Collect the precipitate by centrifugation at 10,000 to 50,000 × *g* for 30 min–1 h at 4 °C. Note that a solution that is highly saturated in ammonium sulfate is often dense and quite difficult to pellet.
6. Separate the pellet from supernatant and proceed to the next required percent saturation, if supernatant contains the target protein.
7. Resuspend recovered pellets in a volume of buffer that is equal to the volume of the extract. To get rid of excess ammonium ions, dialysis (as described below) can be carried out using the desired buffer before proceeding to the next stage of protein purification or prior to protein storage.

The process of ammonium sulfate precipitation offers an easy and relatively inexpensive way to concentrate and purify proteins. However, it is important to consider the associated drawbacks while selecting this method of separation. Most importantly, it is essential to get rid of the salt from the protein sample, so the downstream processing steps in the form of either dialysis or chromatography will be required.

5.4.4 Salting-in

Protein solubility is affected by the ionic strength of the solution. If a protein is placed in aqueous solution like water, the only ionic component in the solution will be protein molecules. Although water is polar, it only slightly ionizes and therefore proteins tend to aggregate based on the protein-protein ionic interactions [42, 43]. This interaction between the proteins is more favored than protein-water interactions resulting in an irreversible precipitation. At low ionic concentrations of salts such as NaCl, the presence of other ionic species now can compete with the ionic protein-protein interactions. These ions in the solution tend to shield the protein molecules from the charge of the other protein molecules. The decrease in the electrostatic interaction between the protein molecules eventually increases the solubility of the protein, referred to as "salting-in" (Fig. 5.1). However, at a point when the ionic strength starts getting too high, it imparts a negative effect on the protein solubility, resulting in "salting-out," as discussed in the previous section.

Salting-in of the proteins occurs generally near its isoelectric point (pI) [13]. In addition to the electrostatic effect, the limited charge on the surface of the protein affects the water associated with the protein [44]. All in all, the pairing of salt ions with the charged groups on protein molecules increases protein solubility resulting in "salting-in" of the proteins.

5.4.5 Dealing with Proteins in the Inclusion Bodies

Often high expression of the recombinant proteins in *E. coli* results in the formation of insoluble and aggregated proteins referred to as "inclusion bodies." In addition to being insoluble, the inclusion bodies have a non-native structure and therefore require solubilization and refolding of the target protein to its native structure [30]. This generally requires a lot of optimization and is also time-consuming. In addition, one is expected to face significant loss in the protein amount during the process of refolding. The formation of the inclusion bodies in *E. coli* can be avoided by modifying several conditions during protein expression. Most commonly, use of lower temperature such as 18 °C and < 0.5 mM IPTG (isopropyl thiogalactoside) can be used to tackle the problem of inclusion body formation [2]. In the presence of higher concentrations of IPTG and/or at high temperatures (37 °C), the protein expression takes place at a high translational rate, which often exhausts the protein quality control system of *E. coli* resulting in partially folded or misfolded protein aggregates in the form of inclusion bodies [45]. However, in certain conditions, wherein expression of large amounts of proteins would otherwise be toxic to the host cell, expression in the form of inclusion bodies allows its large-scale preparation. In addition, the sequestration of proteins in the inclusion bodies prevents it from proteolysis by the cytosolic proteases [30].

For the isolation of inclusion bodies, the cells expressing the target protein are harvested, lysed by mechanical lysis, and then centrifuged at high speed (20,000 × *g*), 15 min at 4 °C. The inclusion bodies are obtained as a pellet. The pellet is then washed with detergents such as 1% Triton-X, followed by denaturation of the protein using chemical denaturants, viz.8 M Urea or 6 M Guanidine-Hydrochloride [30]. The denaturation step can be performed multiple times to achieve maximum recovery. Often, more than 90% purity is obtained by following these steps. In case, the target protein contains impurities, it can be subjected to affinity purification or salting-out precipitation to increase its purity. The denatured protein is then refolded by gradual removal of the denaturant either using dilution or dialysis method. In the dilution method, a drop-by-drop solution of the denatured protein is added in a buffer solution (100× the volume of the denatured protein) [46]. In the case of dialysis, the denatured protein is placed in the dialysis tubing of a particular molecular weight cutoff (depending on the size of the protein), and is suspended in the refolding buffer solution. Some of the critical parameters during refolding include temperature, pH, presence of reducing agents (such as DTT, TCEP), and additives (often in combination) [47]. However, most importantly, screening of multiple conditions and subsequent optimization is required for successful refolding. Often, the success rate for refolding of proteins is not very encouraging. Therefore, a large number of refolding conditions should be tested such that a biologically active form of the protein is obtained in large amounts at the highest purity level. Apart from dilution or dialysis methods, the denaturing agents in the unfolded protein solution can also be removed using different chromatographic techniques. Here, the protein is either slowly allowed to migrate through a column (e.g., gel filtration on a Superdex 75 column) or is bound to the matrix (e.g., affinity chromatography using Ni-NTA agarose) and then eluted in a buffer with decreasing concentration of the denaturant.

5.5 Overview of Chromatography

Column chromatography, most commonly liquid chromatography (LC), is used in purification of the recombinant proteins. In this method of chromatography, the stationary phase is packed in a column and the liquid mobile phase is allowed to pass through the column using a pump or under gravity flow. The sample mixture is introduced at one end, followed by elution with the mobile phase at the other end of the column. The separation of the components in the mixture depends on partitioning of the molecules between the mobile and the stationary phases, which is based on the differences in their molecular weights. For recombinant protein purification, affinity chromatography, ion-exchange chromatography, and size exclusion chromatography are commonly applied to achieve satisfactory purity and homogeneity of the target protein [48].

5.5.1 Affinity Chromatography

There are numerous ways in which the affinity-based method is employed for purification of the recombinant proteins. The most common strategy for

Table 5.4 The list of fusion tags commonly selected for purification of recombinant proteins [14]

Tag	Size (no. of amino acids or kDa)	Ligand	Separation method
Polyhistidine	5–15 a.a.	Ni^{+2} or Cu^{+2}	AC
Glutathione S-transferase	26 kDa	Glutathione	AC
Maltose binding domain	40 kDa	Amylose	AC
FLAG	8 a.a.	mAb based	AC
Strep tag I	9 a.a.	Streptavidin	AC
Strep tag II	8 a.a.	Streptactin	AC
T7-tag	11–16 a.a	mAb based	AC
c-myc	10 a.a.	mAb based	AC
S-tag peptide	15 a.a.	S-protein	AC
Polyaspartic acid	5–16 a.a.	Ion-exchange or precipitation	IEX
VSV tag	11 a.a.	mAb based	AC
Calmodulin binding peptide	26 a.a.	Calmodulin	AC

AC Affinity chromatography, *IEX* Ion-exchange chromatography, *a.a.* Amino acid, *kDa* kilodalton

affinity-based purification involves use of fusion tags (i.e., amino acid sequences attached to the recombinant proteins) that have affinity for ligands immobilized on the column. Some of the commonly used tags are outlined in Table 5.4.

Particularly, the histidine (His) tag (a sequence of six or more histidine residues) is added to either N- or C-terminus of the recombinant protein, and is frequently used for purification of the target protein from a mixture of proteins in the cellular lysate. The His-tagged protein shows affinity toward divalent metal ions such as Ni^{+2} or Zn^{+2}, and therefore the chromatographic technique is known as immobilized metal affinity chromatography (IMAC). IMAC uses matrices such as iminodiacetic acid (IDA) and nitrilotriacetic acid (NTA) to chelate transition metals through three or four coordination sites, respectively [49]. Since the metal ion is only weakly bound via three-coordination sites in IDA compared to four sites in case of NTA, often metal leaching from the IDA matrix occurs during purification resulting in lower yield and impure proteins. Therefore, advancements in NTA-based chemistry have facilitated development of matrices, which securely coordinate metal ions with four-coordination sites while leaving two of the transition metal coordination sites exposed to interact with the His-tag [49]. Due to the presence of the electron donor groups on histidine, the His-tag promotes strong interaction with the immobilized transition metal and gets retained on the IMAC column. After subsequent washes of the matrix to remove nonspecifically bound proteins, the His-tagged recombinant protein gets eluted either by adjusting the pH of the elution buffer or by adding histidine analog, imidazole (concentration of 100–500 mM) [50]. At lower pH (4.5–5.3), the imidazole nitrogen atom of the histidine residue (pKa 6.0) gets protonated that disturbs the coordination bond between histidine and the transition metal [49], while imidazole at concentrations >100 mM acts as a competitor resulting in the elution of the bound his-tagged protein. The major advantage of the IMAC system is that it can tolerate a wide range of buffer conditions, including the presence of additives like detergent and chemical denaturants. In addition, these resins can be regenerated and reused several times, thus enabling development of economical purification strategies in both academia and industrial settings. A schematic overview of Ni-NTA affinity chromatography is provided in Fig. 5.2.

One of the major disadvantages of the polyhistidine tag is the nonspecific binding of the proteins to the IMAC column. Nevertheless, the His-tag offers several advantages due to its small size, which rarely affects protein function. Also in most cases, one can achieve purity up to 90–95% in a single step of purification. The IMAC resin remains unaffected by protease or nuclease activities in the cell lysate making it suitable for purification with crude lysates. One of the advantages of the His-tag is that it can be combined with other affinity tags (listed in Table 5.4) to the same protein to provide great flexibility during the purification process. Overall, IMAC offers a rapid and an inexpensive purification method compared to other affinity-based purification methods.

Similar to polyhistidine tag, alternative affinity tags such as MBP or GST are also used frequently; however, due to large size, these tags need to be removed via a proteolytic cleavage at specific sites introduced between the tag and target protein. Due to the additional processing step involved in the large affinity tags, the use of these tags increases the downstream processing cost for the purification of the recombinant proteins [14, 51].

5.5.2 Ion Exchange Chromatography (IEX)

IEX is based on the electrostatic interactions between the charged groups of the protein with the matrix of the column. Separation of the proteins in IEX depends on surface charge of the proteins, pH, and salt concentration of the mobile phase [52]. IEX is most frequently applied for purification of recombinant proteins that are untagged. Since proteins are amphoteric molecules (act as both acids and bases), any protein can bind the ion-exchange matrix once the pH of the solution is optimized for its binding. The chosen matrix has an ion-load opposite to that of the protein to be separated. A positively charged ion-exchange matrix (i.e., anion-exchanger) adsorbs negatively charged proteins, while a negatively charged matrix (cation-exchanger) adsorbs the positively charged molecules. As a rule, the selected mobile phase should have low to medium salt concentration (i.e., low to medium conductivity). Moreover the chosen pH should lie between the isoelectric point (pI) *aka* acid dissociation constants (pKa) of the charged molecule, and that of the ion on the matrix [53]. For example, in cation-exchange chromatography, where the pKa of matrix is 1.2, the mobile phase of pH 6.0 should be ideally used for a protein with a pI of 8.2. On the other hand, in anion-exchange chromatography where the

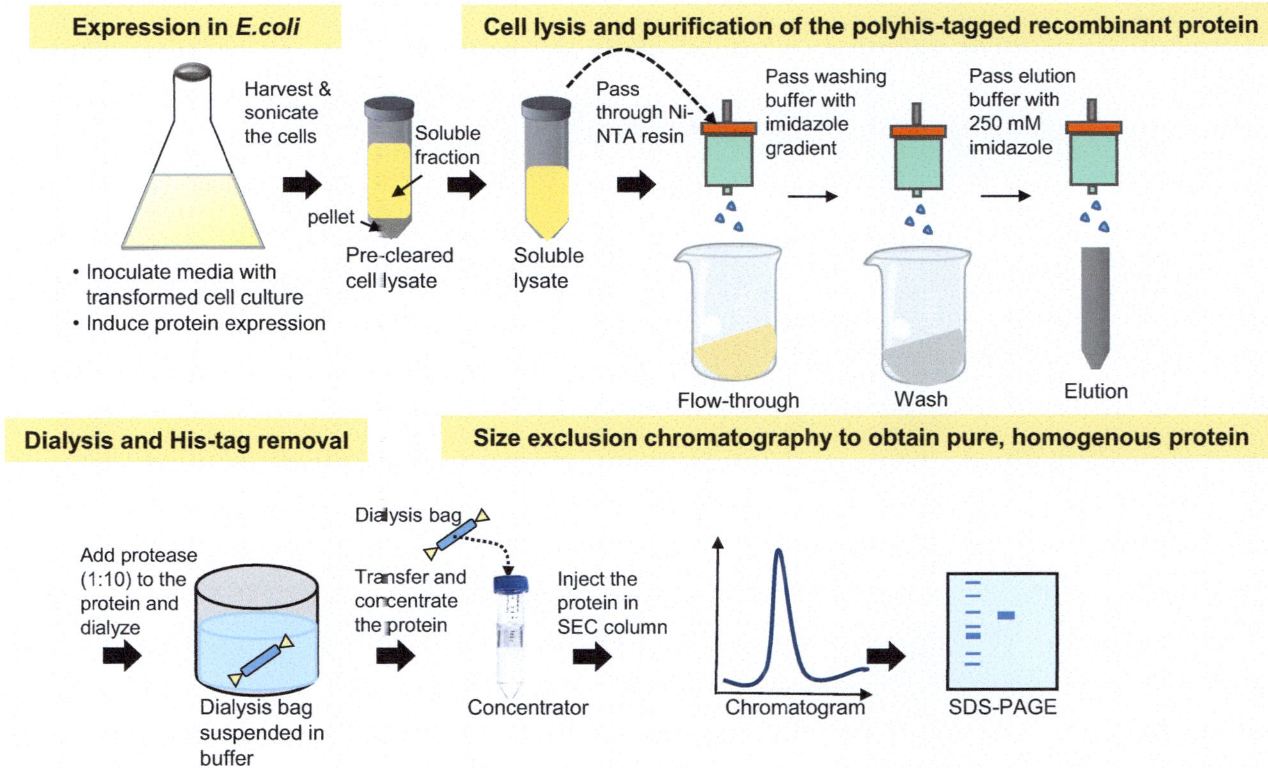

Fig. 5.2 A schematic representation for polyhis-tagged protein purification using Ni-NTA affinity chromatography. The 6X-His-tagged recombinant protein is expressed in the host, such as *E. coli*. Cells over-expressing the protein are harvested, resuspended in the lysis buffer followed by cell lysis using sonication. To remove the cellular debris, the cell lysate is centrifuged at high speed to separate the soluble fraction and pellet. The pre-cleared lysate is then passed through a Ni-NTA affinity column, and the flow-through is collected. The column is given several washes with washing buffer containing low concentrations of imidazole. The desired protein is then eluted using $\geq$250 mM imidazole buffer. The purified protein is treated with site-specific proteases, such as thrombin or TEV

Fig. 5.2 (continued) protease to remove the His-tag, and then dialyzed overnight in the desired buffer (optional step). In order to obtain highly pure and homogenous form of the recombinant protein, the dialyzed protein is concentrated using a concentrator (such as Centricon), and then subjected to size exclusion chromatography (SEC). The eluted fractions are analyzed on SDS-PAGE to check for purity and confirmation of the molecular weight

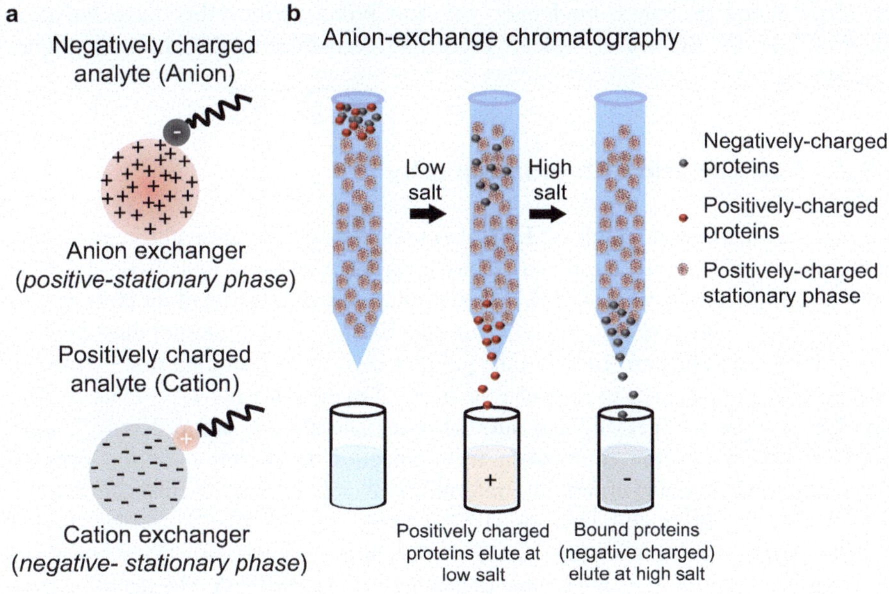

Fig. 5.3 Schematic representation of ion-exchange chromatography. (**a**) Anion and cation exchanger bind to negatively and positively charged molecules, respectively. (**b**) Flow-scheme of anion-exchange chromatography. The positively charged stationary phase of the anion-exchanger allow binding of negatively charged proteins at low ionic concentrations of the mobile phase. In the presence of low salt, the positively charged molecule in the protein mixture elutes first. As the concentration of the salt increases, the bound negatively charged proteins are exchanged by the salt ions and gets eluted from the column

matrix pKa is 10.3, a mobile phase of pH 8.0 should be used for protein molecules with a pI of 6.8.

Binding of the protein onto the column depends on the surface charge of the protein and charge on the matrix (Fig. 5.3a). This interaction, which is reversible, is then disrupted using a linear gradient of salt or varying pH to elute the bound proteins [54]. For instance, the negatively charged proteins can be displaced using negatively charged salt ions. This is because the negatively charged salt ions competitively displace the negatively charged proteins from the functional groups of the matrix. The elution profile of the protein therefore follows a low-to-high salt concentration gradient (Fig. 5.3b). Alternatively, varying pH can also be used to separate the proteins. In cation-exchange chromatography, increasing the buffer pH makes the protein less protonated (i.e., less positively charged), thus disrupting its ionic interaction with the negatively charged matrix and subsequent elution. While in the anion-exchange chromatography, decrease in the pH of the mobile phase results in protonation of the protein (i.e., more positively charged) which promotes protein elution [48].

Under mild conditions, IEX provides high binding capacity and resolution for separation of recombinant proteins. In addition, its ability to scale-up (particularly for recombinant proteins), moderate cost, and broad applicability have led it to become one of the most widely used and versatile liquid chromatography techniques.

5.5.3 Size Exclusion Chromatography

Size exclusion chromatography (SEC), also known as gel filtration chromatography (GFC), separates the macromolecules based on differences in their hydrodynamic volume and molecular weight [55]. The stationary phase consists of an inert spherical bead or gel with pores of a specific size distribution. Proteins larger than the pore size of the gel cannot permeate into the gel particles, thereby eluting first by rapidly passing through the space in-between the beads. On the other hand, proteins smaller than the gel pore size get diffused into the pores and elute at proportionally longer retention times (i.e., the time taken from injection to detection of the protein) [56]. Hence, SEC is also commonly referred to as gel permeation chromatography. Unlike affinity chromatography or IEX, in SEC, the protein molecules do not directly interact with the mobile phase, so the buffer composition does not affect the column resolution (i.e., the degree of separation of the peaks). The retention time of larger proteins are shorter than the smaller proteins, which enables separation of these proteins [54, 57]. Apart from the gel pore size, the column resolution is also influenced by bed height, flow rate, volume of the sample, and the molecular weight of the protein [58–60]. Generally, the highest possible resolution can be obtained with slow to moderate flow rate, long and narrow columns, small pore-size gel and sample volumes (1–5% of the total column volume). A general layout of the size exclusion chromatography is given in Fig. 5.4.

SEC is widely used during the final polishing steps in recombinant protein purification due to its excellent desalting properties. This is usually done when the volume of the sample has been reduced and the major goal remains is to remove aggregates or change the protein buffer (also known as desalting). For desalting purpose, since the difference in the molecular mass of the protein and salt is typically very large, the peaks are very less likely to overlap. Therefore, even sample volumes as much as 30% of the bed volume can be safely applied to the column. In addition to the purification purpose, SEC can also be used for analytical use. With proper molecular weight standards, one can calculate the molecular weight of the unknown protein [61]. In this method, a few proteins of known molecular weights are initially used for calibrating the column. The calibration standard also contains a very large molecular weight protein (such as blue dextran (1000 kDa) to determine the void volume (V_o) of the column, which refers to the excluded volume, i.e., the space between the particles. One can then determine the elution volume (V_e) of the individual standards to determine the molecular weight of the unknown protein. A standard plot can be generated by dividing "V_e" of the standards by the "V_0" (V_e/V_0), and plotting this value against the log of the molecular weights (log M) of the standard. The molecular weight of the unknown protein can then be extrapolated from the standard plot. However, care should be taken while interpreting the data, as

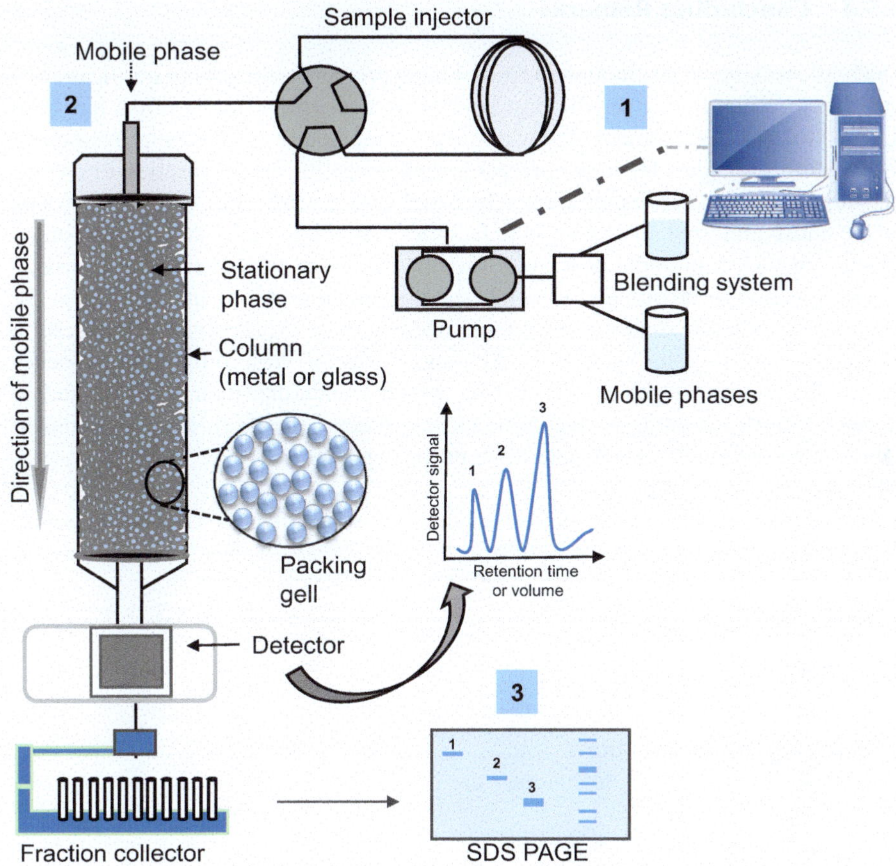

Fig. 5.4 Schematic overview of size exclusion chromatography (SEC). SEC or gel filtration chromatography mostly involves an automated system constituting a pump, sample injector, column, and fraction collector coordinated through a software. With the help of a pump, the mobile phase is passed through the column at a particular flow rate. The protein mixture is injected through a sample injector, and separation is achieved based on the pore size of the packing gel and proteins molecular weight. Protein with higher molecular weight elutes first followed by the smaller proteins. The elution of the protein is monitored in real time in the form of chromatogram, generated by the detector. The eluted fractions are collected either manually or with the help of fraction collector. The collected fractions are further analyzed on SDS-PAGE for purity and confirmation of the molecular weight of the separated proteins

SEC is quite accurate for globular proteins, while it is less accurate for flattened or extended protein molecules.

5.5.4 Concluding Remarks

The bottleneck in the production of recombinant proteins is the cost of purification of the protein. Therefore, over the past several decades, tremendous effort has been directed toward development of new purification methods as well as improvement in the existing strategies. Here, we present general steps for the purification of recombinant proteins that are expressed in a bacterial system such as *E. coli*, which is one of the most preferred microbial cell factories. Although it is suitable for stably expressing folded, globular proteins, difficulties are often encountered in expression and purification of membrane or membrane-associated proteins. In this chapter, we describe the possible routes to meet those challenges of expression and purification of recombinant proteins in *E. coli*. Although many of these outlined approaches might fail at several stages, one has to find ways to standardize their protocols and do extensive troubleshooting to overcome those obstacles since expression and purification of recombinant proteins are often protein specific.

Acknowledgments NS gratefully acknowledges financial support from the DBT-RA Program in Biotechnology and Life Sciences for completion of the manuscript. The authors acknowledge Ms. Chanda Baisane, Bose Lab, ACTREC for formatting the manuscript.

Problems

Multiple Choice Questions

1. Protein purification techniques are based on the following properties except:
   (a) Solubility of the protein
   (b) Charge on the protein
   (c) Viscosity of the protein
   (d) Specific binding affinity of the protein
2. Salting-out refers to
   (a) Precipitation of proteins using ammonium sulfate
   (b) Precipitation of proteins using copper sulfate
   (c) Precipitation of proteins using sodium chloride
   (d) Both (a) and (c)
3. You find that your protein sample shows lots of additional bands of lower molecular weight apart from the desired protein. What can you do about this?
   (a) Add an additional purification step
   (b) Use a protease inhibitor during lysis and purification
   (c) Perform each step as quickly as possible, in a cold-room
   (d) All of the above

Subjective Questions

1. To estimate the molecular mass of an unknown protein, you decide to run a size exclusion chromatography. Next, you run a series of proteins with the known molecular mass and the unknown protein on a Sephadex G-200 column. Below are the elution volumes (V_e) for each protein. The measured void volume (V_0) of the column is 36 mL. Using these values, calculate the molecular mass of the unknown protein.

Protein	Molecular weight (kDa)	V_e (mL)
Lysozyme	14	100
Ovalbumin	45	79
Serum albumin	66	70
Aldolase	150	51
Urease	489	26
Unknown	?	88

2. To separate a mixture of proteins with different pI values, you try anion exchange chromatography using DEAE-cellulose column. For this, you first equilibrate your column with phosphate buffer, pH 6.5. Then you pass the following mixture of proteins, AS (pI – 4.6), BS (pI – 5.0), and CS (pI – 7.0). The proteins are eluted first with weak ionic strength buffer (50 mM NaCl, pH 6.5) and then in the same buffer but with increasing NaCl concentration. Explain what order will the proteins elute?

References

1. Sumner JB. The isolation and crystallization of the enzyme urease preliminary paper. J Biol Chem. 1926;69:435–41.
2. Rosano GL, Ceccarelli EA. Recombinant protein expression in Escherichia coli: advances and challenges. Front Microbiol. 2014;5:172.
3. Structural Genomics, C., C. China Structural Genomics, C. Northeast Structural Genomics, Graslund S, Nordlund P, Weigelt J, Hallberg BM, Bray J, Gileadi O, Knapp S, Oppermann U, Arrowsmith C, Hui R, Ming J, dhe-Paganon S, Park HW, Savchenko A, Yee A, Edwards A, Vincentelli R, Cambillau C, Kim R, Kim SH, Rao Z, Shi Y, Terwilliger TC, Kim CY, Hung LW, Waldo GS, Peleg Y, Albeck S, Unger T, Dym O, Prilusky J, Sussman JL, Stevens RC, Lesley SA, Wilson IA, Joachimiak A, Collart F, Dementieva I, Donnelly MI, Eschenfeldt WH, Kim Y, Stols L, Wu R, Zhou M, Burley SK, Emtage JS, Sauder JM, Thompson D, Bain K, Luz J, Gheyi T, Zhang F, Atwell S, Almo SC, Bonanno JB, Fiser A, Swaminathan S, Studier FW, Chance MR, Sali A, Acton TB, Xiao R, Zhao L, Ma LC, Hunt JF, Tong L, Cunningham K, Inouye M, Anderson S, Janjua H, Shastry R, Ho CK, Wang D, Wang H, Jiang M, Montelione GT, Stuart DI, Owens RJ, Daenke S, Schutz A, Heinemann U, Yokoyama S, Bussow K, Gunsalus KC. Protein production and purification. Nat Methods. 2008;5:135–46.
4. Arnau J, Lauritzen C, Petersen GE, Pedersen J. Current strategies for the use of affinity tags and tag removal for the purification of recombinant proteins. Protein Expr Purif. 2006;48:1–13.

5. Tripathi NK, Shrivastava A. Recent developments in bioprocessing of recombinant proteins: expression hosts and process development. Front Bioeng Biotechnol. 2019;7:420.
6. Wingfield PT. Overview of the purification of recombinant proteins. Curr Protoc Protein Sci. 2015;80:6.1.1–6.1.35.
7. Rath A, Glibowicka M, Nadeau VG, Chen G, Deber CM. Detergent binding explains anomalous SDS-PAGE migration of membrane proteins. Proc Natl Acad Sci U S A. 2009;106:1760–5.
8. Cole JL, Lary JW, Moody TP, Laue TM. Analytical ultracentrifugation: sedimentation velocity and sedimentation equilibrium. Methods Cell Biol. 2008;84:143–79.
9. Park ES, Wallace WE, Guttman CM, Flynn KM, Richardson MC, Holmes GA. A general method for quantitative measurement of molecular mass distribution by mass spectrometry. J Am Soc Mass Spectrom. 2009;20:1638–44.
10. Hong P, Koza S, Bouvier ES. Size-exclusion chromatography for the analysis of protein biotherapeutics and their aggregates. J Liq Chromatogr Relat Technol. 2012;35:2923–50.
11. Audain E, Ramos Y, Hermjakob H, Flower DR, Perez-Riverol Y. Accurate estimation of isoelectric point of protein and peptide based on amino acid sequences. Bioinformatics. 2016;32:821–7.
12. Ahamed T, Chilamkurthi S, Nfor BK, Verhaert PD, van Dedem GW, van der Wielen LA, Eppink MH, van de Sandt EJ, Ottens M. Selection of pH-related parameters in ion-exchange chromatography using pH-gradient operations. J Chromatogr A. 2008;1194:22–9.
13. Shaw KL, Grimsley GR, Yakovlev GI, Makarov AA, Pace CN. The effect of net charge on the solubility, activity, and stability of ribonuclease Sa. Protein Sci. 2001;10:1206–15.
14. Saraswat M, Musante L, Ravida A, Shortt B, Byrne B, Holthofer H. Preparative purification of recombinant proteins: current status and future trends. Biomed Res Int. 2013;2013:312709.
15. Lakowicz JR. Principles of fluorescence spectroscopy. Switzerland: Springer Science & Business Media; 2013.
16. Gasteiger E, Hoogland C, Gattiker A, Wilkins MR, Appel RD, Bairoch A. Protein identification and analysis tools on the ExPASy server. In: The proteomics protocols handbook. New York: Springer; 2005. p. 571–607.
17. Gill SC, Von Hippel PH. Calculation of protein extinction coefficients from amino acid sequence data. Anal Biochem. 1989;182:319–26.
18. Guruprasad K, Reddy BB, Pandit MW. Correlation between stability of a protein and its dipeptide composition: a novel approach for predicting in vivo stability of a protein from its primary sequence. Protein Eng Des Sel. 1990a;4:155–61.
19. Varshavsky A. The N-end rule pathway of protein degradation. Genes Cells. 1997;2:13–28.
20. Mogk A, Schmidt R, Bukau B. The N-end rule pathway for regulated proteolysis: prokaryotic and eukaryotic strategies. Trends Cell Biol. 2007;17:165–72.
21. Guruprasad K, Reddy BV, Pandit MW. Correlation between stability of a protein and its dipeptide composition: a novel approach for predicting in vivo stability of a protein from its primary sequence. Protein Eng. 1990b;4:155–61.
22. Kyte J, Doolittle RF. A simple method for displaying the hydropathic character of a protein. J Mol Biol. 1982;157:105–32.
23. Waegeman H, Soetaert W. Increasing recombinant protein production in Escherichia coli through metabolic and genetic engineering. J Ind Microbiol Biotechnol. 2011;38:1891–910.
24. Vieira Gomes AM, Souza Carmo T, Silva Carvalho L, Mendonca Bahia F, Parachin NS. Comparison of yeasts as hosts for recombinant protein production. Microorganisms. 2018;6
25. Wurm DJ, Slouka C, Bosilj T, Herwig C, Spadiut O. How to trigger periplasmic release in recombinant Escherichia coli: a comparative analysis. Eng Life Sci. 2017;17:215–22.
26. Brady PN, Macnaughtan MA. Evaluation of colorimetric assays for analyzing reductively methylated proteins: biases and mechanistic insights. Anal Biochem. 2015;491:43–51.
27. Seddon AM, Curnow P, Booth PJ. Membrane proteins, lipids and detergents: not just a soap opera. Biochim Biophys Acta. 2004;1666:105–17.

28. Ersson B, Rydén L, Janson JCJPPP. High resolution methods, and applications. In: Introduction to protein purification. Hoboken, NJ: John Wiley & Sons, Inc.; 2011. p. 1–22.
29. Otzen DE. Protein unfolding in detergents: effect of micelle structure, ionic strength, pH, and temperature. Biophys J. 2002;83:2219–30.
30. Singh A, Upadhyay V, Upadhyay AK, Singh SM, Panda AK. Protein recovery from inclusion bodies of Escherichia coli using mild solubilization process. Microb Cell Factories. 2015;14:41.
31. Poole LB. The basics of thiols and cysteines in redox biology and chemistry. Free Radic Biol Med. 2015;80:148–57.
32. Lebendiker M, Maes M, Friedler A. A screening methodology for purifying proteins with aggregation problems. Methods Mol Biol. 2015;1258:261–81.
33. Trivedi MV, Laurence JS, Siahaan TJ. The role of thiols and disulfides on protein stability. Curr Protein Pept Sci. 2009;10:614–25.
34. Middelberg AP. Preparative protein refolding. Trends Biotechnol. 2002;20:437–43.
35. Singh S, Singh J. Effect of polyols on the conformational stability and biological activity of a model protein lysozyme. AAPS PharmSciTech. 2003;4:E42.
36. Lin MZ, Tsien RY. TimeSTAMP tagging of newly synthesized proteins. Curr Protoc Protein Sci. Chapter 26. 2010;Unit 26:25.
37. Wingfield PT. Preparation of soluble proteins from Escherichia coli. Curr Protoc Protein Sci. 2014;78:6.2.1–6.2.22.
38. Cvjetkovic A, Lotvall J, Lasser C. The influence of rotor type and centrifugation time on the yield and purity of extracellular vesicles. J Extracell Vesicles. 2014;3
39. Lodish H, Berk A, Zipursky SL, Matsudaira P, Baltimore D, Darnell J. Hierarchical structure of proteins. In: Molecular cell biology. 4th ed. Bethesda, MD: WH Freeman; 2000.
40. Wingfield P. Protein precipitation using ammonium sulfate. Curr Protoc Protein Sci. 2001; Appendix 3:Appendix 3F.
41. Hyde AM, Zultanski SL, Waldman JH, Zhong Y-L, Shevlin M, Peng F. General principles and strategies for salting-out informed by the hofmeister series. Org Process Res Dev. 2017;21:1355–70.
42. Damodaran S. Amino acids, peptides and proteins. Boca Raton, FL: CRC Press; 2008.
43. Mattos C. Protein–water interactions in a dynamic world. Trends Biochem Sci. 2002;27:203–8.
44. Kramer RM, Shende VR, Motl N, Pace CN, Scholtz JM. Toward a molecular understanding of protein solubility: increased negative surface charge correlates with increased solubility. Biophys J. 2012;102:1907–15.
45. Carrio MM, Villaverde A. Localization of chaperones DnaK and GroEL in bacterial inclusion bodies. J Bacteriol. 2005;187:3599–601.
46. Yamaguchi H, Miyazaki M. Refolding techniques for recovering biologically active recombinant proteins from inclusion bodies. Biomol Ther. 2014;4:235 51.
47. Willis MS, Hogan JK, Prabhakar P, Liu X, Tsai K, Wei Y, Fox T. Investigation of protein refolding using a fractional factorial screen: a study of reagent effects and interactions. Protein Sci. 2005;14:1818–26.
48. Coskun O. Separation techniques: chromatography. North Clin Istanb. 2016;3:156–60.
49. Bornhorst JA, Falke JJ. Purification of proteins using polyhistidine affinity tags. Methods Enzymol. 2000;326:245–54.
50. Porath J. Immobilized metal ion affinity chromatography. Protein Expr Purif. 1992;3:263–81.
51. Mahmoodi S, Pourhassan-Moghaddam M, Wood DW, Majdi H, Zarghami NJCB. Current affinity approaches for purification of recombinant proteins. Cogent Biol. 2019;5(1):1665406.
52. Jungbauer A, Hahn R. Ion-exchange chromatography. In: Methods in enzymology, vol. 463. Netherland: Elsevier; 2009. p. 349–71.
53. Hedhammar M, Karlström AE, Hober SJS. Chromatographic methods for protein purification. Seattle, WA: Semantic Scholar; 2006. p. 1–31.
54. Mori S, Barth HG. Size exclusion chromatography. Singapore: Springer Science & Business Media; 2013.

55. Sun T, Chance RR, Graessley WW, Lohse DJ. A study of the separation principle in size exclusion chromatography. Macromolecules. 2004;37:4304–12.
56. Fekete S, Beck A, Veuthey J-L, Guillarme D. Theory and practice of size exclusion chromatography for the analysis of protein aggregates. J Pharm Biomed Anal. 2014;101:161–73.
57. Burgess RR. A brief practical review of size exclusion chromatography: rules of thumb, limitations, and troubleshooting. Protein Expr Purif. 2018;150:81–5.
58. Afeyan N, Gordon N, Mazsaroff I, Varady L, Fulton S, Yang Y, Regnier FJ. Flow-through particles for the high-performance liquid chromatographic separation of biomolecules: perfusion chromatography. J Chromatogr A. 1990;519:1–29.
59. Granath KA, Kvist BE. Molecular weight distribution analysis by gel chromatography on sephadex. J Chromatogr A. 1967;28:69–81.
60. Striegel A, Yau WW, Kirkland JJ, Bly DD. Modern size-exclusion liquid chromatography: practice of gel permeation and gel filtration chromatography. Hoboken, NJ: John Wiley & Sons; 2009.
61. Erickson HP. Size and shape of protein molecules at the nanometer level determined by sedimentation, gel filtration, and electron microscopy. Biol Proced Online. 2009;11:32.

Protein Purification by Affinity Chromatography

6

Shubhankar Dutta and Kakoli Bose

Abstract

Affinity chromatography involves targeted purification of biological macromolecules from a crude mixture on the basis of highly specific interaction between the macromolecule and a tag protein or peptide. The interaction is typically reversible and purification is implemented by keeping one of the molecules (the affinity ligand or fusion tag) immobilized to the support matrix (containing respective binding resin for interaction with the tag) while its partner (the target protein) is in a mobile phase as part of the crude mixture. In this chapter we will be discussing recombinant protein purification using different affinity tags that are routinely used in a laboratory setup that include polyhistidine, GST (glutathione-S-transferase), maltose-binding protein (MBP) and Strep-tag. As affinity chromatography is a sophisticated purification method that requires significant expertise, the protocol and the problem-solving approaches described in this chapter will act as essential guides to the protein biochemists.

Keywords

Affinity chromatography · Affinity tag · Support matrix · MBP · GST · Strep-tag

S. Dutta
Integrated Biophysics and Structural Biology Laboratory, Advanced Centre for Treatment, Research and Education in Cancer (ACTREC), Tata Memorial Centre, Mumbai, India

K. Bose (✉)
Integrated Biophysics and Structural Biology Lab, Advanced Centre for Treatment, Research and Education in Cancer (ACTREC), Tata Memorial Centre, Mumbai, India

Homi Bhabha National Institute, BARC Training School Complex, Mumbai, India
e-mail: kbose@actrec.gov.in

K. Bose (ed.), *Textbook on Cloning, Expression and Purification of Recombinant Proteins*, https://doi.org/10.1007/978-981-16-4987-5_6

6.1 Introduction

As discussed in the earlier chapter, purification of recombinant proteins came into picture when Sumner et al. purified enzymatic proteins from prokaryotic system in the year 1926 [1]. Since its inception, protein purification process has evolved considerably, giving rise to different types of chromatographic techniques that often in combination give proteins of very high purity (>95%). Affinity chromatography is one such technique that was first introduced by Campbell and co-workers in 1951, when they isolated rabbit anti-bovine serum albumin antibodies using bovine serum albumin as the affinity ligand [2]. In subsequent years, the application of this technique went beyond just antigen-antibody separation and researchers slowly adopted its principle for separation of different types of protein molecules [3]. The salient features that give this process an extra edge over other purifying techniques are high-precision specificity, ease of handling and high yield. A classic affinity chromatography process mainly involves following three steps:

- Incubating the given sample (containing mixture of proteins and other molecules) along with the support matrix (containing affinity ligand) that allow the target protein in the sample to bind with the affinity (immobilized) ligand.
- Binding of the target protein and the affinity ligand depends on the buffer conditions. An optimum buffer condition ensures that the target molecule interacts effectively with the ligand and is retained by the affinity medium as all other molecules wash through the column.
- Post-washing, elution (dissociation from the ligand and recovering the protein) of the target protein from the ligand is done by changing the buffer conditions in a way conducive to disruption of the binding interactions between the target and the ligand.

Generally, elution of the desired protein using affinity-based purification depends on reversible interactions between the target protein and the immobilized ligand bound to support or chromatographic matrix [4]. Most proteins have their designated binding sites that are taken into consideration while selecting the suitable affinity ligand. However, it is necessary that the binding interaction between the target protein and the chosen ligand is reversible as well as specific. In case of recombinant proteins purification, a reversible but specific binding with the ligand is achieved through fusion tags. Fusion tags are generally proteins or peptide molecules that are capable of facilitating expression of the target protein in the expression system by providing resistance to proteolytic degradation [4]. A well-characterized fusion tag enables affinity chromatography of the tagged protein (target protein with the fusion tag attached) using immobilized version of the respective affinity ligand. His_6, MBP (maltose-binding protein) and GST (glutathione-S-transferase) are some of the fusion tags that are generally used in affinity purification [5]. Though in recombinant protein purification fusion tags have their own benefit, choosing a suitable tag, using it for affinity-based protein purification and separating it post-purification are a challenging process. Hence, in this chapter, with detailed protocols for various types of affinity chromatography techniques, we will elucidate its role in purification of recombinant proteins.

6.2 Types of Tags

Selecting suitable fusion tags for the target proteins is an important aspect of recombinant protein purification process. Generally, underlying physiochemical properties of the target protein determine the type of tags to be used for purification. Some of the widely used tags are given below in detail.

6.2.1 Polyhistidine Tag

The polyhistidine affinity tag, *aka* His_6-tag, normally comprises six consecutive histidine (His) residues, but can vary in length from two to ten histidine residues [6]. Protein purification using His_6-tag employs the ability of His to interact with transition metal ions, such as Cu^{2+}, Co^{2+} and Ni^{2+}. His_6-tag is non-toxic, and due to its small size, it induces no effect that would alter the physiochemical properties of the target protein [7]. Usage of His_6-tag is advantageous for protein purification involving prokaryotic expression systems where different vectors such as pET-16b, pET-28a or pET-28b are used [8]. However, in case of eukaryotic systems, where the percentage of His residues is high, protein purification becomes complicated due to non-specific binding of the His residues with the affinity ligands (metal ions). This non-specific background binding is generally circumvented through application of stringent wash conditions [9].

6.2.2 Glutathione-S-Transferase (GST) Tag

GST is a 26 kDa (kilo Dalton) fusion tag belonging to the cytosolic family of eukaryotic and prokaryotic enzymes that catalyse the fusion of the reduced form of glutathione (GSH) to external chemical substrates present within organisms for the purpose of detoxification [7]. It is one of the most extensively used tags for protein purification in prokaryotic expression systems that uses multiple cloning sites of pGEX vectors for generating the fusion tag protein [10]. Due to its ability to enhance tagged-protein solubility, GST affinity tag is also considered as a solubility tag [11]. Since GST-tag has high affinity towards glutathione, glutathione-coated beads are coupled to the chromatographic support matrix during purification of the fused protein of interest [12]. GST-tag is normally attached to the N- or C-terminal region of the target protein, depending on the structural properties of the protein. Owing to its slow binding kinetics with the glutathione ligand, high-scale protein purification using GST-tag sometime becomes time-consuming [13].

6.2.3 Maltose-Binding Protein (MBP) Tag

Similar to GST, MBP also enhance recombinant protein solubility in bacterial expression system [14]. pMAL vector in the *E. coli* expression system is generally used to generate the protein of interest with MBP attached at the N- or C-terminus [15]. One of the advantages of MBP-based purification over polyhistidine tag is the ability to generate high yield of difficult-to-purify proteins [16]. An MBP-fusion protein binds to the cross-linked amylose resin present in the chromatographic matrix [14]. The target protein is eluted by running maltose in the elution buffer solution. MBP-based affinity purification is considerably resistant to denaturing and reducing agents. However, any amylase activity in the crude sample (cell lysate) substantially reduces the efficiency of the amylose resin that is used as the affinity ligand in the support matrix [7].

6.2.4 Calmodulin-Binding Peptide (CBP) Tag

The calmodulin-binding peptide (CBP) is a 26 amino acid fragment that belongs to the C-terminal region of the muscle myosin kinase protein [17]. CBP utilizes its mild binding affinity ($K_d = 10^{-9}$ M) towards calmodulin (affinity ligand) in order to purify the protein of interest from the bacterial system [17]. The mild binding and elution conditions of this purification system enable the fusion protein to maintain its native form post-purification. Expression vectors such as pCAL-c or pCAL-n are used for the expression of CBP-tagged target proteins, where crude cell lysate (containing the fused protein) is passed through the calmodulin affinity resin for its purification and subsequent elution [18, 19]. The 4 kDa size of the CBP tag itself is relatively small and is much less likely to affect the properties of the protein of interest, thus making this an appealing affinity tag in comparison to tags of larger sizes.

6.2.5 Streptavidin-Binding Peptide (SBP) Tag

There are two different versions of SBPs—Strep-tag I and Strep-tag II—that are generally used to immobilize fusion proteins on a streptavidin matrix [20]. In terms of binding specificity, Strep-tag II exhibits stronger interactions with the streptavidin core (StrepTactin Sepharose medium) as compared to Strep-tag I [20]. Strep-tag II is a very small tag of 1 kDa that comprises only eight amino acid residues (WSHPQFEK) [21]. Small size and chemically balanced amino acid composition make it an appropriate tag for purifying functional proteins as it does not affect the protein structure or folding mechanisms. Generally, pASK-IBA or pASG-IBA vector is used to strategically fuse Strep-tag II onto the N- or C-terminus of the target protein for expression in the bacterial system [22]. The affinity core, StrepTactin, is a derivative of streptavidin and exhibits high stability in the presence of various proteases and SDS [23]. This property of streptavidin ensures long-lasting affinity columns that can be re-used several times. Apart from Strep-tag II's small size, its specific and fast binding kinetics with StrepTactin affinity medium make this purification system user-friendly and straightforward.

6.3 Types of Affinity Chromatography

In the subsequent sections, we will be discussing three major types of affinity-based purification system that utilizes following types of tags:

1. A tandem repeat fusion tag (polyhistidine) that binds to the affinity matrix comprising metal ions
2. A larger-sized solubility tag such as GST or MBP that can act as chaperone during protein expression and purification process
3. A smaller size binding peptide (Strep-tag II) that is faster and does not cause any structural perturbation in the protein of interest

6.3.1 Purification of Polyhistidine Tag Protein

As mentioned in the last chapter, IMAC (immobilized metal-affinity chromatography) is a widely employed method to purify recombinant proteins containing a short affinity tag of consecutive histidine residues. The basic principle behind IMAC is interactions between an immovable transition metal ion (Co^{2+}, Ni^{2+}, Cu^{2+}, Zn^{2+}) bound to the matrix and side chain of a specific amino acid such as Histidine [24]. Histidine residue displays strong affinity towards metal ion matrices, due to the presence of imidazole ring that acts as an electron donor group facilitating coordinate bond formation with the transition metal ions. This interaction is reversed during the elution of the target protein by adjusting the concentration of imidazole in the elution buffer or altering the pH of the buffer.

6.3.1.1 Binding with the Polyhistidine Tag

A consecutive stretch of six polyhistidine residues (His_6-tag) is the most commonly used tag in IMAC [6]. Though tags of six histidine residues are generally long enough to yield high-affinity interactions with the matrix, various studies have exhibited successful implementation of either shorter or longer affinity tags [6]. Usage of longer His_6-tag might increase the purity of the final product in some cases [25, 26]. However, it is advisable to use the smallest number of histidine residues wherever applicable, so that structural and functional perturbation of the protein can be minimized [25, 26]. In general, for purification of a novel recombinant proteins whose structure and functions have not been studied extensively, a six-histidine tag is always the most suitable choice [25]. Placement of the polyhistidine tags is done at either extremities of the N- or the C-terminal region, depending upon the physicochemical nature of the protein. For unknown proteins, it has been observed that changing the location (moving the tag to opposite terminus) of affinity tag often resolves the issues related to low yield of the target protein in the expression vector [27]. Moreover, the choice of N- or C-terminal tags also depends upon the downstream studies designed with the purified protein. For example, if protein-protein interaction studies through C-terminal domain are envisaged post-purification, the tag is usually attached to the N-terminus of the protein of interest, keeping the other end free [28].

The main advantage of polyhistidine affinity tag is their small size and non-toxic nature. Owing to their tiny size, they are simple to handle and can be incorporated easily into a variety of expression vectors. Normally, His-tags are introduced into the target DNA (Deoxyribonucleic Acid), encoding the protein of interest, by site-directed mutagenesis [29]. In addition, polymerase chain reaction (PCR) methods can also be applied using primers that have tandem histidine codons (CAC or CAT) [30]. For six histidine tags, eighteen bases encoding the histidine residues are inserted either after the start codon or before the stop codon in the DNA fragments encoding the desired protein [29]. Apart from synthetically preparing the His-tag in laboratory conditions, there are commercially available cloning vectors too that are used for the generation and expression of fusion tag recombinant proteins in various prokaryotic expression systems such as *E. coli.*

6.3.1.2 Components of the Chromatographic Matrix

Among the commercially available metal matrices, nickel-nitrilotriacetic acid (Ni^{2+}-NTA) and Co^{2+}–carboxyl methyl aspartate (Co^{2+}-CMA) are the two widely used immobilized ligands for IMAC [31, 32]. Both Ni^{2+} and Co^{2+} ions have six coordination sites, out of which four sites bind with the IMAC matrix resin and the two remaining sites coupled with the polyhistidine affinity tags [33]. Mechanistic models of their interactions are given in Fig. 6.1. IMAC matrices are generally robust in nature and have the ability to withstand broad range of variable conditions, such as physiological pH, stringent washing procedure as well as varieties of protein denaturants [33]. In terms of binding affinity, Ni^{2+}–NTA matrix has higher affinity towards histidine residues when compared to the Co^{2+}–CMA matrix [35]. At pH 8.0, Ni^{2+}–NTA exhibits binding affinity of $K_d = 10^{-13}$ M with an overall binding capacity exceeding 5 mg protein/ml of matrix resin, making it a more efficient IMAC matrix for purification of polyhistidine-tagged proteins [35].

6.3.1.3 Purification Under Different Conditions

Highly purified fraction of the polyhistidine-tagged recombinant proteins can be obtained under two conditions—native and denaturing. Proteins that are highly soluble in the cytoplasm are normally purified in native condition. Purifying the target protein under native condition helps in preserving its biological activity as this condition rarely alters the structural aspect of the protein [33]. Moreover, native purification with high protein yield is even feasible in mild buffer and imidazole concentration [7]. However, purification under native conditions might not be possible if the protein of interest is insoluble in cell cytoplasm, possesses a tertiary structure that disrupts the bound polyhistidine affinity tag or have the tendency to form aggregates [36]. These issues can be circumvented using denaturing agents

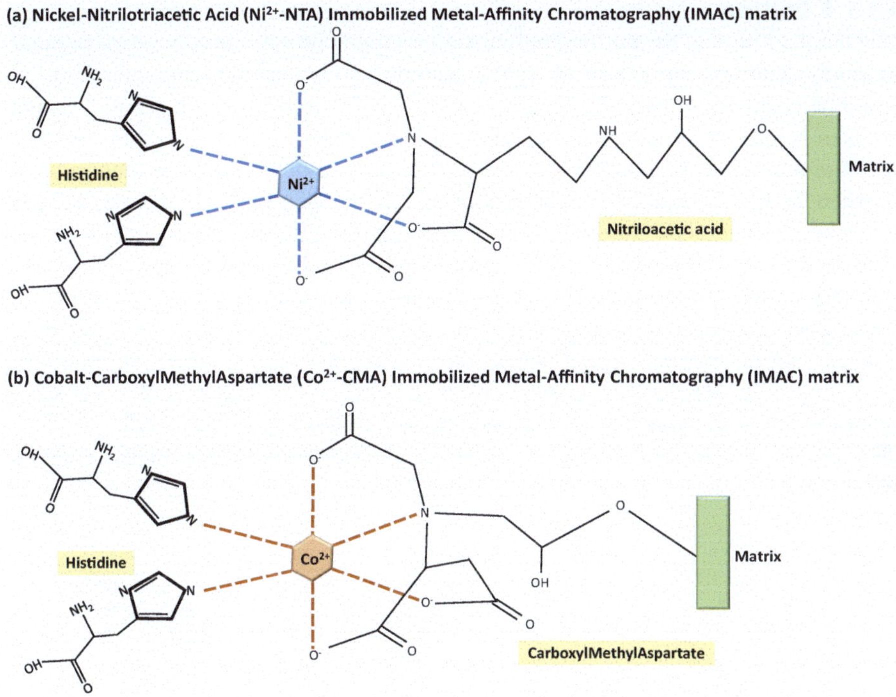

Fig. 6.1 Mechanistic models of interactions between his-tag and IMAC (Immobilized Metal Affinity Chromatography) matrices. (**a**) The nickel–nitrilotriacetic acid matrix (Ni^{2+}-NTA) [34] and (**b**) the cobalt–carboxyl methyl aspartate matrix (Co^{2+}-CMA) [34] are shown here. In both, the metal ion exhibits octahedral coordination by four matrix ligands and two histidine side chains, the latter provided by the polyhistidine affinity tag

including 6 *M* guanidinium hydrochloride or 8 *M* urea during the purification under denaturing conditions [36]. Since polyhistidine tag is an oligopeptide comprising series of tandem histidine repeats, it does not require any specific structural conformation for its proper functioning [37]. Thus, denaturing agents have no adverse effect on the interaction between the matrix resin and the His-tag, which further makes the purification process under denaturing conditions more effective. Between the two agents, denaturation using urea is more preferable for recombinant protein purification as it does not precipitate in the presence of SDS (sodium dodecyl sulphate) unlike guanidinium hydrochloride [37]. Since, SDS–PAGE (sodium dodecyl sulphate–polyacrylamide gel electrophoresis) analysis is one of the standard steps involved in detecting purity of recombinant proteins, formation of aggregates in contact with the SDS should be avoided. Under denaturing conditions, the obtained purified proteins are generally misfolded with reduced or no activity. Hence, the subsequent refolding of the protein into its biologically active state is achieved by washing away the denaturants stepwise through dialysis [38].

6.3.1.4 Elution

In general, elution of the purified polyhistidine-tagged proteins involves two popular methods—altering the pH level of the elution buffer and inducing imidazole at different concentrations [33]. In case of Ni^{2+}–NTA, reduction of pH to a range between 5.3 and 4.5 results in protonation of the nitrogen atom of the imidazole ring in histidine residue (pK_a 6.0) [34]. This subsequently impedes the coordination bond between the histidine and the transition metal. On the other hand, imidazole that acts as a histidine analogue can be used to competitively elute the bound polyhistidine residues by increasing the imidazole concentrations to 100 m*M* or higher [33]. If the tagged protein forms oligomers, more stringent conditions such as lower pH or higher concentrations of imidazole might be required to elute the protein. While both of these elution methods are effective, the use of imidazole is often preferable as exposure to low pH may damage the protein of interest [7]. The overall mechanism behind binding and elution of His-tagged protein is illustrated in Fig. 6.2.

A detailed protocol of IMAC involving Ni^{2+}–NTA as the support matrix is given below [34]:

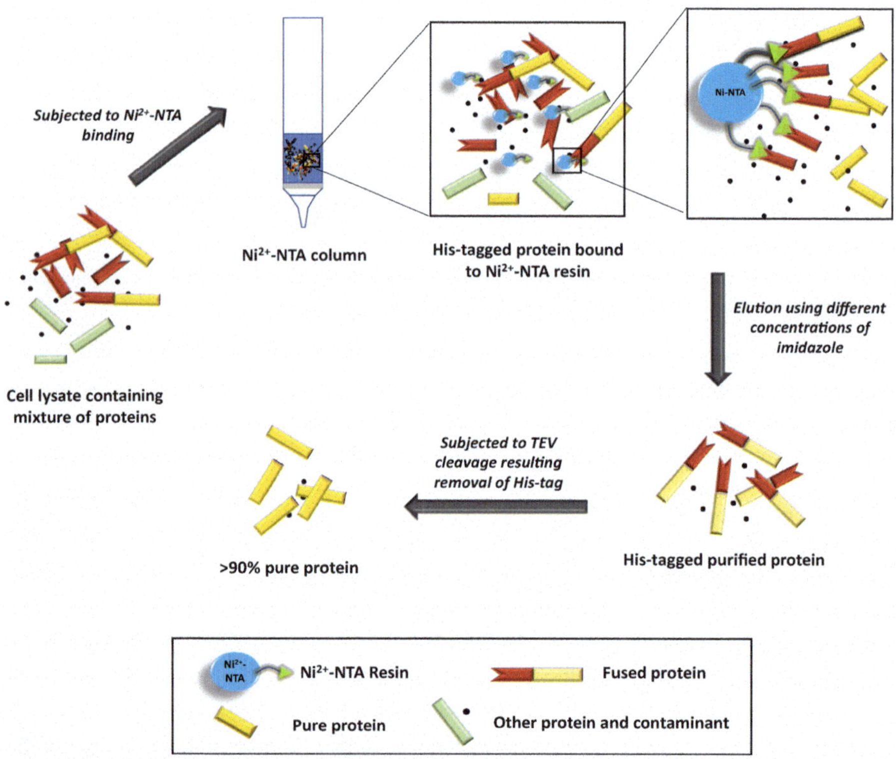

Fig. 6.2 An overall mechanism showing His-tag based affinity purification using Ni^{2+}–NTA. Cell lysate containing mixture of proteins along with the one with His-tag is passed through Ni^{+2}–NTA IMAC column. The fused protein binds to the column and gets eluted subsequently. His-tag removal is carried out using TEV protease, resulting in generation of the pure protein

- Lyse the *E. coli* cells expressing His-tagged recombinant protein suspended in the lysis/loading buffer by sonication on ice. Approximately 3–5 ml of loading buffer should be used per gram (wet weight) of cell pellet. It is important to keep the lysate under cold conditions to avoid any possible proteolysis.
- Centrifuge the cell lysate at 30,000 × *g* for 30 min at 4 °C.
- Add the clear lysate supernatant in the Ni^{2+}–NTA resin pre-equilibrated with an ice-cold loading buffer. Generally, 5–10 mg of protein binds per ml of the resin. Incubate the sample at 4 °C for 1 h on a rocker.
- Load the resin onto a column, followed by washing of the resin with 20 column volumes of a loading buffer containing 10 mM imidazole at 4 °C.
- Elute with 10 column volume gradient of 10–250 mM imidazole prepared in the loading buffer, pH 8.0 and collect the fractions.
- The purified protein is treated with site-specific proteases, such as TEV (Tobacco Etch Virus protease), to cleave the His-tag, followed by overnight dialysis (optional). Usually, most of the proteins are purified with the tag since it is too small and most of the time non-interfering.
- To obtain pure homogenous recombinant protein, the dialysed protein is concentrated using concentrators like Centricon [39], and subjected to SDS-PAGE to check for the purity and expected molecular weight of the recombinant protein.
- If the purification is not >95%, a second round of purification using gel filtration chromatography (discussed in detail in Chap. 8) is performed.

6.3.1.5 Troubleshooting

Problems	Possible reasons	Potential solutions
Purified fraction containing very low yield of the target protein.	Less stringent elution condition might result His-tag to still be bound with the protein.	Buffer condition can be made more stringent by increasing imidazole concentration or decreasing pH.
	Non-specific binding, especially hydrophobic interactions.	Add a non-ionic detergent to the elution buffer (e.g. 0.2% Tween-20).
	Undue protein precipitation in the column or wells.	Reducing protein concentration by eluting with slowly rising imidazole concentration gradient; avoiding steep rise and also decrease the amount of sample in the first place.
Impurity in the eluted protein fraction (SDS-PAGE analysis showing multiple bands).	Contaminants have affinity towards the tagged protein.	During protein expression in the bacterial system, adding detergent and/or reducing agents before sonicating cells might reduce contaminants. Moreover, increasing detergent levels (e.g. up to 2% Tween 20),

(continued)

Problems	Possible reasons	Potential solutions
		or adding glycerol (up to 20%) to the wash might further help in preventing non-specific interactions.
	Insufficient removal of the unbound materials during the washing step.	After application of the sample, repeat the wash step once or twice till highly pure target protein is obtained.
His-tag protein is getting eluted prematurely during loading or washing.	The tag is not adequately exposed.	Usage of denaturing agents such as guanidinium hydrochloride (Gdn-HCL) or urea during purification can be done to verify whether the tag is sufficiently exposed. Alternatively, application of longer polyhistidine tag or addition of linker between the tag and the target protein can significantly increase tag exposure.
	Incubation period is insufficient.	Increasing the incubation time of the crude lysates in the well or decreasing the speed of the centrifugation. Flow rate also must be regulated for efficient his-tag binding to the matrix.
	Imidazole concentration in the loading and/ or binding buffer is above limit.	Imidazole concentration must be lowered. Alternatively, usage of sophisticated affinity matrices such as Ni-Sepharose excel or TALON Superflow might be useful as these matrices do not require imidazole in the loading or binding buffer [40].

6.3.2 Purification of GST-Tagged Protein

GST tag system is a robust mode of affinity purification that involves protein expression and purification at high level in prokaryotic system. GST is a naturally occurring eukaryotic protein of 211 amino acid residues, which exhibits strong affinity towards glutathione [13]. Salient components of GST-tag-based purification are a pGEX plasmid vector, a GST moiety attached at the N- or C-terminus of the protein of interest, and an affinity ligand comprising series of immobilized glutathione beads bound to the chromatographic matrix [12]. One of the striking characteristics that gives GST-based purification an edge over IMAC is the ability of GST to act as a chaperone [41]. Moreover, application of GST enhances protein solubility and avoids its expression in the inclusion bodies [13]. The fusion protein is then captured by immobilized glutathione and impurities are washed away. Elution of the protein is performed under mild, non-denaturing conditions using reduced glutathione. If desired, the removal of the GST affinity tag is accomplished by using a site-specific protease recognition sequence located between the GST moiety and the target protein [7].

6.3.2.1 pGEX Vectors and Their Gene Fusion Construct

GST-tagged proteins are generally constructed by inserting a gene or gene fragment into the multiple cloning sites of pGEX vectors [10]. There are 13 types of commercially available pGEX vectors each containing a tac promoter that is capable of high-level protein expression, if induced by IPTG (Isopropyl ß-D-1-thiogalactopyranoside) [11]. An internal laqI^q gene facilitates tight control over expression of the gene insert by binding to the tac promoter until IPTG is applied. All these vectors possess cleavage site so that GST can be separated from the protein of interest after purification, using any of the three enzymes, namely, Thrombin, PreScission Protease and Factor Xa [11, 42]. Generally, the pGEX-T series, pGEX-X series and pGEX-P series contain protease cleavage sites for Thrombin, Factor Xa and PreScission protease, respectively, as represented in Fig. 6.3 [11, 42]. A suitable vector is determined on the basis of the future application of the target protein. While selecting the protease cleavage site, it is to be noted that the target protein must not possess an internal recognition sequence for this protease. Out of the three proteases, thrombin is the most cost-effective, as relatively small amounts of thrombin and short incubation times at 37 °C are sufficient to cleave the protein with high efficiency [43]. On the other hand, Factor Xa has very high specificity, but is expensive and generally requires high enzyme-to-substrate ratios for efficient cleavage [11, 43]. PreScission protease, in particular, has several advantageous characteristics, namely, it is effective at low temperature (5 °C) [11].

6.3.2.2 Expression of the Fused Protein

Generally, it is very difficult to predict whether the protein of interest will remain soluble in the cytoplasm or will accumulate in inclusion bodies. With increase in the protein size, structural complexity increases resulting requirement of intricate series of protein folding to avoid accumulation in the inclusion bodies. Though GST's chaperone-like behaviour ensures proper folding of the GST fusion protein, high-scale protein expression also requires maintaining of optimum culture conditions such as appropriate growth temperature and proper aeration [44]. Yields of GST fusion proteins using prokaryotic expression system like *E. coli* are highly variable, often ranging from 10 to 50 mg/l, but could potentially be much lower in cases where the fusion protein is toxic to the cells or unstable [11, 44]. Since expression levels are typically high, adequate amounts of protein usually can be obtained from several litres or less amount of the bacterial culture. The expressed protein can further be analysed using SDS-PAGE analysis. A detailed protocol explaining the cloning and expression of the fused protein is given below [11, 44, 45]:

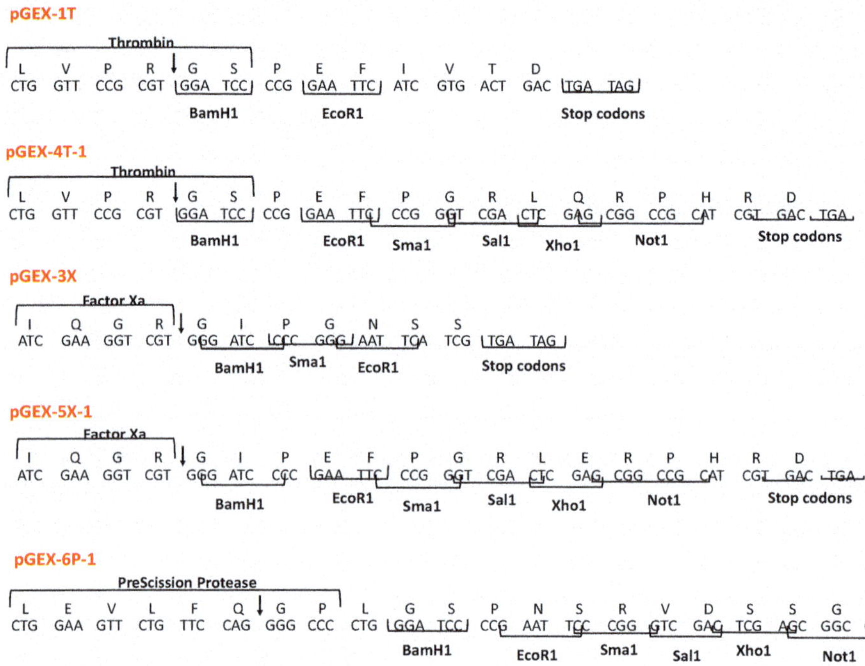

Fig. 6.3 An illustration showing various forms of pGEX vectors along with their proteolytic cleavage sites. T, X and P series of pGEX vectors are shown, which contain enzymatic cleavage sites for proteases such as Thrombin, Factor Xa and PreScission Protease, respectively. The vectors also contain different restriction sites represented by restriction endonucleases including BamH1 (from *Bacillus amyloliquefaciens*), EcoR1 (from *Escherichia coli*), Sma1 (from *Serratia marcescens*), Xho1 (from *Xanthomonas holcicola*), Not1 (from *Nocardia otitidis-caviarum*) and Sal1 (from *Streptomyces albus*)

- PCR amplification of the target gene sequence (DNA sequence of the target protein) containing the restriction sites at the ends of the target gene fragment that are in-frame and complementary to the vector are chosen.
- Digestion of the PCR product using relevant restriction enzymes followed by cleaning using gel electrophoresis.
- Transformation into the *E. coli* system, and growing the transformants on LB agar plates at 37 °C overnight.
- Screening the colonies for verifying that the gene insert is oriented properly and the reading frame is in correct order. Colonies are then transferred to separate tubes consisting PCR beads with 10 picomole each of pGEX primer being added.
- The colonies showing positive PCR are selected and an individual mini-culture of 5 ml is grown for each transformant for further screening of the expression of the protein of interest.
- Glycerol stocks can further be prepared using the exponential phase of the bacterial culture and 70% glycerol for storage purposes. Integrity of the target sequence is analysed using DNA sequencing.

- For expression procedure, some of the glycerol culture containing the isolated colonies are transferred to 100 ml LB (Luria-Bertani broth) with 100 μg/ml ampicillin (antibiotic) added as supplement.
- The inoculated culture is incubated overnight at 37 °C, inside an incubator shaker with 250–300 rpm, that is followed by extraction of 1 ml aliquot of the semi-opaque culture (based on the optical density, OD_{600}) the next morning.
- The obtained culture is termed as the starting culture, which is further diluted to 1:20 ratio using 600 ml fresh LB supplemented with 100 μg/ml ampicillin.
- Incubate the culture at 37 °C at 250–300 rpm until the optical density measured at 600 nm wavelength (OD_{600}) is within the range of 0.5 to 0.7 (log phase) [45].
- Prior to IPTG induction, 1 ml aliquot of the uninduced culture is kept for SDS-page analysis.
- Post-induction, the culture is incubated at 37 °C at 250–300 rpm for an additional 3 h, while monitoring the growth at OD_{600}. At saturation, they will stop dividing.
- Harvest cells by centrifugation at 4000 × *g* for 20 min at 4 °C.
- Carefully decant the supernatant, leaving ~15–50 ml in the centrifugation bottle.
- Resuspend the cells and transfer to a 50 ml centrifuge tube.
- Centrifuge for 20 min at 4000 × *g*, 4 °C.
- Decant the supernatant.
- Analyse un-induced and induced samples by SDS-PAGE to check protein expression levels.

6.3.2.3 Affinity-Based Purification of the GST-Fused Protein

After attaining the desired expression level, the GST fusion protein is subjected to purification from the bacterial cell lysate using immobilized glutathione coupled to a Sepharose column. There are different types of glutathione chromatography resin commercially available for the purification process. One of them is Sepharose 4B resin that is poured into the chromatographic column with an attached peristaltic pump that can control flow rates of the cell lysates [5, 44]. A detailed protocol describing the affinity purification using glutathione Sepharose 4B column is given below [11, 44, 45]:

- Prior to addition of the cell lysate, the glutathione Sepharose 4B column is washed thoroughly with 5–10 bed volumes of phosphate-buffered saline (PBS) for removal of any contaminants such as ethanol solution.
- Pelleted *E. coli* cells are lysed by sonication on ice for 10 s with 1 min interval between the bursts. This must be done at least 10 times with a time-gap between every burst to avoid sample heating. 50 μl of the resultant cell lysate is kept for SDS-PAGE analysis and the rest is subjected to centrifugation at 48,000 × *g* for 20 min at 4 °C.
- Supernatant is decanted into a sterile 50 ml centrifuge tube and kept for further gel analysis.
- 5–10 μl each of both the cell lysate and the decanted supernatant are run on SDS-PAGE gel to assess whether the fusion protein is in the decanted

supernatant. Upon verifying the presence of the fused protein in the supernatant, the fraction is subjected to affinity ligand binding.

- Keeping a flow rate of 0.1 ml/min (flow rate is kept low due to the slow binding kinetics between glutathione and GST), the supernatant fraction is loaded onto the Sepharose 4B column for binding with the glutathione resin. To verify whether the target protein with the GST-tag has bound with the glutathione, SDS-PAGE gel analysis is carried out.
- At this point, the obtained protein should be more than 90% pure. For the removal of the GST affinity tag, the purified fraction is further subjected to enzymatic cleavage through digestion by the serine proteases.

6.3.2.4 Elution and Removal of the GST Tag

On the basis of the vector chosen during cloning, different proteases such as thrombin, factor Xa or PreScission are used for cleavage of the GST affinity tag. These proteases are added in the glutathione buffer for inducing cleavage; however, it is to be ensured that the buffer must not contain any protease inhibitors. Post-cleavage, the target protein is separated from the GST moiety through re-chromatography using the glutathione Sepharose column. The detailed protocol of the elution process is given below [11, 44, 45]:

- Sufficient amount of thrombin or factor Xa or PreScission is added to the purified GST-tagged protein and kept for incubation at appropriate temperature (37 °C for thrombin, 25 °C for factor Xa and up to 5 °C for PreScission proteases).
- Enzymatic cleavage is carried out for a suitable length of time before being subjected to a non-specific protease inhibitor, called Phenylmethylsulphonyl Fluoride (PMSF). 0.3 mM of PMSF is added to terminate the cleavage process.
- The resultant fraction is given for dialysis, at least twice using 2 l of PBS or EDTA (Ethylenediamine Tetra-acetic Acid) for a minimum of 4 h per dialysis.
- The dialysed sample is subjected to centrifugation for 20 min at 4000 × *g*. This step ensures removal of unwanted precipitated materials that might have been developed during dialysis or pre-dialysis stages.
- Re-chromatography of the dialysed sample is done by loading the sample onto the Sepharose column at a flow rate of 0.1 ml/min. This fraction is collected for further gel analysis.
- The column is further washed using 2–3 bed volumes of PBS or EDTA at 1.5 ml/min.
- Finally, the bound GST and un-cleaved fusion protein are eluted using a reduced glutathione buffer at 0.3 ml/min for 5 bed volumes. All these fractions are analysed using SDS-PAGE, and are subsequently pooled to obtain the protein of interest.
- It is to be noted that the sample might contain some residual GST that did not rebind to the Sepharose column during re-chromatography. Moreover, there can be other contaminants too, such as proteolytic fragments, precipitates and aggregates. Hence, for further polishing, ion-exchange chromatography or gel

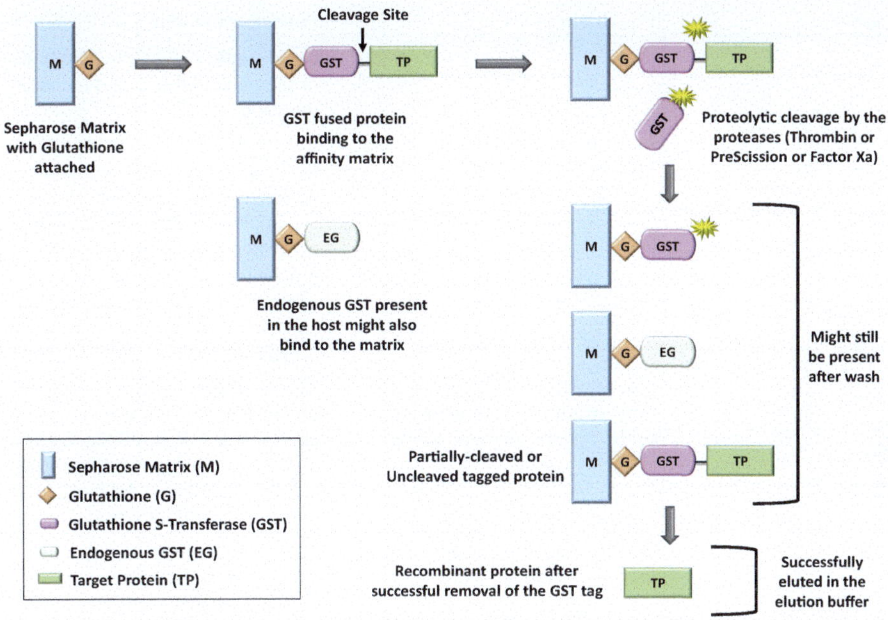

Fig. 6.4 Schematic representation of purification of GST-tagged protein. The target protein fused with GST gets attached to the Glutathione matrix, which is followed by the sequential purification steps as described in the text. The GST tag is subsequently removed by the proteolytic actions of proteases like Thrombin, Factor Xa or Precision Protease

filtration chromatography can be used. These procedures are discussed in detail in the subsequent chapters.

An illustration elucidating the affinity purification of GST-tagged protein is given in Fig. 6.4.

6.3.2.5 Troubleshooting

Problems	Possible reasons	Potential solutions
GST-tag fusion protein poorly binds to the affinity matrix column.	The flow rate used during sample loading is too high.	The binding kinetics between GST and glutathione is very slow. Thus, it is important to keep the flow rate low during sample loading onto the column. This ensures proper binding and increases the binding capacity of the column.
	The tagged protein has already aggregated into the sample during expression.	Prior to sonication of the cells, addition of reducing agents like DTT (Dithiothreitol) may

(continued)

Problems	Possible reasons	Potential solutions
		significantly reduce precipitation of the fusion protein and increase the yield.
	Low concentration of the tagged protein.	The binding kinetics between GST and glutathione Sepharose medium is directly proportional to sample concentration. A highly expressed protein binds more efficiently than the one with low expression. Hence, increasing the concentration of the tagged protein might improve binding.
	The Sepharose column has not been equilibrated properly prior to the fusion protein loading.	Before adding the cell lysates to the Sepharose medium, the column must be equilibrated with 5–10 bed volumes of buffers such as PBS (ensure that the buffer pH is between 6.5 and 8.0).
	The glutathione medium has been used too many times.	Time to time change of the Sepharose medium and use of freshly prepared medium considerably increases the binding efficiency of the affinity matrix column.
GST-tagged protein is not eluted efficiently.	The pH and ionic strength of the elution buffer is low.	Sometimes, increasing the basicity of the elution buffer or adding NaCl salt (0.1–0.3 M) might improve the elution yield of the target protein
	Concentration of glutathione is low.	Normally, the presence of 10 mM glutathione in the elution buffer is sufficient. However, for some tricky proteins, increasing the concentration of glutathione to a range of 20–40 mM might improve elution process.
	Non-specific interactions.	Solubility of the fusion proteins might get affected by non-specific hydrophobic interactions with the affinity medium, resulting aggregation. Addition of detergent molecules such as 0.1% Tween-20 or 2% n-octylglucoside into the elution buffer might prevent non-specific binding
Multiple bands are seen in SDS-PAGE analysis.	Partial degradation of the tagged protein during enzymatic cleavage using thrombin, factor Xa or PreScission.	Adding protease inhibitors like PMSF to the cell lysate solutions might improve elution. AEBSF or 4-(2 aminoethyl)-

(continued)

Problems	Possible reasons	Potential solutions
		benzenesulphonyl fluoride hydrochloride is another alternative to PMSF [46].
	Cell disruption during cell lysis process.	Decreasing lysis duration as over-lysis generally leads to the co-purification of the host cell (*E. coli*) proteins along with the GST-tagged one. Moreover, addition of lysozyme solution before the commencement of mechanical lysis prevents frothing and denaturation of the tagged protein
	Antibodies that react with various *E. coli* proteins may be present in the tagged protein sample.	Sometimes, the commercially available anti-GST might contain antibodies that can interact with *E. coli* proteins, resulting in non-specific background binding. This can be avoided using anti-GST, which has already been cross-adsorbed against all kinds of *E. coli* proteins.
Attached GST-tag is not cleaved properly.	Insufficient enzyme concentration and/or less incubation time might result in partial or no detachment of the GST-tag from the target protein.	Normally increasing the incubation time to 24 h or more facilitates efficient cleavage of the tag. However, higher reaction time might result in target protein degradation, which can be alternatively circumvented by increasing the enzyme concentration of the particular protease (PreScission Protease, thrombin, or Factor Xa) [11, 44, 45].
	During cloning of the fused protein, specific sites for protease cleavage might get altered.	Nucleotide sequence of the generated construct must be checked thoroughly to verify the presence of cleavage sites. Comparative alignment of the resultant sequence with a known sequence might help in identifying the altered site.
	Presence of endogenous protease inhibitors might interfere with the enzymatic action of the proteases.	In case of PreScission Protease, the inhibitors can be removed by dialysing the fused protein against 50 mM Tris–HCl, 150 mM NaCl, 1 mM EDTA, 1 mM DTT at pH 7.5 [11, 44, 45]. For factor Xa, fused protein can be subjected to buffer exchange on a desalting column, or dialysis

(continued)

Problems	Possible reasons	Potential solutions
		against 50 mM Tris–HCl, 150 mM NaCl, 1 mM $CaCl_2$ at pH 7.5 [11, 44, 45].

6.3.3 Purification of MBP-Tag Recombinant Proteins

MBP, encoded by malE gene, is a part of the maltose/maltodextrin system of *E. coli* that can be expressed in both secreted and non-secreted forms in the prokaryotic system [47]. Usage of MBP enhances solubility of the target protein by facilitating proper folding that further results in high yield [14, 48]. Fostering protein-folding ability of MBP lies in its ability to function as a chaperone magnet that enable it to recruit chaperones at the vicinity of the fused protein [49]. Although MBP exerts a high-metabolic burden on the expression system due to its large size (42 kDa), its exquisite ability to circumvent problems associated with heterologous protein expression overshadows its limitations and makes it a popular affinity tag [50]. It also enables segregation of the fused protein from its heterogeneous forms (resulting due to posttranslational modifications), and expedites its expression in the host system [14, 50]. Generally, in MBP-based affinity purification, a series of pMAL vectors can be used for aiding fusion of the MBP-tag with the target protein. Followed by successful attachment of the MBP-tag, the fused protein is subjected to binding with the amylose resin (affinity chromatographic matrix). Similar to GST, MBP is also a highly efficient fusion system due to its smooth purification methods, mild elution criteria and high compatibility with almost all downstream applications. A detailed methodology depicting the purification of high-quality MBP-fused protein is discussed below.

6.3.3.1 Expression of MBP-Tag Protein Using pMAL Vector

E. coli competent cells are subjected to transformation using pMAL vectors for expression of the MBP-tag protein. These vectors enable the expression of the secreted or cytosolic form of MBP, fused to the target protein, under the regulation of an IPTG-induced *tac* promoter. The use of this promoter allows pMAL vectors to be used in a wide variety of bacterial hosts, since the *tac* promoter utilizes the bacterial RNA polymerase for transcription. A detailed protocol describing the expression process is given below [14, 48]:

- *E. coli* competent cells are transformed with 10 ng of pMAL vector and plated on LB agar plates containing 100 mg/ml ampicillin and incubated for 16 h at 37 °C. Generally, 1 l of the LB medium is prepared by 10 g bacto-tryptone [51], 5 g bacto-yeast [51] extract, 10 g NaCl, 166 mL NaOH (10 N) and 10 ml $MgSO_4$ (1 M).

- A single colony is used to inoculate a tube containing 10 ml LB with 100 mg/ml ampicillin.
- The cells are grown in a shaker incubator for 16 h, before being transferred into 1 L of LB medium where the inoculum to medium ratio is kept at 1:100. The cells are placed in an incubator shaker having temperature 37 °C.
- When the OD_{600} reaches around 0.6, IPTG is added to a final concentration of 0.3 mM.
- The cells are harvested after 6 h of incubation at 30 °C and the pellets are stored at −80 °C if not immediately used for purification.

6.3.3.2 Binding to the Amylose Affinity Column and Purification

The main component of the chromatographic matrix for purification of the MBP-fused protein is amylose-agarose resin. For generation of the matrix, 25 ml of Sepharose 6B is washed with water in a Sinter glass and with 1 M sodium carbonate kept at pH 11 [47]. It is then allowed to react with 5 ml vinyl sulphonic acid for 70 min at room temperature. After washing with 500 ml of water, the resin is resuspended in a 25 ml solution of 2.6 g amylose in 1 M sodium carbonate at pH 11, with continuous stirring overnight. The resultant column is rigorously washed and kept in a solution of 20% ethanol and 80% water at 4 °C for proper maintenance. When the fused protein is passed through the column, MBP binds to amylose primarily via hydrogen bonds [52]. Addition of high ionic salts such as NaCl (1 M) can be used in order to reduce non-specific adsorption of proteins onto the resin [52]. The overall procedure of the purification method is discussed here:

- The frozen cell pellet from a 1 l culture is thawed on ice and resuspended in 70 ml of buffer. The cells are then mechanically lysed using a Micro-fluidizer, where the pressure is kept at 21,000 psi (pounds/inch^2).
- Post-lysis, the lysate is subjected to centrifugation at 4 °C for 20 min (15,000 × *g*), followed by filtration for removal of insoluble cell debris.
- Pre-equilibrated amylose-agarose column is loaded with filtered lysate at 1.7 ml/min and washed with buffer A at 2.5 ml/min up to low optical density (~5 CV, column volume).
- Protein is eluted with elution buffer (Binding buffer comprising 20 mM Tris–HCl, 200 mM NaCl, 1 mM EDTA at pH of 7.4 and 10 mM maltose) at 1.5 ml/min, collecting fractions of 9 ml during 4 CV.
- Samples from each fraction are analysed for protein content by SDS–PAGE.

6.3.3.3 Removal of the MBP-Tag Through Proteolytic Cleavage

Presence of maltose in the elution buffer allows the binding of the MBP-fused protein with the maltose and subsequent elution. However, the generated eluate still contains MBP tag fused with the protein of interest, which is required to be removed to obtain pure protein. Post-purification, removal of MBP-tag is done by the application of proteases such as Factor Xa, Thrombin, Enterokinase or TEV proteases [14, 53]. The MBP-containing vectors consist of designated proteolytic cleavage site for these proteases at the junction between the MBP and the target protein. Several studies have shown that usage of proteases like Factor Xa and Thrombin might result in non-specific digestion of the target protein; however, other proteases such as Enterokinase, Rhinovirus 3C protease and TEV are more specific [53–55]. Among them, TEV protease is the most predominantly used protease for MBP-based purification of the target protein [53]. TEV remains active even at 4 °C, whereas the other proteases usually require higher temperatures and a long period of incubation, which further increase the risk of aggregation and inactivation of the cleaved protein [53]. Another advantage of using this protease is its resistance to detergents that are often essential in the preparation of membrane proteins and other hydrophobic proteins [56]. After successful removal of the tag, the purified protein (70–90%) is further polished using ion-exchange chromatography or gel filtration chromatography that are discussed in detail in the subsequent chapters. The purification process is culminated by the regeneration of the affinity column, where the amylose–agarose resins are regenerated with 0.1% SDS at room temperature, followed by water wash, and maintained in 20% ethanol at 4 °C.

6.3.3.4 Troubleshooting

Problems	Possible reasons	Potential solutions
Inadequate binding of MBP-tagged proteins to the amylose resin.	Presence of endogenous amylases in the bacterial system might competitively inhibit binding to the amylose column.	This can be prevented by the application of 0.2% glucose in the growth medium that would substantially reduce the expression of endogenous amylase [52].
	The presence of non-ionic detergents.	Occasionally, the presence of detergents such as triton X-100 and Tween-20 can interfere with amylose and MBP binding. Concentration of the detergents can be reduced to 0.05% or less in order to solubilize the extract [52, 53]. Moreover, binding can be improved by screening alternative detergents.
	Oligomeric property of the target protein also can affect its binding to the amylose column.	Due to oligomerization, soluble aggregates are formed, which can be detected by gel filtration. The formation of oligomers can be reduced by changing the expression conditions or the purification procedure and by screening different buffers and additives [52].
Protein is not eluted efficiently from the affinity column.	Elution kinetics is very slow.	In case of slow elution kinetics, following changes might improve the elution process [57]:

(continued)

Problems	Possible reasons	Potential solutions
		(a) Decreasing the elution flow rate. (b) Overnight incubation in the elution buffer, when performing batch purification. (c) Increasing the concentration of maltose in the elution buffer using varied concentration of maltose (20–100 mM).
	SDS-PAGE analysis showing multiple bands post-elution.	Non-specific proteolysis can be identified by Western blot analysis. Undue proteolysis or degradation of the target protein can be prevented by conducting all purification steps at 4 °C. Moreover, use of protease inhibitors during the cell lysis process, can also help in reducing the chance of proteolysis [14, 57].
	Contaminants are non-covalently linked to the recombinant protein.	Increase ionic strength in all buffers for cell lysis and purification (up to 1 M NaCl) or add mild detergents, 0.1% Tween, 0.1% CHAPS) [14].
Column has clogged.	Top filter is clogged.	Top filter can be changed in case of clogging. Moreover, amylose column can be replaced if repetitive column regeneration has exhausted the column capacity.
	Cell debris in the lysate may have clogged the column.	Centrifuge and/or filter the sample through a 0.22 μm or a 0.45 μm filter or otherwise optimize sample pre-treatment before loading the next sample [52, 57].

6.3.4 Purification of Strep-Tag II Recombinant Proteins

The Strep-tag II is a short peptide that is capable of high selective binding with StrepTactin [21, 58]. StrepTactin is an engineered streptavidin, which in comparison to streptavidin, possesses 100-fold higher binding affinity ($K_d = 1$ μM) with Strep-tag II [58]. One of the striking advantages of this purification system is that it facilitates one-step purification of a wide range of recombinant proteins under natural physiological conditions, thus preserving its bioactivity [22]. Moreover, unlike GST or MBP tags, the Strep-tag II does not disturb the structural aspect of the protein of interest due to its small size. The Strep-tag II purification system can be implemented in various expression systems, such as bacterial, mammalian and insect; however, in this section we will mainly focus on its application using the bacterial expression system.

6.3.4.1 Expression of the Strep-Tag II Fused Protein

There are two types of vector systems used for expressing the Strep-tag II fused protein in the bacterial cells such as *E. coli*. One of them is pASK-IBA/pASG-IBA vector system that carries an inducible tetracycline promoter/operator (*tet*-promoter) responsible for fusion of the Strep-tag II at the N- or C-terminus of the recombinant protein [59]. pPR-IBA/pPSG-IBA vector system, on the other hand, utilizes T7 promoter and T7 RNA polymerase for high-level expression of the target protein in BL21 strain of *E. coli* [59]. Generally, *tet* system is preferred to T7 system, as it is independent of *E. coli* strain unlike T7 system, which needs a definite source of the T7 RNA polymerase (present in BL21 strain of *E. coli*) recombinant protein expression [60]. Hence, the following protocol describes the generation of cell lysates for the expression of Strep-tag II fusion proteins using *tet* system [20, 61, 62]:

- LB medium is prepared using 10 g/l tryptone, 5 g/l yeast extract and 5 g/l NaCl. 100 μg/ml of Ampicillin is added to the LB medium. Upon addition, 100 ml of the resultant medium is inoculated with a fresh bacterial colony containing the pASK-IBA expression plasmid and shake overnight (200 rpm) at 37 °C.
- Optical density of the sample is measured at 600 nm (OD_{600}) wavelength. When OD_{600} reaches between 0.5 and 0.6 range, 1 ml of the sample is taken out to use as uninduced control and then subjected to induction. The sample is induced by adding 10 μl stock solution of tetracycline antibiotic derivative such as anhydrotetracycline (prepared by 2 mg/ml of anhydrotetracycline in Dimethylformamid or DMF).
- The induced cell sample is subjected to shaking for 3 h at 200 rpm. However, for some proteins overnight induction might increase its yield (standardization with small amount is preferred prior to large scale purification). Culture is harvested and pellets are prepared by centrifugation at 4500 × *g* for 12–15 min.
- Washing buffer is prepared by 100 mM Tris/HCl (pH 8.0), 150 mM NaCl and 1 mM of EDTA. The pellet generated from the culture is resuspended in 1 ml of chilled washing buffer (at 4 °C).
- 10 μl of the resultant buffer content is given for SDS-PAGE analysis and rest of the residual suspension is subjected to sonication.
- The suspension is then centrifuged at 13000 rpm (microfuge) for 15 min at 4 °C resulting in sedimentation of the insoluble cell components. The clear supernatant is carefully separated from the sediments and collected in a tube.

6.3.4.2 Purification and Elution of the Fused Protein Using StrepTactin Affinity Column

Purification of Strep-tag II fused protein predominantly depends on the efficiency of binding between Strep-tag II and StrepTactin. The StrepTactin Sepharose column is made up of miniscule StrepTactin beads (average 30 μm), having high-resolution separating capacity, which is capable of generating target proteins in highly pure form [62]. Once the Strep-tag II bound protein binds with the StrepTactin resin, the unbound proteins in the cell lysates are washed away by the wash buffer (100 mM Tris/HCl at pH 8.0, 150 mM NaCl and 1 mM EDTA), prepared to maintain physiological condition [60]. The StrepTactin bound Strep-tag II protein is then eluted by addition of the elution buffer (wash buffer supplemented with 2.5 mM desthiobiotin) [20, 61, 62]. The desthiobiotin is a competitive inhibitor of Strep-tag II, which results in competitive binding at the biotin binding pocket and leads to recovery of the target protein [20, 22]. A complete protocol discussing the purification and elution procedure is given below [20, 61, 62]:

- Using 2 column bed volumes (CVs) of wash buffer, the StrepTactin column is equilibrated. During equilibration, addition of the wash buffer is done by removing the top cap of the column first, followed by the outlet cap. This order is maintained to avoid the column from becoming dry.
- If the storage solution of the column is 20% ethanol, it is recommended that the column is washed with at least 5 CVs of the wash buffer or distilled water. The flow rate during equilibration is set between 50 and 100 cm/h.
- After equilibration, clear supernatant of the cell lysates (cell or protein extract) generated during protein expression is added onto the column. The volume of the cell extract can be in the range of 0.5 to 10 CVs. It is to be noted that higher the concentration of the extract, more is the yield. Generally, the concentration within the range of 50–100 nanomolar is recommended for generating good yield.
- Once the cell extract has completely entered the column, wash the column 5 times with 1 CV of wash buffer. These washings will remove all the unbounded host proteins. Eluate is collected in fractions of 1 CV and each fraction is subjected to SDS-PAGE analysis.
- Elution buffer, containing desthiobiotin [63], is added for 6 times in 0.5 CV. Eluate is collected in fractions of 0.5 CV. 20 μl of each fraction is then given for SDS-PAGE analysis to verify whether the target protein has been successfully recovered. It is to be noted that addition of desthiobiotin results in its binding with StrepTactin followed by release of the target protein.
- Post-elution, the column is needed to be regenerated through removal of the desthiobiotin. The column is cleaned with 3 CVs of distilled water and 3 CVs of 0.5 M NaOH, followed by another wash by 3 CVs of distilled water.
- Subsequently, the column is re-equilibrated using 5 CVs of wash buffer prior to next purification. Alternatively, column re-equilibration can also be done using 15 CVs (large amount) of 1 mM HABA (2-[4′-hydroxy-benzeneazo] benzoic acid) in the wash buffer. When HABA is added in excess, it binds to the biotin by displacing desthiobiotin from the binding pocket. This binding results in colour-change of the medium from yellow to red, indicating that the column regeneration has been successfully accomplished.

An illustration elucidating the affinity purification of recombinant proteins using Strep-tag II is given in Fig. 6.5.

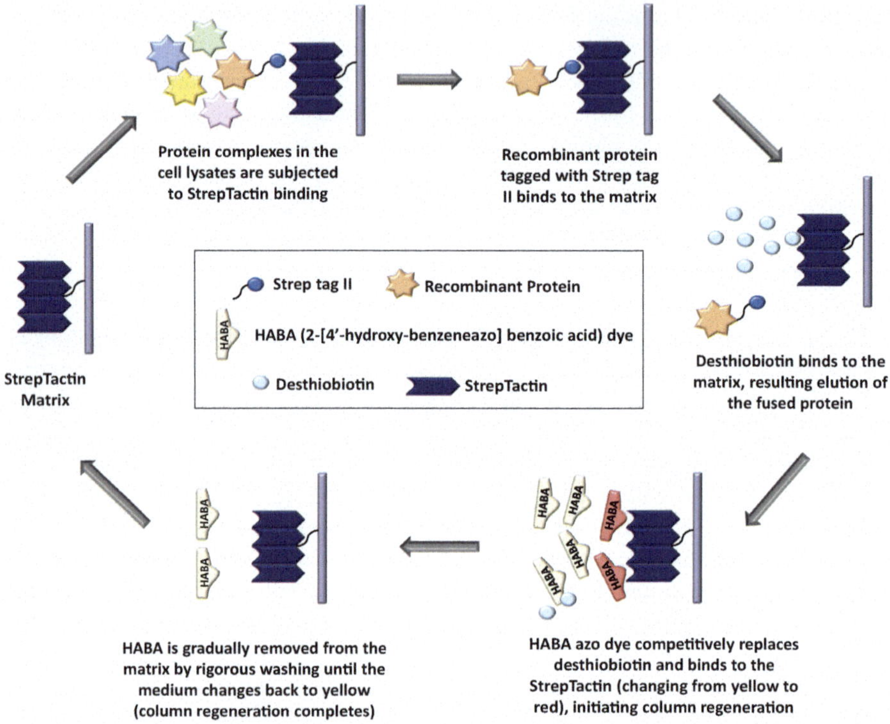

Fig. 6.5 A model delineating the mechanism behind purification of Strep-tag II-fused protein. Cell lysates containing the cocktail of proteins is subjected to binding to StrepTactin column. Target protein containing the Strep-tag II gets attached to the affinity column. Addition of desthiobiotin leads to its competitive binding with the affinity matrix resulting releasing of the fused target protein. The StrepTactin column is regenerated by implementation of HABA dye in excess amount, which displaces desthiobiotin, followed by rigorous washing

6.3.4.3 Troubleshooting

Problems	Possible reasons	Potential solutions
Increased column pressure in the opposite direction.	Unclarified cell lysates and highly viscous solution.	Dilution of the cell extract prior to sonication might reduce the concentration of host nucleic acid. Sometimes, bringing the cell extract at room temperature (from 4 °C) can reduce viscosity.
	Incomplete sonication.	Increasing the sonication time can ensure efficient cell disruption. Moreover, sonication must be conducted in ice as overheating might result in froth formation and denaturation of the target protein.

(continued)

Problems	Possible reasons	Potential solutions
Poor binding with the column matrix.	Strep-tag II is missing or inaccessible.	Proteases that are capable of disrupting the 8-amino acid tag might be present in the *E. coli* strain used for cell lysates. Implementation of proper protease inhibitor can avoid this degradation of the Strep-tag II.
	Biotinylated proteins present in the cell extract might competitively block the ligand.	The host cell proteins might contain biotin in significant concentration that interferes with the StrepTactin binding of Strep-tag II. Addition of avidin (a biotin blocking buffer) facilitates Strep-tag II binding [62].
	Protein has already formed inclusion bodies in the column.	Very high protein concentration leads to precipitation. Decreasing the flow rate and amount of sample load can block protein.
Protein contaminants	Contaminants are generally referred to as the shorter fragments of the fused protein generated during elution.	Uncontrolled proteolysis can be prevented either by using protease-deficient *E. coli* strains or by inducing protease inhibitors post cell lysis [64]. Sometimes, reversing the fusion terminus of Strep-tag II and addition of chelating agents like EDTA can improve protein purity by inhibiting protease activity.
	Formation of covalent and electrostatic interactions between the contaminants and the tagged protein.	Covalent interactions such as disulphide bonds can be disrupted by adding reducing agents such as DTT during cell lysis step. Electrostatic bond formation can be prevented by either increasing the ionic strength up to 1 M NaCl of the wash buffer or by adding weak detergents like 0.1% Tween and 0.1% CHAPS (3-[(3-cholamidopropyl) dimethylammonio]-1-propanesulphonate).

6.4 Role of Affinity Tags in Identifying Protein-Protein Interactions

Apart from column-based protein purification, affinity tags also play a crucial role in identifying protein-protein interactions [28, 65, 66]. Generally, for validating interactions between two proteins, pull-down assay is one of the most common methods, which involves segregation of a protein complex with the help of

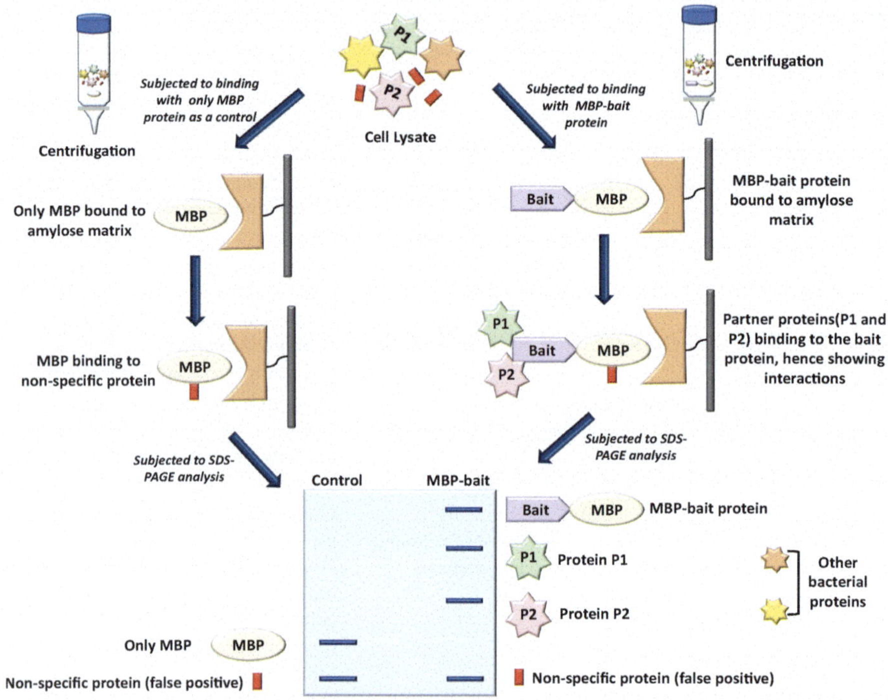

Fig. 6.6 Outline of an MBP-based pull-down assay. The procedure on the left shows the control experiment. In this example, two proteins (P1 and P2) are identified by SDS-PAGE as interacting partners with the bait protein (right lane) bound to the MBP tag. One additional non-specific protein (red) was pulled down as a false positive (MBP binder) by the control (left lane)

immobilized beads [67]. The beads are made of specific ligand resins that bind to the affinity tags such as histidine, GST or MBP attached to the either extremities of the protein complexes [67, 68]. Besides, identifying the interaction between two binding partners, affinity pull-down assays are also used for isolation of low amount (in µg) of complexes, mainly to recognize discrete domains or subunits [52, 67].

One such affinity pull-down assay is MBP pull-down assay that involves affinity purification of one or several unknown proteins from a biological sample using an MBP-tagged bait protein. The MBP-tagged bait protein is attached to the immobilized amylose matrix, which acts as the affinity ligand. The basic principle is that the MBP-tagged bait protein binds to its interacting partner/s forming a complex that gets captured by the affinity matrix. A control is also included to identify non-specific false positive bindings where only MBP is kept without the bait protein. The control can either be lysate from separately transformed cells that express MBP (not the bait fusion protein) or lysate from non-transformed cells to which MBP is added. An illustration explaining the overall process of MBP-based affinity pull-down assay is given in Fig. 6.6. Similar principle can also be used for pull-down using other tags like GST, histidine or Strep-tag II.

6.5 Conclusion

In this chapter, we have discussed about affinity tags where their reversible interactions with the target protein are utilized in the purification process. Though affinity-based chromatography is one of the most sophisticated chromatographic techniques, it has its own limitations. As it relies on the specific interactions of the affinity ligand with the protein of interest, preparation of these ligands sometimes becomes too expensive. The shelf-life of these ligands and the support matrix further adds to the problem, as they require frequent regeneration of the column and regular monitoring. Moreover, choice of tags, media selection, maintaining optimum flow rate and preparation of precise washing and elution buffer require extensive analytical skills of the users for successful execution. Troubleshooting is a key aspect of the affinity chromatographic technique and this chapter has vividly explained how in different scenarios these problems can be dealt with.

Nevertheless, it is a powerful technique that can be utilized at its highest potential with prior knowledge about the structure and function of the target protein. Although affinity chromatography alone is sufficient for purification of a wide variety of recombinant proteins, sometime it might require combination of two or more techniques for purification of few specific proteins to attain highest level of purity for subsequent biophysical or structural studies. In the forthcoming chapters, we will further discuss other chromatographic techniques and purification of proteins which are difficult to purify using so-called conventional methods.

Acknowledgements The authors thank Advanced Centre for Treatment, Research and Education in Cancer (ACTREC) for providing necessary infrastructure and resources for successful completion of the chapter. The authors acknowledge Ms. Chanda Baisane, Bose Lab, ACTREC for formatting the manuscript.

Problems

Multiple choice questions

1. The specific biological interaction that is not used in affinity chromatography purification is:
   (a) Receptor-ligand
   (b) Antigen-antibody
   (c) Cations-anions
   (d) Enzyme-substrate
2. The first step of affinity chromatography purification process is:
   (a) Addition of affinity ligand into the matrix
   (b) Precipitation
   (c) Elution
   (d) Binding of the ligand with the tag

(continued)

3. The property of an ideal affinity chromatography matrix is:
   (a) The matrix materials should be polymeric and organic
   (b) The matrix should be based on inorganic compounds
   (c) The matrix should be mechanically stable and exhibit good flow property
   (d) The matrix should form reversible but specific interaction with the affinity tag

Subjective questions

1. A 50 kDa $His_{6\times}$-tagged protein was adequately expressed in *E. coli* BL21 (DE3) host strain and subsequently purified using Ni-NTA column. Buffer conditions were as follows:
   Binding buffer: 50 mM NaH_2PO_4, 150 mM NaCl and 10 mM Imidazole, pH 7.4
   Elution buffer: 50 mM NaH_2PO_4, 150 mM NaCl and 100 mM imidazole, pH 7.4
   On purification, the protein co-elutes with chaperones and non-specific bands. In addition, white precipitates were observed in the eluted fractions. List a few strategies that can be employed to obtain better yield and purity of target protein.
2. To assess the interactions between protein A and protein B, an MBP pull-down assay was performed. Protein A (22 kDa) was tagged with MBP and considered the bait protein, whereas protein B (48 kDa) was kept untagged. The proteins were incubated in following buffer conditions:
   Binding buffer: 20 mM Tris–HCl, 200 mM NaCl, 1 mM EDTA, pH 7.4
   Elution buffer: 10 mM Maltose, 20 mM Tris–HCl, 200 mM NaCl, 1 mM EDTA, pH 7.4
   Post-elution, the pull-down eluates were subjected to SDS-PAGE analysis keeping only MBP as the control. The resultant gel analysis showed an unexpected band at 70 kDa which neither corresponds to protein A nor B. State a reason that can explain the band and also provide a strategy that can be implemented to avoid such bands.

References

1. Sumner JB. The isolation and crystallization of the enzyme urease: preliminary paper. J Biol Chem. 1926;69(2):435–41.
2. Campbell DH, Luescher E, Lerman LS. Immunologic adsorbents: I. isolation of antibody by means of a cellulose-protein antigen. Proc Natl Acad Sci U S A. 1951;37(9):575–8.

3. Gräslund S, et al. Protein production and purification. Nat Methods. 2008;5(2):135–46.
4. Gómez-Arribas LN, et al. Tag-specific affinity purification of recombinant proteins by using molecularly imprinted polymers. Anal Chem. 2019;91(6):4100–6.
5. Mahmoodi S, et al. Current affinity approaches for purification of recombinant proteins. Cogent Biol. 2019;5(1):1665406.
6. Spriestersbach A, et al. Chapter One - Purification of his-tagged proteins. In: Lorsch JR, editor. Methods in enzymology. Cambridge, MA: Academic Press; 2015. p. 1–15.
7. Kimple ME, Brill AL, Pasker RL. Overview of affinity tags for protein purification. Curr Protoc Protein Sci. 2013;73:9.9.1–9.9.23.
8. Trigoso YD, et al. Cloning, expression, and purification of histidine-tagged *Escherichia coli* dihydrodipicolinate reductase. PLoS One. 2016;11(1):e0146525.
9. Schoonen L, et al. Alternative application of an affinity purification tag: hexahistidines in ester hydrolysis. Sci Rep. 2017;7(1):14772.
10. Aatsinki JT, Rajaniemi HJ. An alternative use of basic pGEX vectors for producing both N- and C-terminal fusion proteins for production and affinity purification of antibodies. Protein Expr Purif. 2005;40(2):287–91.
11. Harper S, Speicher DW. Purification of proteins fused to glutathione S-transferase. Methods Molecul Biol (Clifton, NJ). 2011;681:259–80.
12. Young CL, Britton ZT, Robinson AS. Recombinant protein expression and purification: a comprehensive review of affinity tags and microbial applications. Biotechnol J. 2012;7 (5):620–34.
13. Schäfer F, et al. Purification of GST-tagged proteins. Methods Enzymol. 2015;559:127–39.
14. Fox JD, Waugh DS. Maltose-binding protein as a solubility enhancer. In: Vaillancourt PE, editor. *E. coli* gene expression protocols. Totowa, NJ: Humana Press; 2003. p. 99–117.
15. Maina CV, et al. An Escherichia coli vector to express and purify foreign proteins by fusion to and separation from maltose-binding protein. Gene. 1988;74(2):365–73.
16. Duong-Ly KC, Gabelli SB. Affinity purification of a recombinant protein expressed as a fusion with the maltose-binding protein (MBP) tag. Methods Enzymol. 2015;559:17–26.
17. Klein W. Calmodulin-binding peptide as a removable affinity tag for protein purification. In: Vaillancourt PE, editor. *E. coli* gene expression protocols. Totowa, NJ: Humana Press; 2003. p. 79–97.
18. Zhao X, Li G, Liang S. Several affinity tags commonly used in chromatographic purification. J Anal Methods Chem. 2013;2013:581093.
19. Fraseur JG, Kinzer-Ursem TL. Next generation calmodulin affinity purification: clickable calmodulin facilitates improved protein purification. PLoS One. 2018;13(6):e0197120.
20. Schmidt TGM, et al. Development of the twin-strep-tag® and its application for purification of recombinant proteins from cell culture supernatants. Protein Expr Purif. 2013;92(1):54–61.
21. Rai J, et al. Strep-tag II and twin-strep based cassettes for protein tagging by homologous recombination and characterization of endogenous macromolecular assemblies in *Saccharomyces cerevisiae*. Mol Biotechnol. 2014;56(11):992–1003.
22. Ivanov KI, et al. One-step purification of twin-strep-tagged proteins and their complexes on strep-tactin resin cross-linked with bis(sulfosuccinimidyl) suberate (BS3). JoVE. 2014;86:51536.
23. De Giuseppe A, et al. Purification by strep-Tactin affinity chromatography of a delete envelope gp51 protein of bovine Leukaemia virus expressed in Sf21 insect cells. Protein J. 2010;29 (3):153–60.
24. Cheung RCF, Wong JH, Ng TB. Immobilized metal ion affinity chromatography: a review on its applications. Appl Microbiol Biotechnol. 2012;96(6):1411–20.
25. Mohanty AK, Wiener MC. Membrane protein expression and production: effects of polyhistidine tag length and position. Protein Expr Purif. 2004;33(2):311–25.
26. Hemdan ES, et al. Surface topography of histidine residues: a facile probe by immobilized metal ion affinity chromatography. Proc Natl Acad Sci U S A. 1989;86(6):1811–5.

27. Booth WT, et al. Impact of an N-terminal Polyhistidine tag on protein thermal stability. ACS Omega. 2018;3(1):760–8.
28. Chaganti LK, Kuppili RR, Bose K. Intricate structural coordination and domain plasticity regulate activity of serine protease HtrA2. FASEB J. 2013;27(8):3054–66.
29. Singh MI, Jain V. Tagging the expressed protein with 6 histidines: rapid cloning of an amplicon with three options. PLoS One. 2013;8(5):–e63922.
30. Goh HC, et al. Going native: complete removal of protein purification affinity tags by simple modification of existing tags and proteases. Protein Expr Purif. 2017;129:18–24.
31. Hochuli E, Döbeli H, Schacher A. New metal chelate adsorbent selective for proteins and peptides containing neighbouring histidine residues. J Chromatogr A. 1987;411:177–84.
32. Chaga G, Hopp J, Nelson P. Immobilized metal ion affinity chromatography on Co_2^+–carboxymethylaspartate–agarose Superflow, as demonstrated by one-step purification of lactate dehydrogenase from chicken breast muscle. Biotechnol Appl Biochem. 1999;29 (1):19–24.
33. Bornhorst JA, Falke JJ. Purification of proteins using polyhistidine affinity tags. Methods Enzymol. 2000;326:245–54.
34. Crowe J, et al. 6xffis-Ni-NTA chromatography as a superior technique in recombinant protein expression/purification. In: Harwood AJ, editor. Protocols for gene analysis. Totowa, NJ: Humana Press; 1994. p. 371–87.
35. Schmitt J, Hess H, Stunnenberg HG. Affinity purification of histidine-tagged proteins. Mol Biol Rep. 1993;18(3):223–30.
36. Schlager B, Straessle A, Hafen E. Use of anionic denaturing detergents to purify insoluble proteins after overexpression. BMC Biotechnol. 2012;12(1):95.
37. Haneskog L. Purification of histidine-tagged proteins under denaturing conditions using IMAC. Cold Spring Harb Protoc. 2006;1:pdb.prot4221.
38. Sinha D, Bakhshi M, Vora R. Ligand binding assays with recombinant proteins refolded on an affinity matrix. BioTechniques. 1994;17(3):509–12. 514
39. Richard V, Viallon J, Cao-Lormeau V-M. Use of centrifugal filter devices to concentrate dengue virus in mosquito per os infection experiments. PLoS One. 2015;10(9):e0138161.
40. Antaloae AV, et al. Optimisation of recombinant production of active human cardiac SERCA2a ATPase. PLoS One. 2013;8(8):e71842.
41. Maity R, et al. GST-His purification: a two-step affinity purification protocol yielding full-length purified proteins. JoVE. 2013;80:e50320.
42. Dobrowolski P. Short protocols in molecular biology. In: Ausubel FM, et al., editors. A compendium of methods from "current protocols in molecular biology", Acta biotechnologica, vol. 13(1). 2nd ed. Hoboken, NJ: John Wiley & Sons; 1993. p. 88. ISBN: 0-471-57735-9.
43. Kumar Purbey P, et al. GST fusion vector with caspase-6 cleavage site for removal of fusion tag during column purification. BioTechniques. 2005;38(3):360–6.
44. Harper S, Speicher DW. Expression and purification of GST fusion proteins. Curr Protoc Protein Sci. 1997;9(1):6.6.1–6.6.21.
45. Harper S, Speicher DW. Expression and purification of GST fusion proteins. Curr Protoc Protein Sci. 2008;52(1):6.6.1–6.6.26.
46. James GT. Inactivation of the protease inhibitor phenylmethylsulfonyl fluoride in buffers. Anal Biochem. 1978;86(2):574–9.
47. Nikaido H. Maltose transport system of *Escherichia coli*: an ABC-type transporter. FEBS Lett. 1994;346(1):55–8.
48. Baneyx F, Mujacic M. Recombinant protein folding and misfolding in *Escherichia coli*. Nat Biotechnol. 2004;22(11):1399–408.
49. Randall LL, et al. SeeB: A chaperone from *Escherichia coli*. In: Methods in enzymology. Cambridge, MA: Academic Press; 1998. p. 444–59.
50. Sachdev D, Chirgwin JM. Order of fusions between bacterial and mammalian proteins can determine solubility in *Escherichia coli*. Biochem Biophys Res Commun. 1998;244(3):933–7.

51. Rosano GL, Ceccarelli EA. Recombinant protein expression in *Escherichia coli*: advances and challenges. Front Microbiol. 2014;5:172.
52. Riggs P. Expression and purification of recombinant proteins by fusion to maltose-binding protein. Mol Biotechnol. 2000;15(1):51–63.
53. Kapust RB, Waugh DS. Controlled intracellular processing of fusion proteins by TEV protease. Protein Expr Purif. 2000;19(2):312–8.
54. Nallamsetty S, Waugh DS. A generic protocol for the expression and purification of recombinant proteins in *Escherichia coli* using a combinatorial His6-maltose binding protein fusion tag. Nat Protoc. 2007;2(2):383–91.
55. Sheffield P, Garrard S, Derewenda Z. Overcoming expression and purification problems of RhoGDI using a family of "parallel" expression vectors. Protein Expr Purif. 1999;15(1):34–9.
56. Mohanty AK, Simmons CR, Wiener MC. Inhibition of tobacco etch virus protease activity by detergents. Protein Expr Purif. 2003;27(1):109–14.
57. Lebendiker M, Danieli T. Purification of proteins fused to maltose-binding protein. In: Walls D, Loughran ST, editors. Protein chromatography: methods and protocols. New York, NY: Springer New York; 2017. p. 257–73.
58. Yeliseev A, Zoubak L, Schmidt TGM. Application of strep-Tactin XT for affinity purification of twin-strep-tagged CB(2), a G protein-coupled cannabinoid receptor. Protein Expr Purif. 2017;131:109–18.
59. Johar SS, Talbert JN. Strep-tag II fusion technology for the modification and immobilization of lipase B from Candida antarctica (CALB). J Genet Eng Biotechnol. 2017;15(2):359–67.
60. Meyer-Ficca ML, et al. Comparative analysis of inducible expression systems in transient transfection studies. Anal Biochem. 2004;334(1):9–19.
61. Schmidt TGM, Skerra A. The strep-tag system for one-step purification and high-affinity detection or capturing of proteins. Nat Protoc. 2007;2(6):1528–35.
62. Maertens B, et al. Chapter Five - Strep-tagged protein purification. In: Lorsch JR, editor. Methods in enzymology. Academic Press; 2015. p. 53–69.
63. Kresoja-Rakic J, Felley-Bosco E. Desthiobiotin-streptavidin-affinity mediated purification of RNA-interacting proteins in mesothelioma cells. JoVE. 2018;134:57516.
64. Meerman HJ, Georgiou G. Construction and characterization of a set of E. coli strains deficient in all known loci affecting the proteolytic stability of secreted recombinant proteins. Bio/Technology. 1994;12(11):1107–10.
65. Singh N, et al. Dual regulatory switch confers tighter control on HtrA2 proteolytic activity. FEBS J. 2014;281(10):2456–70.
66. Kummari R, et al. Discerning the mechanism of action of HtrA4: a serine protease implicated in the cell death pathway. Biochem J. 2019;476(10):1445–63.
67. Bell MR, et al. To fuse or not to fuse: what is your purpose? Protein Sci. 2013;22(11):1466–77.
68. Kummari R, et al. Elucidating the role of GRIM-19 as a substrate and allosteric activator of pro-apoptotic serine protease HtrA2. Biochem J. 2021;478(6):1241–59.

Protein Purification by Ion Exchange Chromatography

7

Ayon Chakraborty, Rashmi Puja, and Kakoli Bose

Abstract

Separation of similar biomolecules and proteins with little or no differences in molecular weight or without tags can be difficult with chromatographic techniques such as affinity or size exclusion. To circumvent this problem, distinct physicochemical properties of protein molecules have been harnessed for their separation. Since proteins carry overall electrical charges due to their chemical composition; ion exchange chromatography (IEX) uses this property to separate positively or negatively charged molecules via interaction with charged ion exchange resins as stationary media. Charged proteins bind to the resins in normal buffering conditions and can be gradually eluted with increasing salt concentration or by changing the pH of the mobile phase. Depending on the protein's isoelectric point (pI) value, cation or anion exchange chromatography media can be used. If the pH environment of a protein is lower than its pI, it will carry a positive surface charge and strongly bind the cation exchange resins, while proteins with the negative surface charge will bind to the anion exchange counterpart. The purpose of this chapter on ion exchange chromatography is to describe its basic principle, protocols, applications in protein purification as well as provide troubleshooting tips.

Both the authors contributed equally to the work.

A. Chakraborty
Integrated Biophysics and Structural Biology Laboratory, Advanced Centre for Treatment, Research and Education in Cancer (ACTREC), Tata Memorial Centre, Mumbai, India

R. Puja · K. Bose (✉)
Integrated Biophysics and Structural Biology Lab, Advanced Centre for Treatment, Research and Education in Cancer (ACTREC), Tata Memorial Centre, Mumbai, India

Homi Bhabha National Institute, BARC Training School Complex, Mumbai, India
e-mail: kbose@actrec.gov.in

K. Bose (ed.), *Textbook on Cloning, Expression and Purification of Recombinant Proteins*, https://doi.org/10.1007/978-981-16-4987-5_7

Keywords

Ion exchange chromatography · Cation exchange · Anion exchange · Isoelectric point · Resins · Surface charge

7.1 Introduction

Besides affinity chromatography, one of the most useful liquid chromatographic techniques of protein purification is ion exchange chromatography (IEX). Affinity chromatography has limitations by its requirement of fusion tags during protein purification as discussed in the earlier chapter. Additionally, separation and purification of proteins having similar molecular mass (size) are challenging while working with size exclusion chromatography. In such conditions, IEX is an excellent choice as it separates molecules based on the interaction between the charged solute molecules and the complementarily charged matrix of the column. Moreover, due to high resolving power and capacity, this technique is often used for the isolation and purification of proteins. The mechanism of separation using ion exchange relies on two main factors such as competitive binding (attraction) between opposite charged ions and repulsion between similarly charged ions fixed on the matrix of the column. In addition to the charge variations, a couple of other considerations such as charge density and charge distribution on their surfaces play important roles in determining the degree of separation. The inclusion of these factors enables IEX to efficiently separate proteins even with a single charged residue difference. Thus, IEX is considered one of the most powerful protein purification techniques [1–3].

The concept of ion exchange emerged with the introduction of polystyrene matrix by D'Alelio in 1944, which accelerated the usage of ion exchange as an analytical tool [4]. However, later in 1956, IEX of protein was first successfully demonstrated by Peterson and Sober using cellulose matrix [5]. Thereafter, a series of research endeavors for almost two decades led to the development of a robust ion exchange chromatography system by Bauman and coworkers in 1975 [6]. Within the next 5 years, the latest anion and cation exchange chromatographic tools were also developed leading to a breakthrough in biochemical and industrial research [7–9]. Since then, ion exchange has been considered as one of the most important preparative and analytical tools for the separation, purification, and characterization of nucleic acids, proteins, peptides, and other charged biomolecules [10–12]. This chapter will discuss the basic principles and applications of IEX in recombinant protein purification.

7.2 Basic Principles of Ion Exchange Chromatography

Ion exchange is a process in which ions attached to a matrix are exchanged for other ions in solution [13, 14]. The basic process of ion exchange for a negatively charged resin (i.e., cation exchanger) is provided by Eq. (7.1):

$$(\mathrm{Res}^-)\mathrm{B}^+ + \mathrm{C}^+\ (\mathrm{soln}) \Leftrightarrow (\mathrm{Res}^-)\mathrm{C}^+ + \mathrm{B}^+(\mathrm{soln}) \tag{7.1}$$

where Res^- represents the negatively charged ion exchange resin, i.e., cation exchanger, B^+ is the counterion of the opposite charge associated with the exchanger matrix, C^+ is the charged molecule (bearing the same charge as the counterion) in the sample to be separated. The positively charged molecule (C^+) can interchange with the counterion (B^+) to find appropriate binding sites on the negatively charged resin. All the other neutral or negatively charged ions present in the buffer do not interact with the exchanger. Bound ions, C^+, can now be eluted from the resin in two ways. The first way of elution includes passing of solvent through the resin with increasing concentrations of B^+. This method increases the possibility that B^+ will replace C^+ by substitution in the above-stated equilibrium due to the presence of B^+ in high concentration. The quantity of charge possessed by C^+ determines the concentration of B^+ required for elution. The greater the charge carried by C^+, the higher the concentration of B^+ required for eluting it. Another way of eluting the bound C^+ ions from the resin is by altering the pH of the solvent in a way such that C^+ is converted to an uncharged moiety. In this case, p*K*a of C^+ is the determining factor to the requirement of the pH of the solvent. The higher the p*K*a value of C^+, the higher pH is required for the elution [3].

Similarly, the method of ion exchange for a positively charged resin (i.e., anion exchanger) is presented by Eq. (7.2):

$$(\mathrm{Res}^+)\mathrm{X}^- + \mathrm{Y}^-\ (\mathrm{soln}) \Leftrightarrow (\mathrm{Res}^+)\mathrm{Y}^- + \mathrm{X}^-(\mathrm{soln}) \tag{7.2}$$

where Res^+ represents the positively charged ion exchange resin, i.e., anion exchanger, X^- is the counterion associated with the exchanger matrix, Y^- is the charged molecule (bearing the same charge as the counterion) in the sample to be separated. To find the appropriate binding sites on the positively charged resin, the negatively charged residues (Y^-) from the solution can interchange with the counterion (X^-). The remaining positively charged and neutral residues will pass through the resin without interacting with the exchanger. Elution of the bound Y^- ions from the resin can be done by using higher concentrations of X^- or by increasing the pH of the solvent. Therefore, the interchange of similarly charged ions between the solvent and the resin through substitution forms the basic principle of ion exchange chromatography (Fig. 7.1) [1–3].

Based on the ion exchange mechanism, IEX separates molecules depending on their net charges [3, 14]. It is a unique type of adsorption chromatography in which ionic solutes (mobile phase) interact with a charged resin (stationary phase) in a reversible electrostatic interaction. The ion exchange resin is composed of an inert,

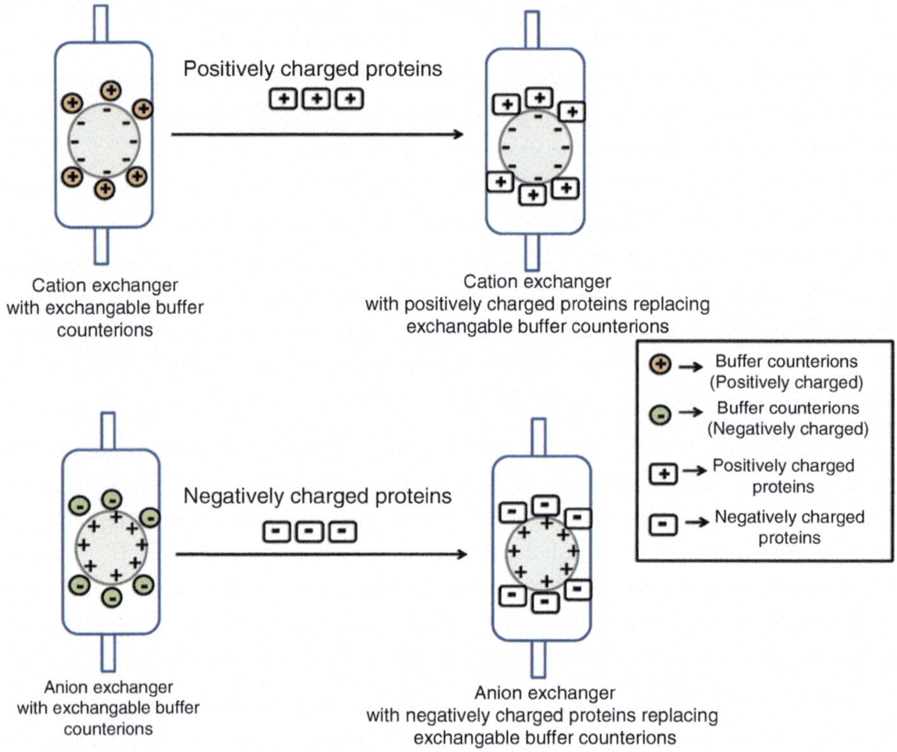

Fig. 7.1 Working principle of cation and anion exchangers. The cation exchangers carry negatively charged functional groups and attract positive charge counterions from the surrounding buffer solution. Due to the presence of positively charged protein in the buffer, these buffer counterions are replaced by positively charged proteins. On the other hand, the anion exchangers carry positively charged functional groups and attract negative charge counterions from the buffer solution. Upon the addition of protein (negatively charged), these buffer counterions are replaced by negatively charged proteins

porous matrix covalently bound to a significant number of specific ionic functional groups [15]. These specific ions are stabilized by equivalent and oppositely charged ions from the buffer specified as counterions. For the purification of proteins, both **cation** and **anion exchangers** are used [15]. As mentioned earlier, **cation exchangers** possess negatively charged groups and therefore attract positively charged cations from the proteins that have an overall positive charge at the pH in which the experiment is conducted. Since their negative charges arise from the ionization of acidic groups, these exchangers are also known as acidic ion exchangers. Conversely, **anion exchangers** that have positively charged groups attract negatively charged anions. Since positive charges on the resin result from the binding of protons with basic groups, these exchangers are often named basic ion exchangers [3, 10, 15]. The working principle of cation and anion exchangers is shown in Fig. 7.1.

Proteins carry many ionizable groups in the charged polar side chains of amino acid residues. The free amino group (α-NH_2) at one end and the free carboxylic group (α-COOH) at the other end of the peptide chain or protein may also ionize. Depending on the pH, the α-NH_2 group and the side-chain groups such as NH_2, imidazole, and guanidino groups may accept protons (H^+) to form cations, while the α-COOH and side-chain COOH groups may donate H^+ to form anions. Thus, a protein shows its amphoteric nature (property to act equally as an acid and a base) by consequently acting both as a donor and an acceptor of H^+. At a specific pH known as the isoelectric pH (pI), the amino acid exists as a dipolar ion or zwitterions carrying equal numbers of positive and negative charges on its ionizable groups so that the net charge is minimum or zero. In a buffered solution, if the pH of the solution is lower than the pI of a protein, the protein exists as a cation by accepting H^+ from the acidic solution and binds to the negatively charged functional groups of a cation exchange resin [2, 16]. On the other hand, if the pH of the buffer solution exceeds the pI of the same protein, it forms an anion by donating H^+ to the alkaline solution and binds to the positively charged functional groups of an anion exchange resin [2, 16]. In the subsequent sections, we will illustrate various components, specific conditions, and protein purification methods using IEX.

7.3 Components and Factors of Ion Exchange Chromatography

7.3.1 Ion Exchange Resins

Ion exchange resins act as the stationary phase of IEX having two major components—a matrix and functional ionic groups. The matrix is an inert, three-dimensional, porous material to which charged groups are bound covalently. It is composed of cross-linked polymers that are typically polystyrene-, cellulose- or agarose-based. The extent of cross-linking of the resin should be sufficient to provide required mechanical stiffness, insolubility, and porosity [17]. However, smaller pore sizes are not conducive to working with macromolecules. Therefore, polystyrene, which is hydrophobic and with less porosity, is suitable for inorganic ions and small molecules [18, 19], while cellulose and agarose are hydrophilic and useful for larger, biologically essential molecules such as proteins and nucleic acids [20–23].

The other constituents of the ion exchange resin are the functional ionic groups. These functional ionic groups are covalently attached to the matrix specifying the nature and strength of the ion exchanger. Depending on their affinity for either positive or negative ions, functional ionic groups are classified as cationic exchangers and anionic exchangers, respectively. Thus, the charge carried by the exchangeable ion defines whether the material is anionic or cationic.

Depending on the ionization state, ion exchange resins are categorized as strong and weak exchangers. The strong ion exchange resins contain strongly ionized groups like sulfopropyl and diethyl-(2-hydroxyl-propyl)-aminoethyl. These get completely ionized and exist in the charged form except at extreme pH values.

The weak ion exchange materials, on the contrary, contain groups such as carboxylate and diethylaminoethyl, whose ionizations are pH-dependent [2, 24].

The ion exchange process, which involves replacing the resin exchangeable ions (A_r) by the counterions (B_s) from a solution, can be expressed as Eq. (7.3):

$$A_r + B_s \Leftrightarrow B_r + A_s \tag{7.3}$$

In this reversible process, the selectivity coefficient (K) for the charged ions is defined as Eq. (7.4):

$$K = [B_r][A_s]/[A_r][B_s] \tag{7.4}$$

where the terms in parentheses reflect the concentrations of ions A and B in the resin or mobile phase, respectively. Relative affinities of ions for a particular resin can be obtained experimentally by determining the standard values of the selectivity coefficient (K). Thus, if the obtained value of K is greater than 1, the resin indicates more affinity for ion B, whereas if the experimental value of K is lesser than 1, the resin exhibits more affinity towards ion A.

7.3.2 Capacity

Another significant feature of ion exchange resin is its capacity. Capacity is characterized as the potential of an ion exchanger to adsorb its counterions [2]. The total capacity of an ion exchanger is the amount of charged groups per unit weight of a dry exchanger. On the other hand, the amount of counterions that can bind to the ion exchanger under specific experimental conditions is the available capacity of the exchanger. The exchange capacity of a cation exchange resin may be calculated experimentally by determining the amount of sodium (Na^+) ion which are absorbed by 1 g of the dry resin in H^+ form. Similarly, the exchange capacity of a strongly basic anion exchange resin is estimated by quantifying the amount of chloride (Cl^-) ion taken up by 1 g of dry resin in OH^- form. The exchange capacity is usually expressed as millimoles per gram of exchanger. In some cases, the porosity of the resin matrix decreases due to the presence of excessive covalent cross-linking. As a result, large molecules cannot enter the matrix and can bind only with surface-charged groups. Thus, in these cases, the available capacity will be significantly reduced in comparison to the total capacity [2].

7.3.3 Selection of the Ion Exchange Resins

During protein purification, the selection of appropriate ion exchange resin primarily confines the folding and stability of target proteins and their relative molecular mass or size. The following criteria are used for the selection of ion exchange resin.

7.3.3.1 Choice of Anionic and Cationic Exchangers

Usually, a cation exchanger should be used if the target protein shows maximum stability below its pI value. Preparing a protein solution under an appropriate buffer condition at a pH lower than its pI results in a net positive charge on the protein, which is then purified using a cation exchanger [2]. On the other hand, when the target protein is most stable above its pI value (giving it a net negative charge), an anion exchanger is used [2]. Some proteins are stable over a broad pH range both above and below their pI value. Either type of exchanger can be used for the purification of these proteins [2].

7.3.3.2 Choice of Strong and Weak Exchangers

The preference between a strong and a weak exchanger is dependent on the target protein stability over various pH ranges. The expressions weak and strong do not refer to the binding strength of a protein to an ion exchanger, but rather to the degree of ionization of the exchanger as a function of pH. Due to their p*K*a values, strong ion exchangers are fully ionized over [3, 10, 15] a broader pH range compared to weak ion exchangers. A strong exchanger can be used if the pH required for binding of the target protein is very acidic (down to pH 2.0) or basic (up to pH 12.0) provided the target protein retains its stability at the working pH [2]. In this case, the functional groups remain charged over a larger pH range. In contrast, weak exchangers might be more suitable for proteins that do not require an extreme pH for binding. Weak exchangers are advantageous in protein purification because the binding tendency of weakly charged impurities is very low, resulting in improved elution characteristics [2]. The routinely used ion exchangers for protein purification are listed in Table 7.1 [25].

Table 7.1 Commonly used ion exchange resins

Sl. no.	Resin name	Type	Functional group	Matrix	Class	Refs.
1	SP-Sephadex	Cation Exchangers	Sulfopropyl	Dextran	Strong	[26]
2	P-Cellulose		Phosphate	Cellulose	Intermediate	[27]
3	Bio-Rex 70		Carboxylate	Acrylic	Weak	[28]
4	CM-Sephacel		Carboxymethyl	Sephacel		[29]
5	CM-Sephadex		Carboxymethyl	Dextran		[30]
6	QAE-Sephadex	Anion Exchangers	Diethyl-(2-hydroxyl-propyl)-aminoethyl	Dextran	Strong	[31]
7	AG3		Tertiary amine	Polystyrene	Weak	[32]
8	DEAE-Sephacel		Diethylaminoethyl	Sephacel		[33]
9	DEAE-Sephadex		Diethylaminoethyl	Dextran		[34]

7.3.3.3 Choice of Particle Size and Porosity of the Resin Matrix

Flow characteristics and chromatographic resolution of an ion exchanger are primarily determined by the particle size and porosity of the matrix particles. The pore size or degree of cross-linking of the resin matrix does not influence the selectivity but impacts the resolution of separation. Smaller pore size offers better resolution but usually imposes lower flow rates. This type of resin is frequently used while working with high-resolution media/buffer as well as in the last phase of preparative chromatography (for purification and separation) using smaller proteins. Viscous samples such as cell lysates or samples containing glycerol and larger proteins often cannot be isolated using small pore size resins due to the elevated backpressure of resins, which might exceed the column's working pressure level. A larger pore size matrix allows for increased flow rates but results in lower resolution. Large pore size resins are mostly employed for the purification of bigger proteins from the cell lysates when IEX is used in the initial steps of the protein purification process. Altogether, the viscosity and size of the protein define the selectivity of the resin porosity for its efficient purification [1–3].

7.3.4 Buffer

Buffers regulate small changes in pH by maintaining solutions within a constant pH range. A buffer system consists either of a weak acid and its conjugate base or a weak base and its conjugate acid. The pH range of a buffer is based on its p$K_{a,}$ which is defined as the pH at which 50% of the molecules are in their acidic, and 50% in their basic forms [35]. The effective range of a buffer to maintain a constant pH of a solution is when the pH of the solution remains at p$Ka \pm 1$. Commonly used buffers

Table 7.2 Recommended buffer substances for cation exchange chromatography

Sl. no.	Effective pH range	Buffer compounds	Conc. (mM)	Counter-ion	p*K*a (25 °C)	Refs.
1	2.6–3.6	Citric acid	20	Na^+	3.13	[36]
2	3.3–4.3	Lactic acid	50	Na^+	3.86	[37]
3	3.7–4.7	Succinic acid	50	Na^+	4.21	[38]
4	4.3–5.3	Acetic acid	50	Na^+ or Li^+	4.75	[39]
5	5.1–6.1	Succinic acid	50	Na^+	5.64	[40]
6	5.6–6.6	2-(N-morpholino)ethanesulfonic acid (MES)	50	Na^+ or Li^+	6.27	[41]
7	6.7–7.7	Monophosphate	50	Na^+	7.2	[42]
8	7.0–8.0	4-(2-hydroxyethyl)-1-piperazineethanesulfonic acid (HEPES)	50	Na^+ or li^+	7.56	[43]
9	7.8–8.8	N,N-Bis(2-hydroxyethyl)glycine (BICINE)	50	Na^+	8.33	[44]

Table 7.3 Recommended buffer substances for anion exchange chromatography

Sl. no.	Effective pH range	Buffer compounds	Conc. (mM)	Counter-ion	p*K*a (25 °C)	Refs.
1	4.3–5.3	N-Methylpiperazine	20	Cl^-	4.75	[45]
2	4.8–5.8	Piperazine	20	Cl^- or $HCOO^-$	5.33	[46]
3	5.5–6.5	L-Histidine	20	Cl^-	6.04	[47]
4	6.0–7.0	Bis–Tris	20	Cl^-	6.48	[48]
5	6.2–7.2	Bis–Tris propane	20	Cl^-	6.65	[49]
6	7.3–8.3	Triethanolamine	20	Cl^- or CH_3COO^-	7.76	[50]
7	7.6–8.6	Tris	20	Cl^-	8.07	[51]
8	8.0–9.0	N-Methyldiethanolamine	20 or 50	SO_4^{2-} or cl^- or CH_3COO^-	8.52	[52]
9	8.4–9.4	Diethanolamine	20 or 50	Cl^-	8.88	[53]
10	8.6–9.6	bis-Tris propane	20	Cl^-	9.1	[54]
11	9.0–10.0	Ethanolamine	20	Cl^-	9.5	[55]
12	9.2–10.2	Piperazine	20	Cl^-	9.73	[56]
13	10.6–11.6	Piperidine	20	Cl^-	11.12	[57]

for cation exchange and anion exchange chromatography are enlisted in Tables 7.2 and 7.3, respectively [2].

7.3.5 Selection of Buffer

Buffer selection is an important part of the IEX as the electrostatic interactions between the oppositely charged surface and protein depend on buffer conditions. The choice of buffers broadly relies on the following factors:

7.3.5.1 Buffer Substance

Selecting a proper buffer substance is a critical factor for IEX. Buffering ions should have the same charge as the functional groups on the ion exchanger. Therefore, cationic buffers should be used while working with anionic exchangers and vice versa. If the buffering ions bear an opposite charge to that of the functional groups of exchangers, it participates in the ion exchange process by competing with proteins for binding sites. Thus, this process greatly reduces the capacity of the column. Additionally, it can also cause significant pH fluctuations while elution. Therefore it is always advisable to use identical charged buffering ions as the functional groups on the ion exchanger.

7.3.5.2 pH of the Buffer

Another critical factor for choosing a buffer is its working pH range. The foremost thing is the protein to be separated should be stable in the pH range of the buffer. Ensuring strong electrostatic interactions between the surface and protein, the pH of the buffer should be kept in between the p*K*a of the surface functional groups and the pI of the protein molecules. Based on this criterion, the p*K*a of the chosen buffer should at least be within 0.7 units of the working pH; however, the ideal value is 0.3 units [58].

7.3.5.3 Ionic Strength of the Buffer

The ionic strength of the buffer has a major impact on protein adsorption, i.e., its attachment to the matrix. With increased ionic strength, the buffer counterions compete with the adsorbed protein molecules to substitute on the surface of the matrix. The salt ions also exhibit a shielding effect by hindering the interactions between the charged groups of proteins and surface binding sites. At a constant pH and ionic strength, the type of ionic species in the buffer might also interfere with protein binding. Sodium chloride (NaCl) is the most common salt to increase the ionic strength of the eluent while using IEC for proteins. The use of NaCl is advantageous as it does not alter the protein structure and is thus considered a gentle eluent. Nevertheless, NaCl is not always the greatest elution choice. It has been demonstrated that the type of the cations and anions influence the separation and chromatographic resolution [2]. Salts containing monovalent cations (such as Na^+, Li^+, and K^+) and monovalent anions (Cl^-, $HCOO^-$, CH_3COO^-) are preferred in the elution buffer [2, 59, 60]. The presence of multivalent ions (Ba^{2+}, Ca^{2+}, Mg^{2+}, Al^{3+}, Th^{4+}, SO_4^{2-}) in the buffer results in a reduced ion exchange rate as well as a considerable decrease in the surface charge of cation and anion exchangers. Thus, the affinity of proteins to the cationic and anionic surface decreases [2, 59–61].

7.3.5.4 Temperature of the Buffer

The p*K*a value of a buffering substance is dependent on temperature [62]. Consequently, the pH of the buffer varies with an increase or decrease in the temperature. For example, Tris has different p*K*a values at different temperatures. At 0 °C it shows a p*K*a value of 8.85, which is distinctly reduced to 8.06 and 7.72 when the temperature increases to 25 °C and 27 °C, respectively. Therefore, at certain temperatures, the buffer exhibits a very low buffering capacity that might result in a working pH different from the desired pH range for IEX. Thus, to evade this trouble, it is recommended to prepare buffer solutions at the same temperature at which they are supposed to be used.

7.4 Protein Purification Using Ion Exchange Chromatography

Purification of proteins by ion exchange chromatography using either cation exchangers or anion exchangers primarily consists of the following steps which are described below.

7.4.1 Equilibration

The first step of an IEX is the equilibration of the resin matrix, which provides a condition to ensure that the target proteins interact effectively with the matrix. In this step, an equilibration or starter buffer is used so that the charged groups in the matrix are surrounded by buffer exchangeable counterions. As the pH and ionic strength of the buffer play crucial roles, they are selectively chosen to ensure the proper binding of the protein of interest.

7.4.2 Loading of Sample

The given protein sample solution comprising the target proteins of interest and bacterial protein impurities is then loaded onto the resin in the same buffer that is used for equilibration. Under experimental conditions, proteins entering the resin may possess a negative charge, positive charge, or neutral charge. Proteins with the opposite charge as the resin bind tightly yet reversibly to the resin. The strength of binding depends on the charge and charge density (amount of charge per unit volume) of the solute. The greater the charge or the charge density, the stronger is the binding.

7.4.3 Washing

Extensive washing is done afterward to remove the nonspecific interaction of the proteins with the resin. Neutral proteins or those with the same charge as the resin do not show any affinity and washes out of the resin.

7.4.4 Elution

The bound proteins can be eluted from the resin using a buffer of increased ionic strength or pH (salt or pH gradient). As ionic strength increases, the proteins with the lowest net charge elute first from the column at the selected pH. Similarly, at a given pH, the proteins with maximum charge content are retained strongly and elute at the end. Choices of elution using increased ionic strength are of two general types such as linear gradient elution and step elution [63]. A linear gradient elution refers to a gradual increase of ionic strength in the elution buffer where weakly bound proteins elute first followed by stronger binding proteins. Linear gradients are ideal for the purification of an unknown sample or if peak resolution is important. In step elution, the ionic strength over which the target protein will elute is already known and it can be easily eluted using that particular ionic strength. Step elutions are often faster to perform and use less total volume to elute the protein in comparison to linear gradient elutions. The linear gradient elution and step elution are shown in Fig. 7.2.

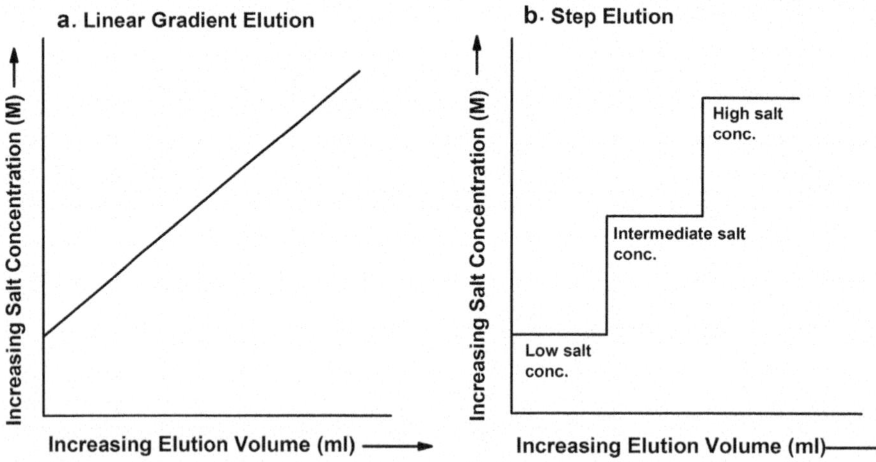

Fig. 7.2 Elution profiles of ion exchange chromatography using increased ionic concentration. Elution of proteins from ion exchange resin can be performed using increased ionic strength or salt concentration. This can be done in two ways such as linear gradient elution (**a**) and step elution (**b**). In linear gradient elution, there is a continuous elution of the proteins from the ion exchanger, whereas the elution of proteins is discontinuous in step elution

7.4.5 Regeneration

In the final step, increased ionic strength or increased pH buffer is used to eliminate any protein still bound to regenerate the resin for further use. Increasing ionic strength of the buffer releases bound proteins by displacement while increasing buffer pH weakens the interaction by lowering the charge on the protein or the resin [64]. All the steps of protein purification using ion exchange chromatography are illustrated in Fig. 7.3.

7.5 Instrumentation for Ion Exchange Chromatography

A typical ion exchange chromatography includes several elements that are schematically presented in Fig. 7.4 and described as follows.

7.5.1 Pump

The pump is used to deliver the mobile phase into the chromatographic system. It ensures a continuous and constant flow of the solvent through the injector, guard column, ion exchange column, and finally to the detector. The recommended flow rate depends on the ion exchange resin and should be included in the supplier's instructions. Usually, the chromatographic steps are carried out at a lower flow rate as compared to the column washing and equilibration steps.

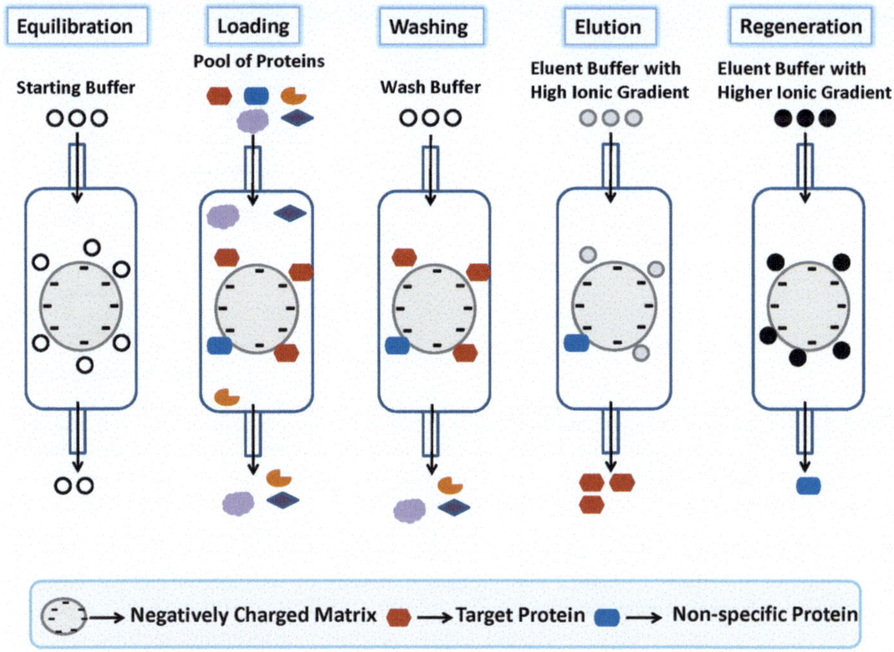

Fig. 7.3 Generic steps of ion exchange chromatography. A schematic model showing different steps involved in ion exchange chromatography. In the first step, equilibrium of the matrix (here, negatively charged) is done followed by loading the pool of protein samples. While the target protein binds to the matrix of the resin, other proteins flow through the column. The washing step ensures the removal of all nonspecific interactions. Elution steps involve deattachment of the target protein from the matrix using different ionic gradients. Finally, the column is regenerated for further use by using a higher ionic strength buffer to release if any proteins are still bound

7.5.2 Injector

An injector introduces the sample into the mobile phase or eluent stream into the column. It is capable of introducing small as well as a large volume of samples depending on the requirement.

7.5.3 Guard Column

A guard column is fitted next to the injector and is placed anterior to the ion exchange column. The role of a guard column is to protect the ion exchange column by filtering impurities and suspended particles that might clog the separation column.

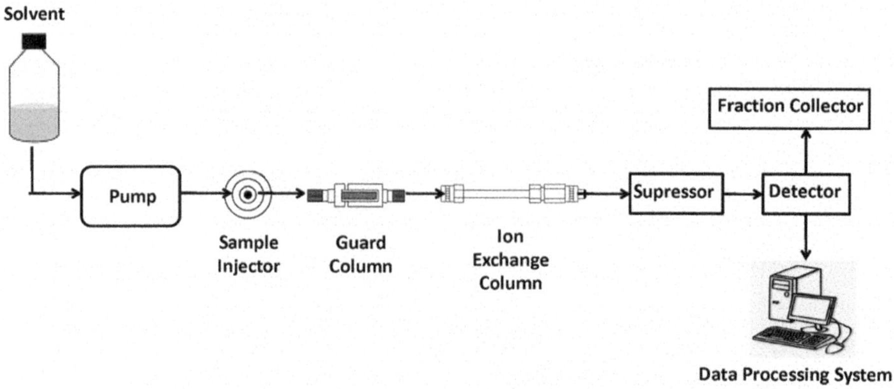

Fig. 7.4 Schematic overview of instrumentation for ion exchange chromatography. Ion exchange chromatography involves an automated system that includes a pump, sample injector, columns, suppressor, fraction collector, and detector coordinated through software. With the help of the pump, the solvent is passed through the guard column and ion exchange column at a certain flow rate. Protein mixture is introduced into the system through a sample injector and is separated by ion exchange columns based on their net charges. A suppressor reduces the conductance of the solvent. The elution of proteins is monitored through data processing by the detector and consequently collected using a fraction collector

7.5.4 Column

The choice of column for IEX depends on the target protein to be separated. If the target protein in the working buffer carries a net positive charge, a cation exchanger should be used, while if the target protein is negatively charged, an anion exchanger is preferred. Similarly, a strong exchanger is selected when the pH required for binding of the target protein is very acidic (down to pH 2.0) or basic (up to pH 12.0). On the contrary, when the pH required for binding of the target protein is moderate (pH 6–8), weak exchangers are chosen. The length-to-diameter ratio of a column varies according to the type of elution. For preparative (purification and separation) purposes using Fast Protein Liquid Chromatography (FPLC) and analytical (characterization) purposes using High Performance Liquid Chromatography (HPLC) most commonly used columns have a length and diameter ratio of 5:1 or lower (Details about FPLC and HPLC have been discussed in Chap. 8) [2]. Frequently used column sizes for preparative purposes of proteins are 10 × 50 mm (vol = 3.9 ml), 16 × 100 mm (vol = 20.1 ml), 26 × 200 mm (vol = 106 ml), or 50 × 300 mm (vol = 589 ml). Smaller columns provide better resolution in step elution whereas longer columns are further appropriate for linear gradient elution [2]. Short columns with a diameter of 5 mm and a length of 50 mm (vol = 1 ml) and 100 mm (vol = 2 ml) are more common for analytical purposes. The column material is usually made up of comparatively inert materials such as stainless steel, titanium, glass, or inert plastics [65].

7.5.5 Suppressor

In IEX, the eluent generally has a high ionic concentration. As a result, the conductivity of eluent increases which decreases the sensitivity of the detector. Conductivity is the measure of the concentration of ions present in the eluent [66]. This is calculated by the ability of the analyte to transmit an electrical current over a defined area. During ion exchange chromatography, a suppressor is installed between the column and the detector to lower the conductivity of the eluent, thus improving the sensitivity of IEX [66].

7.5.6 Detectors

For monitoring the column eluent, UV and conductivity detectors are used. While using UV detectors, two wavelengths are scanned such as 220 nm (for amide peptide bond detection) and 280 nm (for aromatic amino acid residues) [67].

7.5.7 Fraction Collector

An automated fraction collector with tube holders helps in collecting the sample that is eluted from the column.

7.5.8 Data Processing System

Provides a graphical representation of peaks from different fractions, stores data, and provides a graphical representation of the data.

7.6 Protocol for Recombinant Protein Purification

For the purification of a recombinant protein using ion exchange chromatography, the first step is to estimate the isoelectric point (pI) of the target protein to be purified using bioinformatics resources such as the ProtParam feature of ExPASy (Expert Protein Analysis Software) (details mentioned in Chap. 5). The estimated pI of the target protein helps to decide whether the cation or anion exchange chromatography to be used along with the pH of buffers. The detailed manufacturer instructions about the maximum and optimal flow rates of the columns should be carefully followed. A detailed protocol for the purification of recombinant protein using **anion exchange chromatography** is discussed as follows.

7.6.1 Instruments and Materials

- FPLC connected with a pump and a UV detector.
- QAE Sephadex A-25 resin or any other corresponding anion exchange resin.
- Recombinantly expressed protein sample.
- Binding buffer: 20 mM Tris–Cl, pH 7.5.
- Wash buffer: 20 mM Tris–Cl (pH 7.5)/100 mM NaCl.
- Elution buffer: 20 mM Tris–Cl (pH 7.5)/350 mM NaCl.
- Regeneration buffer: 20 mM Tris–Cl (pH 7.5)/2 M NaCl.
- Collection tubes.

7.6.2 Procedure

7.6.2.1 Step 1 (Equilibration of the Column)

- Follow the instructions given by the manufacturer to remove the column storage buffer and place the column in water.
- Equilibrate the resin with a minimum of 2 column volumes (CV) of the binding buffer. Before proceeding, verify that the readings on UV absorption (280 nm and 220 nm) are stable.
- Set the measurement to zero for UV absorption at both wavelengths.

7.6.2.2 Step 2 (Binding of the Protein Sample)

- Before loading, adjust the pH and salt concentration of the protein sample to initial optimal values. Be sure that the buffer composition of the protein sample is the same as the binding buffer used for the equilibration of the column. Else the protein should be dialyzed in the binding buffer.
- A fraction of the sample should be collected before loading into the column for running the SDS-PAGE assay later.
- Load the protein sample into the system by injecting the protein sample through the injector.
- If the pressure of the column goes beyond its maximum pressure, decrease the flow rate.
- A fraction of the flow-through should be collected.

7.6.2.3 Step 3 (Removal of Unbound Proteins)

- Wash the column with a minimum of 2 column volumes of the binding buffer.
- The UV absorbance both at 280 nm and 220 nm should be checked. The washing step should carry on till the absorbance (A_{280} and A_{220}) is zero or close to zero.
- A fraction of the wash flow-through should be collected.

7.6.2.4 Step 4 (Elution of the Bound Protein)

- *Elution with increasing salt concentration:* There are two methods for elution with increasing salt concentration. First, for gradient elution, apply a gradient from 0% (20 mM Tris–Cl, pH 7.5) to 100% (20 mM Tris–Cl, pH 7.5/350 mM

NaCl) of elution buffer over 20 column volumes. Collect each of the fractions. For step elution, apply the elution buffer containing 20 mM Tris–Cl, pH 7.5/350 mM NaCl over 10 column volumes.
- *pH-based elution:* For pH-based elution, the pH of the elution buffer is altered. For an anion exchanger, elution occurs when the pH of the elution buffer is decreased and vice versa. So apply 10 column volumes of low pH eluent buffer for elution of the desired protein.
- For both the elutions, the elution buffer should be run till there is no significant absorbance at 280 nm or 220 nm.
- Analyze the fractions (loaded sample, flow-through, washes, eluted protein) using gel electrophoresis (SDS-PAGE).
- Collect the pooled fractions of eluted proteins containing the purified desired protein.

7.6.2.5 Step 5 (Storage of the Protein)
- Dialyze to remove the excess salt from collected pooled fractions of the desired protein.
- Store it at −80 °C or liquid nitrogen as per the requirement.

7.6.2.6 Step 6 (Regeneration of the Column)
- Wash the column with 2–5 column volumes of the regeneration buffer (20 mM Tris–Cl, pH 7.5/2 M NaCl) to remove any very tightly bound species.
- Run 2 column volumes of binding buffer through the column to re-equilibrate the column with the binding buffer.

The protocol for *cation exchange chromatography* is exactly similar to that of anion exchange chromatography being the only differences are in the buffer composition and pH and the selection of appropriate cation exchanger resin.

7.7 Choice of Different Combination of Chromatographic Techniques

With the aim of achieving the highest level of purity and yield, the protein purification from a crude sample is often performed in a multi-step downstream process using the purification strategy of capture, intermediate purification, and polishing [68]. In the capture step, the desired protein is isolated, concentrated, and stabilized. During the intermediate purification step, the objective is to eliminate most bulk contaminants, like other proteins and nucleic acids. In the polishing stage when most of the impurities have already been removed, the goal is to attain high purity levels (>95%) by eliminating residual impurities or closely related moieties.

Ion exchange chromatography (IEX) is frequently used in most multi-step purification systems. If a specific affinity medium is not available or if the desired protein is little understood, IEX should be considered as the initial step towards any purification. This technique is highly flexible to be applied at any stage of purification, viz., capture, intermediate, or polishing, based on the exact purpose. As IEX offers several selectivities (anion or cation exchangers), and the purification pH may be varied to alter the charge of target proteins, this technique might be used more than once under the same purification schemes. Additionally, IEX can be employed with gradual elution to provide the maximum resolution for a quick capture or with gradient elution in a polishing step.

- *Ion exchange as a capture step*: The goal of using IEX as a capture step is to rapidly adsorb the protein of interest from the crude sample and separate it from critical impurities such as proteases and glycosidases. As a result, the target protein is concentrated and transferred to an environment that will preserve its function. It is usually used in cases where the researcher prefers to purify the protein without a tag and hence affinity chromatography remains no longer an option [69]. This can be followed by hydrophobic interaction chromatography (HIC), reverse phase chromatography (RPC), and/or gel filtration chromatography (GFC) for further purification that has been elaborated in later sections of this book.
- *Ion exchange for intermediate purification*: The aim of using IEX for intermediate purification is to eliminate the majority of the significant impurities such as proteins, nucleic acids, endotoxins, and viruses. The main focus in a typical intermediate purification step is on capacity and resolution to maintain productivity (the amount of target protein processed per column in unit time) and achieve high purity. For example, when a protein after purification through affinity chromatography exhibits impurities comprising proteins of similar molecular weights, then IEX can be used as an intermediate step [70]. Afterward, HIC or RPC can be followed as the final polishing steps if GFC does not work very well.
- *Ion exchange as a polishing step:* Most contaminants are eliminated when IEX is used for polishing, except for trace quantities of closely related moieties such as structural variations of the target protein, nucleic acids, viruses, or endotoxins. The goal of the separation is to minimize these variants and trace impurities to levels that are acceptable for the application. In contrary to capture steps where quick, high-capacity step-elution is usually utilized, a polishing step focuses on producing the maximum possible resolution [71].

Like other chromatographic separation techniques, IEX is rarely sufficient as the sole purification stage for the separation of crude protein mixture samples. Ion exchange is frequently combined with other techniques such as gel filtration chromatography, hydrophobic interaction chromatography, reversed phase chromatography, as well as affinity chromatography for obtaining proteins of very high purity to be used in structural and biophysical studies.

7.8 Advantages and Disadvantages of Ion Exchange Chromatography

Alike other chromatographic techniques, IEX also has some advantages as well as drawbacks. Some of them are discussed below:

7.8.1 Advantages

- It is one of the most effective techniques for separating charged particles and can be employed for nearly any category of charged molecules including large proteins, small nucleotides, and amino acids.
- The purification technique does not require any additional tag and hence is capable of retaining the native architecture of the recombinant protein.
- IEX can be run with high flow rates. The high flow rate results in quick separation and purification of the target protein. Thus, it is crucial for the recovery of active protein.
- A high yield of the desired protein can be achieved through IEX.

7.8.2 Disadvantages

- One main disadvantage of IEX is that it can only separate the charged molecules. Molecules that do not carry any charges cannot be separated through IEX.
- The pH of buffers plays a crucial role in separation through IEX and a little alteration in pH can greatly alter the binding profile of proteins to the ion exchange resin.
- Excessive buffers with different ionic strengths or pH are essentially required while performing IEX which increases the overall cost of protein purification and separation using this technique.
- Since the method requires high salt concentration or high or low pH conditions for elution of desired proteins, the salt concentration and pH must be adjusted before they can be used for biochemical, biophysical, or structural studies.

7.9 Applications of Ion Exchange Chromatography

7.9.1 Purification of Recombinant Proteins

IEX is one of the widely used and efficient techniques for the purification of recombinant proteins. This technique is often combined with other techniques, which separate proteins depending on other parameters such as size (gel filtration), hydrophobicity (hydrophobic interaction chromatography or RPC), or biological activity (affinity chromatography). The protocols of protein purification using cation exchange and anion exchange chromatography have been discussed above. It is a simple technique that includes loading the sample with the desired protein onto the chromatographic system and collection of the eluent with the aid of a fraction collector.

7.9.2 Purification of Enzymes

In the purification of biologically active enzymes, the retrieval of enzymatic activity is very crucial. Purification of enzymes with high activity recovery can be achieved through IEX. Additionally, the purification of isoenzymes can also be done with IEX. The enzyme isoforms usually have around the same molecular weight. This renders separation by gel filtration chromatography difficult. However, with the slight variations in charge owing to the variation of the amino acid composition, isoenzymes are separated by IEX. As an example, for the diagnosis of hematological malignancy, N-Acetyl β-D-glucosaminidases are intensively explored. An isoenzyme called "Intermediate 1 Form" was described in the case of common acute lymphoblastic leukemia [72]. The separation of these isoenzymes is improved by using chromatography with high-resolution ion exchange.

7.9.3 Miscellaneous Applications

Apart from these, IEX is routinely used for industrial purposes for analytical applications such as quality assessment. It is a useful tool in the downstream process to scrutinize the fermentation process of various enzymes [73, 74]. In clinical studies, this method is used for the separation and purification of blood components such as albumin and recombinant growth factors [75, 76].

7.10 Troubleshooting

Sl. no.	Problems	Possible reasons	Possible solutions
1	Decreased or no flow through the column	Outlet closed or leakage or pumps not working	Open the outlet, check for leakage and pumps
2	Poor resolution	Flow rate is either too fast or too slow	Adjust flow rate
		Suboptimal elution condition	Adjust elution conditions by altering the pH and ionic strength of the buffer
		Column overloaded or sample is too viscous	Decrease sample load or dilute as required
3	Protein does not bind or elute	Incomplete column equilibration	Run more equilibration buffer through the column until the conductivity and pH are stabilized

(continued)

Sl. no.	Problems	Possible reasons	Possible solutions
		Protein may be unstabilized in the elution buffer	Adjust the pH and ionic strength of the buffer
		Protein aggregated in the column	Clean the column and adjust buffer conditions for more protein stability
4	Lower protein yield	Protein degradation by proteases	Add protease inhibitor
		Protein aggregated in the column	Adjust buffer conditions for more protein stability
		Nonspecific adsorption or precipitation of the protein	Reduce the ionic strength of the elution buffer to minimize hydrophobic interaction. Add suitable detergent or organic solvent
5	Protein loses activity during the procedure	Protein is destabilized or aggregated in the column	Adjust buffer conditions for more protein stability
		During purification, a necessary cofactor has been removed	Add cofactor

7.11 Conclusion

In this chapter, we have discussed the separation and purification of proteins by IEX. Ion exchange chromatography is a highly adaptable method for protein purification, which is vital for some studies. The technique can be used at any stage of the purification level. Alongside, the availability of a large variety of resins offers a wide range of selectivity which can be tailored according to the experimental conditions or protein of interest. Although IEX is an incredibly versatile method for protein purification, the ability to separate only the charged molecules limits its use. Besides, the selection of appropriate resin and buffers requires extensive analytical skills of the users for the successful execution of this method. We will discuss more other chromatographic techniques for the purification of proteins in the subsequent chapters.

Acknowledgments The authors thank Advanced Centre for Treatment, Research and Education in Cancer (ACTREC) for providing necessary infrastructure and resources for successful completion of the chapter. The authors acknowledge Ms. Chanda Baisane, Bose Lab for formatting the manuscript.

Problems

Multiple Choice Questions

1. The capacity of the resin for ion exchange relies on:
   (a) The cumulative molecular mass of the resin
   (b) Length of the ion exchange resin
   (c) The total number of ion active groups
   (d) Solubility of the ion exchange resins
2. The concept of ion exchange chromatography is based on:
   (a) Electrostatic attraction
   (b) Electrical mobility of ionic species
   (c) Adsorption
   (d) Partition
3. In anion exchange chromatography:
   (a) The column contains negatively charged beads where positively charged proteins bind
   (b) The column contains positively charged beads where negatively charged proteins bind
   (c) The column contains both positive and negatively charged beads where proteins bind
   depending on their net charge
   (d) All of these

Subjective Questions

1. A protein has an isoelectric point (pI) of 5.2. What is the net charge on this protein in BICINE [(N, N-bis(2-hydroxyethyl)glycine] buffer (pH 8.5)? Explain.
2. A crude lysate sample comprising four proteins (1, 2, 3, and -galactosidase) is obtained by a protein biochemist. He wants to purify β-galactosidase using ion exchange chromatography. The respective isoelectric points of these proteins are enlisted below:

Protein	Isoelectric point (pI)
1	3.7
2	6.8
3	9.5
β-Galactosidase	5.3

(continued)

He equilibrated an anion exchange column using a buffer of pH 5.0. (A) At this condition, which protein(s) from the lysate sample will bind to the column? (B) How the bound protein(s) can be eluted from the anion exchange column? He then recognized the fraction containing β-galactosidase from the anion exchange column and opted to purify it using a cation exchange column. (C) Explain how a cation exchange column may be used to separate β-galactosidase from any residual contaminated protein (s).

References

1. Cummins PM, Rochfort KD, O'Connor BF. Ion-exchange chromatography: basic principles and application. In: Walls D, Loughran ST, editors. Protein chromatography: methods and protocols. New York, NY: Springer New York; 2017. p. 209–23.
2. Roos PH. Chapter 1 – Ion exchange chromatography. In: Kastner M, editor. Journal of Chromatography Library. Netherland: Elsevier; 2000. p. 3–88.
3. Fekete S, Beck A, Veuthey JL, Guillarme D. Ion-exchange chromatography for the characterization of biopharmaceuticals. J Pharm Biomed Anal. 2015;113:43–55.
4. Production of synthetic polymeric compositions comprising sulphonated polymerizates of poly-vinyl aryl compounds and treatment of liquid media therewith, Google Patents; 1944.
5. Peterson EA, Sober HA. Chromatography of proteins. I. Cellulose ion-exchange adsorbents. J Am Chem Soc. 1956;78(4):751–5.
6. Small H, Stevens TS, Bauman WC. Novel ion exchange chromatographic method using conductimetric detection. Anal Chem. 1975;47(11):1801–9.
7. Fritz JS, Gjerde DT, Becker RM. Cation chromatography with a conductivity detector. Anal Chem. 1980;52(9):1519–22.
8. Gjerde DT, Fritz JS. Effect of capacity on the behaviour of anion-exchange resins. J Chromatogr A. 1979;176(2):199–206.
9. Gjerde DT, Fritz JS, Schmuckler G. Anion chromatography with low-conductivity eluents. J Chromatogr A. 1979;186:509–19.
10. Duong-Ly KC, Gabelli SB. Using ion exchange chromatography to purify a recombinantly expressed protein. Methods Enzymol. 2014;541:95–103.
11. Singhal RP. Separation and analysis of nucleic acids and their constituents by ion-exclusion and ion-exchange column chromatography. Sep Purif Methods. 1974;3(2):339–98.
12. Dragull K, Beck JJ. Isolation of natural products by ion-exchange methods. Methods Mol Biol. 2012;864:189–219.
13. Ćurković L, Cerjan-Stefanović Š, Filipan T. Metal ion exchange by natural and modified zeolites. Water Res. 1997;31(6):1379–82.
14. Lucy CA. Evolution of ion-exchange: from Moses to the Manhattan project to modern times. J Chromatogr A. 2003;1000(1–2):711–24.
15. Alexandratos SD. Ion-exchange resins: a retrospective from industrial and engineering chemistry research. Ind Eng Chem Res. 2009;48(1):388–98.
16. Edward Bańkowski M, Galewska Z, Gogiel T, Małkowski A, Romanowicz L, Sobolewski K, Wolańska M. Biochemistry workbook. Białystok: Medical University of Białystok.
17. Duncan JK, Chen AJ, Siebert CJ. Performance evaluation of non-porous versus porous ion-exchange packings in the separation of proteins by high-performance liquid chromatography. J Chromatogr. 1987;397:3–12.

18. Toshio Y, Masaharu S. Studies of polystyrene-based ion exchange fiber. I. The preparation and fundamental characteristics of polystyrene-based ion exchange fiber. Bull Chem Soc Jpn. 1983;56(12):3726–9.
19. Hubicki Z, Wawrzkiewicz M, Wójcik G, Kołodyńska D, Wołowicz A. Ion exchange method for removal and separation of noble metal ions. In: Ion exchange-studies and applications. London: IntechOpen; 2015.
20. Ralph Himmelhoch S. [26] Chromatography of proteins on ion-exchange adsorbents. Methods Enzymol., Academic Press. 1971:273–86.
21. Stone MC, Carta G. Protein adsorption and transport in agarose and dextran-grafted agarose media for ion exchange chromatography. J Chromatogr A. 2007;1146(2):202–15.
22. Hughes EN, Engelsberg BN, Billings PC. Purification of nuclear proteins that bind to cisplatin-damaged DNA. Identity with high mobility group proteins 1 and 2. J Biol Chem. 1992;267 (19):13520–7.
23. Levison PR, Badger SE, Hathi P, Davies MJ, Bruce IJ, Grimm V. New approaches to the isolation of DNA by ion-exchange chromatography. J Chromatogr A. 1998;827(2):337–44.
24. Harland CE. Ion exchange: theory and practice. Cambridge, London: Royal Society of Chemistry; 1994.
25. Boyer R. Modern experimental biochemistry. Delhi, India: Pearson Education; 2000.
26. Graziano SL, Huang RC. Chromatographic separation of chick brain chromatin proteins using a SP-sephadez column. Biochemistry. 1971;10(25):4770–7.
27. Groves ML. Preparation of some iron-binding proteins and alpha-Lactalbumin from bovine Milk. Biochim Biophys Acta. 1965;100:154–62.
28. Rombauts WA, Schroeder WA, Morrison M. Bovine lactoperoxidase. Partial characterization of the further purified protein. Biochemistry. 1967;6(10):2965–77.
29. Kent UM. Purification of antibodies using ion-exchange chromatography. Methods Mol Biol. 1999;115:19–22.
30. Fweja LW, Lewis MJ, Grandison AS. Isolation of lactoperoxidase using different cation exchange resins by batch and column procedures. J Dairy Res. 2010;77(3):357–67.
31. MacGillivray AJ, Rickwood D. The heterogeneity of mouse-chromatin nonhistone proteins as evidenced by two-dimensional polyacrylamide-gel electrophoresis and ion-exchange chromatography. Eur J Biochem. 1974;41(1):181–90.
32. Tuchscherer G, Steiner V, Altmann K-H, Mutter M. De novo Design of Proteins. In: Dunn BM, Pennington MW, editors. Peptide analysis protocols. Totowa, NJ: Humana Press; 1994. p. 261–85.
33. Lu CM, Wu YJ, Chen CC, Hsu JL, Chen JC, Chen JY, Huang CH, Ko YC. Identification of low-abundance proteins via fractionation of the urine proteome with weak anion exchange chromatography. Proteome Sci. 2011;9:17.
34. Tjian R, Robbins A. Enzymatic activities associated with a purified simian virus 40 T antigen-related protein. Proc Natl Acad Sci U S A. 1979;76(2):610–4.
35. Stoll VS, Blanchard JS. [4] Buffers: principles and practice. In: Deutscher MP, editor. Methods in enzymology. New York: Academic Press; 1990. p. 24–38.
36. Kang X, Frey DD. High-performance cation-exchange chromatofocusing of proteins. J Chromatogr A. 2003;991(1):117–28.
37. Larre C, Popineau Y, Loisel W. Fractionation of gliadins from common wheat by cation exchange FPLC. J Cereal Sci. 1991;14(3):231–41.
38. Sakakibara Y, Yanagisawa H. Techniques for the separation of proteins by isoelectric point column chromatography. Nat Sci. 2007;56:45–9.
39. Pabst TM, Antos D, Carta G, Ramasubramanyan N, Hunter AK. Protein separations with induced pH gradients using cation-exchange chromatographic columns containing weak acid groups. J Chromatogr A. 2008;1181(1–2):83–94.
40. Perkins M, Theiler R, Lunte S, Jeschke M. Determination of the origin of charge heterogeneity in a murine monoclonal antibody. Pharm Res. 2000;17(9):1110–7.

41. Pabst TM, Carta G. pH transitions in cation exchange chromatographic columns containing weak acid groups. J Chromatogr A. 2007;1142(1):19–31.
42. DePhillips P, Lenhoff AM. Determinants of protein retention characteristics on cation-exchange adsorbents. J Chromatogr A. 2001;933(1–2):57–72.
43. Debelak D, Fisher J, Iuliano S, Sesholtz D, Sloane DL, Atkinson EM. Cation-exchange high-performance liquid chromatography of recombinant adeno-associated virus type 2. J Chromatogr B Biomed Sci Appl. 2000;740(2):195–202.
44. Rindler-Ludwig R, Braunsteiner H. Cationic proteins from human neutrophil granulocytes. Evidence for their chymotrypsin-like properties. Biochim Biophys Acta. 1975;379(2):606–17.
45. Lim T-k, Nakamura N, Matsunaga T. Automated flow immunoassay for detection of food allergen using anion-exchange resin and alkaline phosphatase conjugated immunoglobulin G. Anal Chim Acta. 1998;370(2):207–14.
46. Johansson BL, Belew M, Eriksson S, Glad G, Lind O, Maloisel JL, Norrman N. Preparation and characterization of prototypes for multi-modal separation media aimed for capture of negatively charged biomolecules at high salt conditions. J Chromatogr A. 2003;1016(1):21–33.
47. Clezardin P, Hunter NR, MacGregor IR, MacGregor JL, Pepper DS, Dawes J. Tandem purification of mouse IgM monoclonal antibodies produced in vitro using anion-exchange and gel fast protein liquid chromatography. J Chromatogr. 1986;358(1):209–18.
48. Manji B, Hill A, Kakuda Y, Irvine DM. Rapid separation of milk whey proteins by anion exchange chromatography. J Dairy Sci. 1985;68(12):3176–9.
49. Shan L, Anderson DJ. Effect of buffer concentration on gradient chromatofocusing performance separating proteins on a high-performance DEAE column. J Chromatogr A. 2001;909 (2):191–205.
50. Clezardin P, McGregor JL, Manach M, Robert F, Dechavanne M, Clemetson KJ. Isolation of thrombospondin released from thrombin-stimulated human platelets by fast protein liquid chromatography on an anion-exchange mono-Q column. J Chromatogr. 1984;296:249–56.
51. Tsuneda S, Saito K, Furusaki S, Sugo T. High-throughput processing of proteins using a porous and tentacle anion-exchange membrane. J Chromatogr A. 1995;689(2):211–8.
52. Urabe M, Xin KQ, Obara Y, Nakakura T, Mizukami H, Kume A, Okuda K, Ozawa K. Removal of empty capsids from type 1 adeno-associated virus vector stocks by anion-exchange chromatography potentiates transgene expression. Mol Ther. 2006;13(4):823–8.
53. Meyer DJ, Coles B, Pemble SE, Gilmore KS, Fraser GM, Ketterer B. Theta, a new class of glutathione transferases purified from rat and man. Biochem J. 1991;274(Pt 2):409–14.
54. Lindblom H, Axio-Fredriksson UB, Cooper EH, Turner R. Separation of urine proteins on the anion-exchange resin mono Q. J Chromatogr. 1983;273(1):107–16.
55. Kunz C, Lonnerdal B. Human-milk proteins: analysis of casein and casein subunits by anion-exchange chromatography, gel electrophoresis, and specific staining methods. Am J Clin Nutr. 1990;51(1):37–46.
56. Hearn MT, Hodder AN, Aguilar MI. High-performance liquid chromatography of amino acids, peptides and proteins. LXXXVII. Comparison of retention and bandwidth properties of proteins eluted by gradient and isocratic anion-exchange chromatography. J Chromatogr. 1988;458:27–44.
57. Seely JE, Richey CW. Use of ion-exchange chromatography and hydrophobic interaction chromatography in the preparation and recovery of polyethylene glycol-linked proteins. J Chromatogr A. 2001;908(1–2):235–41.
58. Yamamoto S, Nakanishi K, Matsuno R. Ion-exchange chromatography of proteins. Boca Raton: CRC Press; 1988.
59. Salgin S, Takac S, Ozdamar TH. Effect of ionic environments on the adsorption and diffusion characteristics of serine alkaline protease enzyme in polyethersulfone ultrafiltration membranes. J Colloid Interface Sci. 2006;299(2):806–14.
60. Sata T, Izuo R, Mizutani Y, Yamane R. Transport properties of ion-exchange membranes in the presence of surface active agents. J Colloid Interface Sci. 1972;40(3):317–28.

61. Oliver RT. Cation-exchange behavior of di-, tri-, and tetravalent metal ions in ethylenediammonium salt solutions. Iowa: Iowa State University; 1959.
62. Beynon RJ, Easterby JS. Buffer solutions: the basics. Florida: Taylor & Francis; 2004.
63. Kang K, McCoy BJ. Protein separation by ion exchange chromatography: a model for gradient elution. Biotechnol Bioeng. 1989;33(6):786–90.
64. Xu T. Regeneration of the ion-exchange resin. In: Drioli E, Giorno L, editors. Encyclopedia of membranes. Berlin, Heidelberg: Springer Berlin Heidelberg; 2015. p. 1–3.
65. Williams PA, Hudson MJ. Recent developments in ion exchange: 2. Singapore: Springer Science & Business Media; 2012.
66. Rocklin RD, Pohl CA, Schibler JA. Gradient elution in ion chromatography. J Chromatogr A. 1987;411:107–19.
67. Noble JE, Bailey MJA. Chapter 8: Quantitation of protein. In: Burgess RR, Deutscher MP, editors. Methods in enzymology. New York: Academic Press; 2009. p. 73–95.
68. Kornecki M, Mestmacker F, Zobel-Roos S, Heikaus de Figueiredo L, Schluter H, Strube J. Host cell proteins in biologics manufacturing: the good, the bad, and the ugly. Antibodies. 2017;6(3)
69. Singh RS, Walia AK, Kennedy JF. Purification and characterization of a mitogenic lectin from Penicillium duclauxii. Int J Biol Macromol. 2018;116:426–33.
70. Gunzer G, Hennrich N. Purification of alpha 1-proteinase inhibitor by triazine dye affinity chromatography, ion-exchange chromatography and gel filtration on Fractogel TSK. J Chromatogr. 1984;296:221–9.
71. Liu W, Sun Y, Yu J, Chen Q, Bao Z, Fan X, Liang Y, Peng X, Xian M, Nian R. Advance chromatin extraction improves the performance of electropositive mixed-mode chromatography as a capture step and enables its integration with void-exclusion anion exchange chromatography as a two-column-step purification platform for monoclonal antibody production. Biochem Eng J. 2019;142:145–52.
72. Scott CS, Patel M, Stark AN, Roberts BE. Fast protein liquid chromatography (FPLC) of leukaemic cell N-acetyl beta-D hexosaminidases. Leuk Res. 1987;11(5):437–44.
73. Kamble RD, Jadhav AR. Isolation, purification, and characterization of xylanase produced by a new species of *bacillus* in solid state fermentation. Int J Microbiol. 2012;2012:683193.
74. Bajaj BK, Singh NP. Production of xylanase from an Alkalitolerant Streptomyces sp. 7b under solid-state fermentation, its purification, and characterization. Appl Biochem Biotechnol. 2010;162(6):1804–18.
75. Raoufinia R, Mota A, Keyhanvar N, Safari F, Shamekhi S, Abdolalizadeh J. Overview of albumin and its purification methods. Adv Pharm Bull. 2016;6(4):495–507.
76. Raines EW, Ross R. Purification of human platelet-derived growth factor. Methods Enzymol. 1985;109:749–73.

Gel Filtration Chromatography

8

Raghupathi Kummari and Kakoli Bose

Abstract

Apart from finding multitudes of applications in chemical, medicinal, and pharmaceutical research, gel filtration chromatography (GFC) has also become a routine tool in almost every biomedical research laboratory especially protein biochemistry. With rapid advancement and implementation of recombinant DNA technology in basic research, the requirement of purifying bacterially expressed proteins for further characterization has increased manifold. Gel filtration chromatography is usually adopted in the last or polishing purification step to obtain highly purified proteins that are later used for biophysical and structural studies. The versatility and robustness of GFC lies in the fact that the protein molecules do not adhere to the column during separation unlike ion-exchange and affinity chromatography. This gives GFC a significant advantage as it allows proteins to be eluted in a buffer condition that is conducive toward their future applications or storage. Since GFC separates analytes as a function of their size or molecular weight, it is also used as an analytical tool to determine molecular weights and oligomeric properties of macromolecules. This chapter, which is a sequel to the previous two chapters on chromatographic techniques, describes the theory, instrumentation, and applications of gel filtration chromatography along with elaborate discussions on several important protocols and troubleshooting tips.

R. Kummari · K. Bose (✉)
Integrated Biophysics and Structural Biology Lab, Advanced Centre for Treatment, Research and Education in Cancer (ACTREC), Tata Memorial Centre, Mumbai, India

Homi Bhabha National Institute, BARC Training School Complex, Mumbai, India
e-mail: kbose@actrec.gov.in

K. Bose (ed.), *Textbook on Cloning, Expression and Purification of Recombinant Proteins*, https://doi.org/10.1007/978-981-16-4987-5_8

Keywords

Gel filtration chromatography · Molecular weight determination · Resolution · Matrix · Fast protein liquid chromatography

8.1 Introduction

Gel filtration chromatography (GFC) or size exclusion chromatography (SEC) is a popular protein purification technique that separates the macromolecules based on differences in their hydrodynamic volume or size [1]. The general principle of gel filtration chromatography is fairly simple where the inert gel medium is a porous matrix comprising spherical beads with stable physicochemical properties. When a mixture of analytes with different size distribution is applied to the column, the molecules larger than the pore size are unable to enter into the beads. As a consequence, they pass through the interstitial spaces between beads and elute first. On the other hand, molecules smaller than the gel pore size get diffused into the pores and elute at later time points as a function of their molecular weight [2] (Fig. 8.1).

Although this chromatographic technique is widely used in recombinant protein purification in research laboratories, it is capable of scaling up the process at an industrial scale as well. Based on the type of solvents used in the mobile phase, and other subtle variations, it is also referred to as molecular exclusion, molecular sieve, gel-permeation and gel-exclusion chromatography [3].

One of the distinct features of GFC that separates it from the other two main chromatographic techniques (viz. ion–exchange and affinity chromatography) is that, there is no direct interaction between the protein molecules and the solvent, i.e., the mobile phase. This makes the method more flexible in terms of buffer selection as it has no adverse effect whatsoever on the column resolution. Since GFC is usually used in the last step of protein purification, this unique characteristic allows retention of the protein in a buffer that will be amenable to several structural and functional studies post purification. The other important determinants of column resolution other than the gel pore size are bed height, flow rate, sample volume, and the molecular weight of the protein [4, 5], which will be elaborated in the later sections of this chapter. As mentioned briefly in Chap. 5, the best resolution is usually achieved with an optimum flow rate (slow or medium), long and narrow columns, small pore-size gel and sample volumes (1–5% of the total column volume) [6].

Apart from separation of proteins based on their molecular weights, another important application of GFC is desalting [7]. Since GFC is generally used in the finishing step of protein purification, the loaded protein can be efficiently eluted in a buffer of interest. The salt component present in the loaded sample easily separates from the macromolecule due to their huge size difference. Therefore, sample volumes as much as 30% of the total bed volume can be safely loaded onto the column as chances of peak overlap is unlikely. Furthermore, since the separation

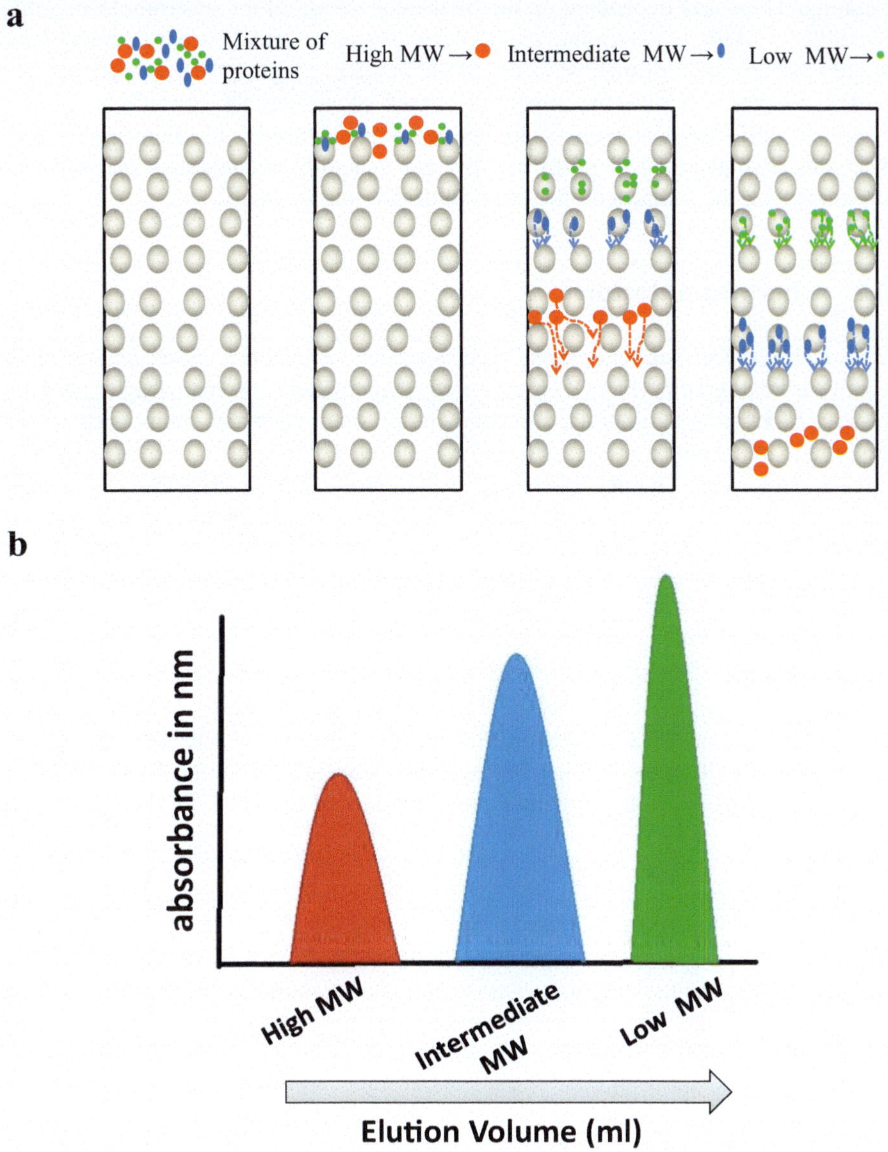

Fig. 8.1 Illustration of the separation principle in gel filtration chromatography. (**a**) The mixture of proteins (shown in different colors) applied to the column. Depending upon their size, protein molecules pass through the gel via different routes. High molecular weight proteins (shown in *red*) that are too large to enter the beads pass through the space between the beads and elute first. Small molecules (shown in *green*) enter the beads, travel slowly, and elute last. (**b**) Elution profile of proteins of different sizes as a function of elution buffer volume

technique is linearly dependent on the molecular weight of the macromolecule, this principle is used to determine the molecular weight and hence the oligomeric property of a protein of interest as described in later sections of the chapter [8].

This chapter, describes the principles of GFC along with its various applications in protein purification as well as studies pertaining to protein biochemistry. It also discusses several practical problems that are generally encountered and ways of circumventing them through efficient troubleshooting methods.

8.2 Instrumentation

Note: Gel filtration chromatography is commercially available as either a fast protein liquid chromatography (FPLC) or high performance liquid chromatography (HPLC) system. HPLC that is often used to characterize small chemical compounds works under high temperature and pressure, which makes it unsuitable for protein purification or molecular weight determination. On the other hand, FPLC system is optimized to purify large biomolecules [9]. Therefore, all protocols will refer to FPLC system in this chapter.

The typical instrumentation of Gel filtration chromatography system is as follows.

8.2.1 Pump

The pump pulls up the solvent from the reservoir and forces it to the column and subsequently to the detector. Operating pressure depends on composition of mobile phase, column dimensions, flow rate, and particle size.

8.2.2 Injector

Introduces the protein solution into the mobile phase.

An efficient injector should have the following properties:

1. Should be capable of introducing small as well as large volumes of samples (depending on applications).
2. Multiple injections of the sample solution should be possible.
3. It should not agitate the mobile phase of the column.

8.2.3 Column

Usually used for separation of protein of interest from its mixtures.

The size and dimensions of columns depend on the type of application.

1. *Analytical:* In analytical chromatography, the main objective is to characterize the sample so as to obtain its properties in a quantitative manner. This is mainly used in synthetic chemistry labs, pharmaceutical industry, etc. Recovery of the loaded sample is not of primary importance and hence the eluent is often removed as waste [10].
2. *Preparative*: This is mainly used for protein isolation and purification for further biochemical and biophysical characterization. Therefore, the eluents are collected in several fractions [11].

Typically, in GFC, analytical columns have 7.5–8 mm and preparative columns have 22–25 mm diameter. Column lengths range between 25 and 60 cm. In protein biochemistry labs, preparative columns are mainly used.

8.2.4 Detector

Regularly utilized detectors are UV-spectroscopy, fluorescence, mass-spectrometric, and electrochemical identifiers. In UV detectors, routinely used wavelengths are 214 or 220 nm (where the amide peptide bond has a strong absorbance) and 280 nm (to detect the presence of other chromophores primarily aromatic amino acids) [12].

8.2.5 Fraction Collector

An automated fraction collector with tube holders helps in collecting the sample eluted from the column.

8.2.6 Data Processing System

Provides data output from different fractions, stores data, and provides a graphical representation of the data. It also helps in column calibration.

A simple layout of the instrument setup is provided as shown in Fig. 8.2 below.

8.3 Principle of Macromolecular Separation Using Gel Filtration Chromatography

The gel filtration columns consist of beads that contain sieves cross-linked with agarose, dextran, polyacrylamide, or their combinations of a particular size, and are designed to suit specific separation tasks. The matrix that acts as a stationary phase consists of two measurable volumes, internal and external. The internal volume consists of liquid within the beads, and external volume, *aka* void volume (V_0), refers to liquid in the space between the beads. The sum of the external and internal volumes makes up the total volume (V_t) (Fig. 8.3) [3].

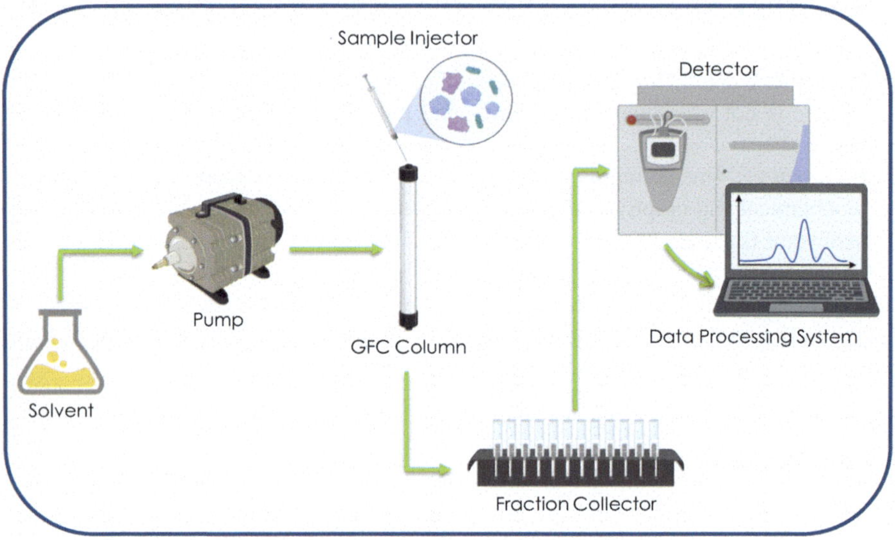

Fig. 8.2 A simplistic representation of gel filtration chromatography instrument setup. All the components of the schematic are labelled and are self-explanatory

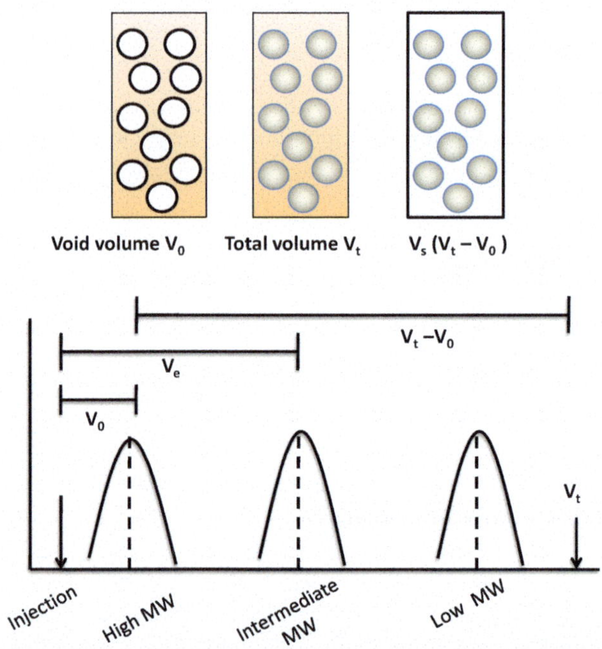

Fig. 8.3 Pictorial and graphical representation of molecular weight determination using gel filtration chromatography. V_t is the total volume of a gel bed, V_0 is the void volume, and V_s is space occupied by solvent inside the medium particles. The colored sections in the three rectangular boxes (representing a GFC column) in the upper panel represent the volume that has been considered, while the white section represents the part of the column volume that is not

When a mixture of proteins of different sizes is loaded onto the column, larger molecules migrate through the interstitial spaces between the stationary phase matrix pores and run down quite rapidly through the column eluting at V_0. Simultaneously, particles smaller than the matrix pores equilibrate with external and internal liquid volumes, causing them to migrate more slowly and eventually elute at a volume (V_e). The elution volume V_e, which is different for different proteins but greater than V_0, is linearly dependent on their size (Fig. 8.3). The elution volume (V_e) of a particular molecule depends on the fraction of the stationary phase available for diffusion. V_t refers to the total column volume. This principle can be mathematically represented by constant K_d or K_{av} (partition or distribution coefficient) as depicted in the following equations [3].

$$V_e = V_0 + K_{av}(V_t - V_0) \tag{8.1}$$

Rearranging the equation:

$$K_{av} = (V_e - V_0)/(V_t - V_0) \tag{8.2}$$

In GFC, K_{av} is represented as a function of its molecular size. When a molecule is adequately large (for example, blue dextran, MW ~ 2000 kDa) so as to bypass the mobile phase between the beads, then $K_{av} = 0$. On the contrary, for an extremely small molecule that enters the innermost mobile phase, then $K = 1$. For other proteins, the K_{av} value is intermediate and varies between 0 and 1. This shifting of the value of K_{av} between 0 and 1 is the fundamental basis for separation of macromolecules as a function of their molecular weight [3].

Other than size and mass, the hydrodynamic radius plays a vital role in determining gel filtration's flow rate. During separation in gel filtration, the general assumption is that molecules have symmetrical shapes with a similar hydrodynamic radius that fall in the resolution range of the matrix used. However, protein molecules that deviate from this assumption (i.e., asymmetric with larger hydrodynamic radius) will be eluted prior to the ones having same molecular weight but are symmetric with lower hydrodynamic radius. For example, fibrous proteins and some protein aggregates are asymmetric with higher hydrodynamic radius compared to their globular counterparts [3].

8.4 Choice of Matrix in Gel Filtration Chromatography

Matrix is a chemically inert but mechanically stable substance with a homogeneous porous structure and pore size. Gel filtration matrices are made from a wide variety of materials including agarose (Sepharose™ series), dextrans (Sephadex™ series), polyacrylamide (Bio-Gel series), polyvinylpyrrolidone, polyvinylethylcarbitol, silica-based materials, and cellulose. These materials have different pore size and particle size distribution that help in separation of molecules belonging to different ranges of molecular weights. Some of the commercially available matrices are a mixture of the above-mentioned substances such as dextran–agarose (Superdex™

Table 8.1 List of various commercially available media for gel filtration chromatography [3]

Sl. no.	Type of media	Commercial name	Molecular weight range (kDa)
1	Agarose	Bio-Gel A-0.5 Bio-Gel A-1.5 Bio-Gel A-5 Sepharose 6B Sepharose 4B	10–500 10–1500 10–5000 10–4000 60–20,000
2	Dextran	Sephadex G-50 Sephadex G-75 Sephadex G-100 Sephadex G-200	1.5–30 3–80 4–150 5–600
3	Polyacrylamide	Bio-Gel P-10 Bio-Gel P-30 Bio-Gel P-100 Bio-Gel P-150 BioGel P-200	1.5–20 2.5–40 5–100 15–150 30–200
4	Dextran-polyacrylamide gels	Sephacryl S-200 Sephacryl S-300 Sephacryl S-400	5–250 10–1500 20–8000

series), dextran-polyacrylamide (Sephacryl™ series), etc. as shown in Table 8.1. Each of them have distinct properties with certain *pros* and *cons*, and are suited for particular types of applications [3].

The type of column appropriate for a particular analyte is determined from "selectivity curve" that plots K_{av} (partition coefficient) versus $\log_{10}$ of MW for a set of standard molecules that is represented by the equations provided below.

$$K_{av} = (V_e - V_0)/(V_t - V_0) \tag{8.3}$$

$$K_{av} = \log\ (MW) + b \tag{8.4}$$

here "b" is the intercept on the Y-axis. The steeper the curve, the higher the resolution is reached [13]. The matrix is chosen for a particular analyte when its MW falls within the range of selectivity curve [13].

The variables such as buffers, organic solvents, pH, and temperature should be considered while choosing the matrix because they should be compatible with molecules or analytes being separated. The matrix should not adsorb the analytes during the separation process. For difficult separation problems, the resolution of the material will be critical and in these instances properties such as bead size, selectivity (given by the pore size distribution), and separation volume (available pore volume) become essential. Another factor to consider is the effect of the sorption properties (physicochemical property that includes both adsorption and absorption) of the matrix under running conditions. Although virtually no support can be expected to be completely free from sorption properties, the nature and degree of these properties vary with the nature of the matrix. Sometimes, these properties have been used to achieve increased separation of the sample component [14, 15].

8.5 Resolution of Gel Filtration Chromatography

Resolution can be defined as the extent of separation between peaks that correspond to MW of different analyte molecules.

Mathematically Resolution (R_s) can be expressed as:

$$R_s = \frac{Vr2 - Vr1}{\frac{W1+W2}{2}} \tag{8.5}$$

where V_{r1} and V_{r2} are the elution volumes of two consecutive peaks, whereas W_1 and W_2 are their peak widths, respectively [16] (Fig. 8.4).

There are different factors that determine resolution as described below. Therefore, by carefully balancing these parameters one can achieve good selectivity and sharp peaks in gel filtration chromatography.

8.5.1 Parameters Affecting Resolution

Several factors influence resolution. Pore size of the particles, particle size and distribution, column dimensions, medium packing density, flow rate, sample volume as well as viscosity of the sample and buffer. The medium selectivity is described by the selectivity curve (as mentioned above), in which a partition coefficient, K_{av}, is plotted against the logarithm of the molecular weight for a set of standard proteins.

The different parameters affecting resolution are discussed below.

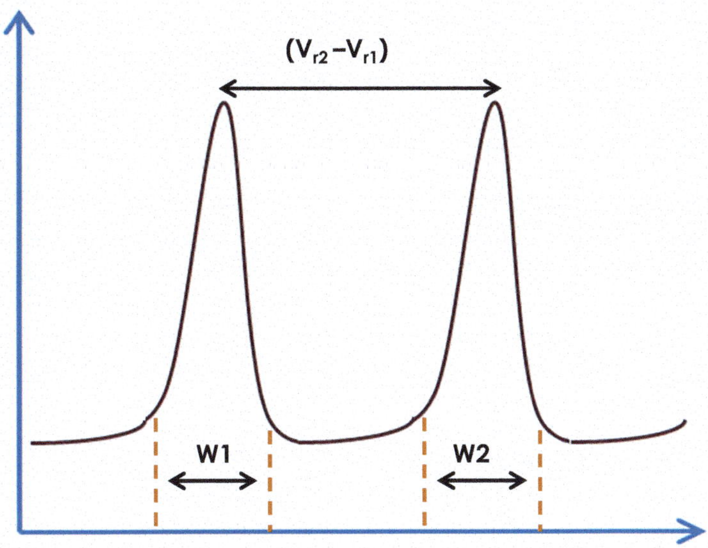

Fig. 8.4 Graphical representation of parameters affecting resolution (R_s) of gel filtration chromatography

8.5.1.1 Column Parameters

In GFC, maximum resolution is obtained with long columns. The ratio of the length of the cylindrical column to its diameter may vary from 20 to 100 [17]. A wide variety of pre-packed columns with different materials and sizes are available. Packing instructions are given below; however, for efficient packing, one needs to strictly follow the product manual's instructions. The loosely packed column with gaps either on top or bottom severely affects the resolution, pressure, etc.

8.5.1.2 Packing the Column

A column needs to be properly packed to obtain good resolution of protein separation through gel filtration chromatography. For proper packing of a column, the protocol to be followed is provided below:

According to the manufacturer's recommendation, add an appropriate amount of double distilled autoclaved water to the gel or dry gel powder. The gel can be pre-swollen at 90 °C for 1–5 h or kept at room temperature for 3–72 h. Never use a magnetic stirrer as it damages the beads. Gently mix the pre-swollen slurry and degas it. Fix the column vertically and pass the buffer to remove air bubbles in the outlet tube. Block the outlet tube and pour the gel slurry to the required height with the help of a glass rod. A gel reservoir can be used for smooth and continuous pouring of the gel. Allow the gel to settle, remove the excess buffer, and put the adaptor. Nowadays, a wide variety of pre-packed columns are commercially available that cater to specific separation requirements, which might be used for obtaining higher resolution [12].

8.5.1.3 Air Bubbles, Uneven Packing, and Cracks

Air bubbles in the column affects its packing and hence care should be taken to prevent air bubbles or remove them if found. The presence of air bubbles can be checked via transmitted light. It is recommended to use colored protein markers such as blue-dextran 200, cytochrome-c, and myoglobin to check the uneven packing, air bubbles, and cracks present in the column [12].

8.5.1.4 Choice of Eluent

For separation of proteins of interest through GFC, a favorable eluent condition is to be maintained. Although the elution profile is independent of the type of buffer used, it is important to provide a stable environment to the macromolecules. This is achieved on a case-to-case basis with the help of additives such as metal ions, cofactors, and protease inhibitors. It is to be noted that high concentrations of salts or solvents that give peaks at 215 nm or 280 nm wavelengths should be avoided [12].

8.5.1.5 Effect of Flow Rate

Since resolution is inversely proportional to the flow rate, a high resolution in GFC is achieved through use of very low flow rate. The optimum flow rate depends on the column and pump systems mentioned by the manufacturer. However, in GFC, the approximate flow rate range of 2 mL/cm^2/h to 30 mL/cm^2/h is preferably used [12].

8.5.1.6 Column Cleaning and Storage

To maintain a standard resolution as well as performance, gel filtration matrices must be cleaned according to the manufacture's instructions. Three basic cleaning protocols are recommended—simple, rigorous, and harsh. Simple is recommended with a low concentration of acetic acid or NaOH when there is an increase in back-pressure due to presence of contaminants. Rigorous cleaning is recommended when the following are observed: an increase in back-pressure; a color change at the top of the column; and a loss of resolution. Harsh cleaning is recommended when the column is contaminated mainly due to microbial growth. The use of 0.02% sodium azide in autoclaved water has proven to be useful to prevent microbial growth and is highly recommended for column storage [3].

8.5.1.7 Sample Preparation

A series of precautionary steps are required during sample preparation to achieve the best possible resolution. A sample or protein concentration of up to 50 mg/mL can be used for a non-viscous protein sample. The sample must be clear and free from any particulate matter, which is usually achieved by filtration using 0.22 μm syringe filter, or centrifugation at 10,000 × *g* for 15–30 min prior to injection. Furthermore, sample stability test can be performed before injecting into the gel filtration column under required salt, pH, and temperature conditions. This will prevent the sample from aggregating or precipitating, which further leads to column clogging and contamination or bacterial growth. High sample viscosities should be avoided. The sample volume should be 3–5% of the total column volume [12].

8.6 Applications of Gel Filtration Chromatography

As discussed earlier in this chapter, gel filtration chromatography has immense applications in protein biochemistry. Apart from serving as the last important step in recombinant protein purification, it is also used for desalting of proteins and determination of their molecular weights, hence oligomeric properties.

Some of the major applications with appropriate protocols are provided below.

8.6.1 Molecular Weight Determination

Gel Filtration Chromatography is one of the simplest tools to determine molecular weight (hence oligomeric property) of a protein of interest. The easy availability and user-friendly attributes of a gel filtration system make it one of the most popular and widely used tools in protein biochemistry laboratory. Since the proteins are run under their native conditions, this technique is also capable of determining the oligomeric property of a protein [18, 19].

An elaborate protocol for molecular weight determination using GFC is provided below.

8.6.1.1 Operating Procedure

The setup is as follows.

GFC System

1. *Chromatography system*: Any commercially available FPLC system with a UV detection system and a fraction collector.
2. *Gel filtration column* (choice should be based on the application and molecular weight of the desired macromolecule—refer to Table 8.1).
3. *Necessary accessories*:
   (a) Vacuum filtration unit (for filtering buffer and protein samples).
   (b) 0.22 μm syringes and filters.

Reagents

1. *Protein of interest* (usually 1 mL of solution with a final concentration of ~2 mg/mL).
2. *Molecular Weight Standards*: Can be made in the lab or purchased; MW range is usually between 12 and 200,000 kDa (for example: 1 mg/mL BSA, 1 mg/mL MBP, 1 mg/mL β-casein, and 1 mg/mL lysozyme).
3. *Blue Dextran* for determination of void volume.
4. *Molecular weight standard buffer:* 50 mM Tris–HCl, 100 mM salt, pH 7.5 (Buffer A).
5. *Sample elution buffer*: 50 mM Tris–HCl, 150 mM salt, pH 7.5. It is advisable to add a reducing agent such as Dithiothreitol or DTT (since the markers and the protein of interest might have multiple cysteine residues). Variation in buffer composition is possible (e.g., 20 mM Na_2HPO_4/NaH_2PO_4 and 200 mM NaCl, 1% (v/v) Glycerol at pH 7.5). Glycerol can be used to maintain stability of the protein of interest. ~200 mM salt is added to the buffer to maintain stability of the protein as well as to prevent it to stick to the column wall.
6. Distilled water (1 L).
7. 20% *Ethanol in distilled water* (1 L)—for cleaning.

Standard Operating Tools

Read the manufacturer's instructions for standard operating procedure and maintenance-related specifications. Every system operating window consists of a system controller, method editor, and evaluation workspace.

1. *System controller*: This wizard is for connecting and running the system manually and to calibrate pumps and detectors.
2. *Method editor*: For creating a new method, defining flow rate, amount of water or buffer to pass, equilibration time, injection time and fraction collection time, etc.
3. *Evaluation*: For analyzing the data, comparing the peaks and exporting data, etc.

The automated system has pump modules, precise UV detectors, conductivity measures, and a pH meter. The system is connected with its supporting software that runs under user-friendly operating systems.

Running the Experiment-Instrumental Setup

1. Turn on the chromatographic system followed by software on the computer and connect it with the system.
2. Create method: Define the parameters such as pressure limit, UV monochromators, flow rate (~0.5 mL/min), volume of the equilibration buffer (~2 column volume), injection time, fraction volume, time of collection, and other required information.
3. It is preferred to inject the protein after equilibration is over to prevent protein aggregation in the loop.
4. Check the tubing connections prior to injecting the protein. System pump is connected to the column directly.
5. Set up the fraction collector: Using the fraction collector, define the type of rack, collection scheme, and fraction size in mL.

Running Analyte Through the GFC Column

Step 1:

1. Column Standardization:
   Before injecting the standards, wash the column with at least 1 column volume of autoclaved double distilled water and equilibrate using 2 column volume of a MW marker buffer. If the protein is being run for the first time, the UV absorbance needs to be monitored at different wavelengths (215 nm, 254 nm, and 280 nm) and then standardized. The flow rate should not exceed ~0.5% of the column volume per minute. The UV monitor needs to be set to "zero" once the reading becomes stable.
2. Obtaining the standard curve:
   (a) Take 1 mg/mL blue dextran (~ 1 mL volume) and inject into the column and elute with Buffer A. Calculate the volume of the buffer required to reach the middle of the peak (as seen on the computer monitor). This measure gives the value of the void volume (V_0).
   (b) Either use commercially available protein standards or use 1 mg/mL BSA, 1 mg/mL MBP, 1 mg/mL β casein, and 1 mg/mL lysozyme and run them individually through the column and elute with 2 column volume of Buffer A. Since the elution volume includes the volume of the protein, the volume can be reset to "zero" prior to each injection. Measure elution volume of each standard macromolecule (V_e).
   Note: If more than three protein markers are used, mixing may cause peak interference because BSA might form a complex with other proteins. It is advisable to prepare an individual or multiple marker groups of 2–3 proteins with significant size differences.

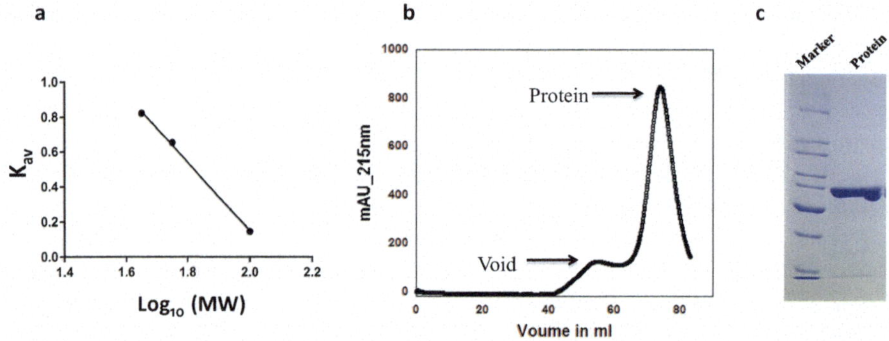

Fig. 8.5 (**a**) Using standard molecular weight markers, a standard curve is plotted and K_{av} values of the marker proteins are determined experimentally. (**b**) Chromatogram of the protein of interest as well as the void volume is shown in the graph obtained from the GFC system. (**c**) The collected fractions were pooled, concentrated, and run on SDS-PAGE. The protein band shows >95% purity. *Marker*: Molecular Weight marker; *Protein*: Protein of interest

(c) Plot K_{av} (Y-axis) versus the logarithm of molecular weight, i.e., $\log_{10}$MW (on the X-axis) for the protein standards (refer to Eqs. (8.1, (8.2, (8.3, and (8.4). Fit the data to a straight line, which is the standard curve (Fig. 8.5a). Once the column is standardized, the values obtained can be used multiple times until there is a major protocol change or the column repacking has been done.

Note: If a problem arises regarding UV signal stabilization, column cleaning is recommended.

Step 2: Determination of the MW of the protein of interest:

1. Wash the column with at least 0.5 to 1 column volume of double distilled water and Buffer A.
2. Equilibrate the column with 2 column volumes of elution buffer. Set the UV reading to "zero" once the signal gets stabilized.
   Note: For an unknown protein, use of multiple wavelengths to measure the peak intensity is advisable (including 215 nm, 260 nm, and 280 nm). This will help choose the best wavelength to use for the particular protein sample.
3. Prior to loading onto the column, concentrate the protein to 1 mL volume with a final concentration not less than 1 mg/mL.
4. Elute the protein with at least 1 column volume of the elution buffer. Calculate V_e/V_o and plot the point on the standard curve to obtain the MW (Fig. 8.5b).
5. The eluted protein (corresponding to the peak observed) is collected in the fraction collector and run on SDS-PAGE for checking its purity (Fig. 8.5c) and further storage.
   Note: If the protein has no tryptophan (that absorbs at 280 nm or 295 nm), set the UV measurement wavelength to either 260 nm (for Tyrosine) or 215 nm (peptide bonds).

Note: The plot of K_{av} versus $\log_{10}$ (MW) yields a straight line. It is important to note that, K_{av} value of the protein of interest should ideally fall between 0 and 1 and can be obtained from the graph. $K_{av} > 1$ indicates some interaction between the sample and the column. On the other hand, $K_{av} < 1$ indicates the formation of channels in the column. In any case, where there is a deviation from linearity, consider replacing the column.

8.6.2 Purification of Recombinant Proteins

Proteins expressed and purified in bacterial system are mainly used for biochemical, biophysical, and structural characterizations. Therefore, large amount of protein with >95% purity is preferred for these studies. Proteins obtained from the bacterial system are usually run through affinity chromatography, and if required followed by ion-exchange chromatography (IEC). This gives the protein of interest a reasonable level of purity (might vary from protein-to-protein). The last polishing step is then achieved by running the protein through an appropriate GFC column.

The protocol for protein purification though similar is much simpler than the one discussed above. The protein of interest is used as the loading sample and the eluent from the peak (as displayed in the chromatogram on the system computer) is collected in the fraction collector. However, for an unknown protein, it is important to have a rough estimate of the elution volume, which is achieved by calculating its elution volume from an already determined standard curve [20, 21].

8.6.3 Desalting and Buffer Exchange

"Desalting" of proteins refers to removal of salt or other undesirable contaminants such as nucleotides and chemicals from a protein solution. On the other hand, "buffer exchange" refers to replacement of the existing buffer with a final buffer that might be required to store the protein under stable and native conformation and/or conducive for certain biochemical, biophysical, and structural analyses [22].

Although the principle behind both the procedures is very similar, there are certain subtle differences in approach as well as applications. For desalting purpose, the GF column is equilibrated with water, while in buffer exchange, the column is equilibrated with the desired buffer solution for the protein. In both the cases, the existing buffer solution within the column is replaced with either water or a different buffer system. Along with the previous buffer, other impurities also pass out of the column.

In both the techniques, the size of the column plays a crucial role and hence standardization of the column size becomes necessary. This is because a very large column might lead to dilution of the protein sample and a smaller one will lead to inadequate separation or exchange.

8.6.4 Miscellaneous Applications

The versatility of gel filtration chromatography has led to a variety of applications that encompass various scientific disciplines and their interfaces. It has also created a niche both in industries and in clinical applications. The various other applications of GFC include group separation that allows fast separation of macromolecules from natural extracts and chemicals [22–24], separation of proteins through PEGylation [25], separation of proteins from peptides [26, 27], separation of viruses [28, 29], and separation of oligonucleotides from contaminants [30, 31]. Furthermore, GFC has found immense application in the fields of enzymology, endocrinology, and clinical pharmacology [32–36] as well.

8.7 Troubleshooting Tips for Running GFC

Different problems encountered while running a gel filtration column and their troubleshooting tips are provided in Table 8.2 below.

Table 8.2 Troubleshooting tips for running various experiments using gel filtration chromatography

Sr. no.	Problem	Recommendation
1.	Column gets dried	Pass 20% methanol followed by water and observe for void volume change and back-pressure. If it changes, then repack the column according to the manufacturer's instructions. The dried column produces channels within the column and perturbs the resolution and hence repacking becomes imperative
2.	Poor resolution	a. Decreasing the flow rate might improve resolution b. Revisit your choice of column and type of matrix for the particular protein of interest since fractionation range and pore-size diameter play important roles in resolution. Therefore, switching to a more appropriate column-matrix combination might help c. Clean the column with appropriate buffers and detergents, organic solvents, denaturing agents (within allowed concentration and tolerable range of the matrix) in an increasing manner as suggested in the users' manual. If there is no improvement, disassemble the column and repack it after washing the beads and other column components
3.	UV lamp is unstable and making noise	Pass at least two column volumes or more buffer by keeping UV "*ON*" while enabling auto-zero option frequently in between. The buffer should not contain components that give signals or noise. Always make fresh buffers if you are adding DTT (reducing agent), which often gets oxidized over time and creates this issue. The column and flow system should be clean to prevent growth of any microorganism

(continued)

Table 8.2 (continued)

Sr. no.	Problem	Recommendation
4.	Air bubbles in column, or in the eluent	Air bubbles are found in the gel bed when eluents are not thoroughly degassed or the column temperature is increased. Since air bubbles affect the resolution to a greater extent, extensive washes with degassed buffer might solve the problem. Consider repacking the column if the problem persists after washes
5.	Very high column pressure	The column and flow system should be rigorously cleaned to eliminate the possibility of growth of microorganisms that sometime lead to high column pressure
6.	Broad peaks	Might be due to loss of column efficiency. To take care of this problem: use mobile phase of lower viscosity; lower the flow rate; pack the column with smaller particle-sized beads; inject smaller sample volume; and elevate column temperature. However, if the problem still persists, then replace the column

8.8 Conclusions

Gel Filtration Chromatography has become an indispensable tool in protein purifications. Due to its simplicity, robustness, and versatility, it has found applications beyond separation of macromolecules and created a place for itself in biomedical research. The huge popularity of GFC has eventually led to significant advancements in its instrumentation as well as broadening of the gamut of its applications at the interface of academia and industry.

Upgradation in the classical GFC technique include recent development of leading-edge tools such as Absolute Size-Exclusion Chromatography (ASEC) and SEC-MALS. The former technique combines principles of dynamic light scattering with that of size exclusion chromatography to obtain absolute size of proteins and other macromolecules. GFC/SEC coupled with DLS is capable of enhancing DLS resolution and is used for aggregation studies [37]. Furthermore, SEC-MALS, which combines the power of size-exclusion chromatography with multi-angle light scattering, offers an advanced and faster analytical technique that bypasses the lengthy column calibration step [37].

Acknowledgments The authors thank Advanced Centre for Treatment, Research and Education in Cancer (ACTREC) for providing necessary infrastructure and resources for successful completion of the chapter. The authors acknowledge Ms. Chanda Baisane, Bose Lab, ACTREC for formatting the manuscript.

Problems

Multiple Choice Question

1. In Gel Filtration Chromatography the best resolution is usually achieved with the following combination:
   (a) Faster flow rate, short and wide columns, small pore-size gel, and less sample volumes (1–5% of the total column volume).
   (b) An optimum flow rate (slow or medium), long and narrow columns, small pore-size gel and sample volumes (1–5% of the total column volume).
   (c) Faster flow rate, short and wide columns, large pore-size gel, and less sample volumes (1–5% of the total column volume).
   (d) An optimum flow rate (slow or medium), long and narrow columns, large pore-size gel, and high sample volumes (10–25% of the total column volume).
2. Which one of the following statements is true?
   (a) K_{av} value of the protein of interest should fall between 0 and 1 and can be obtained empirically by substituting the values (known total, void, and elution volumes) in the standard equation.
   (b) K_{av} value of the protein of interest should fall between 0 and 1 and can be obtained experimentally from the standard graph (known molecular weight markers).
   (c) K_{av} value of the protein of interest should be any integer and can be obtained experimentally from the standard graph (known molecular weight markers).
   (d) K_{av} value of the protein of interest should be any integer and can be obtained empirically by substituting the values (known total, void, and elution volume) in the standard equation.
3. A gel filtration column is filled with Polyacrylamide media (Bio-Gel P-150) with a molecular weight separation range of 15–150 kDa. A mixture of four different proteins: Protein A (MW 55 kDa); Protein B (MW 35 kDa); Protein C has a molecular weight (MW 200 kDa), and Protein D (MW 10 kDa). Protein B has a tendency to form aggregates (>300 kDa) in small proportions. Therefore, the order of their elution time would be:
   (a) Protein D will be eluted first, followed by B, A, C and aggregates will be eluted last.
   (b) Aggregates will be eluted first, Protein C will be the second to elute, followed by A, B, and D.
   (c) Aggregates and Protein C will elute simultaneously in the void volume, followed by D, B, and finally A.
   (d) Aggregates and Protein C elute simultaneously in the void volume, followed by A, B, and finally D.

Subjective Question

1. A researcher was trying to characterize the native oligomeric status of one of the least studied member "**PROTEIN X**" of a member of a serine protease family. "**PROTEIN X**" has been purified using Ni-NTA chromatography and subsequently subjected to gel filtration chromatography using Superdex 200 10/300 HR column ($V_t = 120$ mL) (GE Healthcare, Uppsala, Sweden). The elution volumes with other necessary information are provided in the table below. Blue Dextran was used to determine the void volume of the column. The standard proteins BSA, MBP, and lysozyme were run on the same column and their elution volumes are provided. Answer the following two questions from the following information:
   (a) Calculate the K_{av} of the standards.
   (b) Find the K_{av} values and thereafter calculate the molecular weights and oligomerization status of newly characterized "**PROTEIN X**."

S. no	Protein	Mol. wt.	Elution volume
1	Blue Dextran	2000 kDa	46.5 mL
2	BSA	66.5 kDa	77 mL
3	MBP	45 kDa	86.5 mL
4	Lysozyme	14 kDa	111.5 mL
5	"**PROTEIN X**"	?	Peak 1: 64 mL (major peak) Peak 2: 87 mL (minor peak)

References

1. Sun T, Chance RR, Graessley WW, Lohse DJ. A study of the separation principle in size exclusion chromatography. Macromolecules. 2004;37:4304–12. https://doi.org/10.1021/ma030586k.
2. Fekete S, Beck A, Veuthey JL, Guillarme D. Theory and practice of size exclusion chromatography for the analysis of protein aggregates. J Pharm Biomed Anal. 2014;101:161–73.
3. O'Fágáin C, Cummins PM, O'Connor BF, et al. Gel-filtration chromatography. In: Methods in molecular biology (Clifton, NJ). Amsterdam: Nature Publishing Group; 2011. p. 25–33.
4. Afeyan NB, Fulton SP, Gordon NF, et al. Perfusion chromatography: an approach to purifying biomolecules. Bio/Technology. 1990;8:203–6. https://doi.org/10.1038/nbt0390-203.
5. Granath KA, Kvist BE. Molecular weight distribution analysis by gel chromatography on sephadex. J Chromatogr A. 1967;28:69–81. https://doi.org/10.1016/s0021-9673(01)85930-6.
6. Gritti F, Guiochon G. The current revolution in column technology: how it began, where is it going? J Chromatogr A. 2012;1228:2–19. https://doi.org/10.1016/j.chroma.2011.07.014.
7. Jerker P, Per F. Gel filtration: a method for desalting and group saparation. Nature. 1959;183:1657–9.
8. Erickson HP. Size and shape of protein molecules at the nanometer level determined by sedimentation, gel filtration, and electron microscopy. Biol Proc Online. 2009;11:32–51.
9. Madadlou A, O'Sullivan S, Sheehan D. Fast protein liquid chromatography. Methods Mol Biol. 2011;681:439–47. https://doi.org/10.1007/978-1-60761-913-0_25.

10. Yashin YI, Yashin AY. Analytical chromatography. Methods, instrumentation and applications. Russ Chem Rev. 2006;75:329–40. https://doi.org/10.1070/rc2006v075n04abeh003607.
11. Jungbauer A. Preparative chromatography of biomolecules. J Chromatogr A. 1993;639:3–16.
12. Stanton P. Gel filtration chromatography. In: HPLC of peptides and proteins. Totowa, NJ: Humana Press; 2004. p. 55–74.
13. Bai Y. Detecting protein-protein interactions by gel filtration chromatography. In: Protein-protein interactions: methods and applications. 2nd ed. New York, NY: Humana Press; 2015. p. 223–32.
14. Barth HG. High-performance size-exclusion chromatography of hydrolyzed plant proteins. Anal Biochem. 1982;124:191–200. https://doi.org/10.1016/0003-2697(82)90237-8.
15. Pfannkoch E, Lu KC, Regnier FE, Barth HG. Characterization of some commercial high performance size-exclusion chromatography columns for water-soluble polymers. J Chromatogr Sci. 1980;18:430–41. https://doi.org/10.1093/chromsci/18.9.430.
16. Bruessau R. General characterization of gel-permeation chromatography columns. In: Column handbook for size exclusion chromatography. Netherland: Elsevier; 1999. p. 429–43.
17. Stellwagen E. [25] Gel filtration. Methods Enzymol. 1990;182:317–28. https://doi.org/10.1016/0076-6879(90)82027-Y.
18. Andrews P. Estimation of molecular weights of proteins by gel filtration. Nature. 1962;196:36–9.
19. Andrews P. The gel-filtration behaviour of proteins related to their molecular weights over a wide range. Biochem J. 1965;96:595–606.
20. Chaganti LK, Kuppili RR, Bose K. Intricate structural coordination and domain plasticity regulate activity of serine protease HtrA2. FASEB J. 2013;27:3054–66. https://doi.org/10.1096/fj.13-227256.
21. Kummari R, Dutta S, Chaganti LK, Bose K. Discerning the mechanism of action of HtrA4: a serine protease implicated in the cell death pathway. Biochem J. 2019;476:1445–63. https://doi.org/10.1042/BCJ20190224.
22. Helmerhorst E, Stokes GB. Microcentrifuge desalting: a rapid, quantitative method for desalting small amounts of protein. Anal Biochem. 1980;104:130–5. https://doi.org/10.1016/0003-2697(80)90287-0.
23. Fritz H, Trautschold I, Haendle H, Werle E. Chemistry and biochemistry of proteinase inhibitors from mammalian tissues. Ann N Y Acad Sci. 1968;146:400–12. https://doi.org/10.1111/j.1749-6632.1968.tb20301.x.
24. Bassett EW, Beiser SM, Tanenbaum SW. Purification of antibody to Galactosyl-protein conjugates. Science. 1961;133:1475–6. https://doi.org/10.1126/science.133.3463.1475.
25. Fee CJ. Size-exclusion reaction chromatography (SERC): a new technique for protein PEGylation. Biotechnol Bioeng. 2003;82:200–6. https://doi.org/10.1002/bit.10561.
26. Lindner EB, Elmqvist A, Porath J. Gel filtration as a method for purification of protein-bound peptides exemplified by oxytocin and vasopressin. Nature. 1959;184:1565–6. https://doi.org/10.1038/1841565b0.
27. Flodin P, Killander J. Fractionation of human-serum proteins by gel filtration. Biochim Biophys Acta. 1962;63:403–10. https://doi.org/10.1016/0006-3002(62)90104-X.
28. Kröber T, Wolff MW, Hundt B, et al. Continuous purification of influenza virus using simulated moving bed chromatography. J Chromatogr A. 2013;1307:99–110. https://doi.org/10.1016/J.CHROMA.2013.07.081.
29. Haller W. Chromatography on glass of controlled pore size. Nature. 1965;206:693–6. https://doi.org/10.1038/206693a0.
30. González A, Gómez-Márquez J. Purification of bacteriophage DNA by gel filtration chromatography. Genet Anal Tech Appl. 1990;7:2–4.
31. Kondo T, Mukai M, Kondo Y. Rapid isolation of plasmid DNA by LiCl-ethidium bromide treatment and gel filtration. Anal Biochem. 1991;198:30–5.

32. Van Eijk HG, Monfoort CH, Witte JJ, Westenbrink HGK. Isolation and characterization of some Bence-Jones proteins. Biochim Biophys Acta. 1962;63:537–9. https://doi.org/10.1016/0006-3002(62)90127-0.
33. Epstein WV, Tan M. Gamma globulin interactions in the sera of 2 patients with rheumatoid arthritis studied by gel filtration. J Lab Clin Med. 1962;60:125–37. https://doi.org/10.5555/URI:PII:0022214362900537.
34. Roskes SD, Thompson TE. A simple molecular sieve technique for detecting macroglobulinemia. Clin Chim Acta. 1963;8:489–96. https://doi.org/10.1016/0009-8981(63)90094-9.
35. Osserman EF, Lawlor D. Immunoelectrophoretic characterization of the serum and urinary proteins in plasma cell myeloma and waldenström's macroglobulinemia. Ann N Y Acad Sci. 2006;94:93–109. https://doi.org/10.1111/j.1749-6632.1961.tb35535.x.
36. White WF, Barlow GH, Mozen MM. The isolation and characterization of plasminogen activators (Urokinase) from human urine. Biochemistry. 1966;5:2160–9. https://doi.org/10.1021/bi00871a003.
37. Al-Ghobashy MA, Mostafa MM, Abed HS, et al. Correlation between dynamic light scattering and size exclusion high performance liquid chromatography for monitoring the effect of pH on stability of biopharmaceuticals. J Chromatogr B. 2017;1060:1–9. https://doi.org/10.1016/J.JCHROMB.2017.05.029.

Protein Purification by Reversed Phase Chromatography and Hydrophobic Interaction Chromatography

9

Rucha Kulkarni and Kakoli Bose

Abstract

With the increasing analysis and studies on purified proteins and peptides with respect to their structure and function, the need for high resolution separation of these biomolecules becomes important. Although affinity, ion exchange and gel filtration chromatography techniques are routinely used in protein purification, sometimes the protein (mainly highly hydrophobic) fails to attain the required purity for its further characterization. Reversed Phase Chromatography (RPC) and Hydrophobic Interaction Chromatography (HIC) are two recently developed techniques that make use of protein surface hydrophobicity as a parameter for their separation. RPC and HIC thus aid in the purification of hydrophobic proteins that are otherwise difficult to purify and characterize. These techniques provide great flexibility in the choice of separation buffers, thus helping a large variety of analytes to be separated. RPC and HIC have gained popularity because of the wide variety of nonpolar analytes that can be separated based on the principle of surface adsorption. This chapter elaborates more on the principle of these techniques, their protocols and applications.

Keywords

Reversed phase chromatography · Hydrophobic interaction chromatography · Adsorption · Hydrophobicity · Protein purification

R. Kulkarni · K. Bose (✉)
Integrated Biophysics and Structural Biology Lab, Advanced Centre for Treatment, Research and Education in Cancer (ACTREC), Tata Memorial Centre, Mumbai, India

Homi Bhabha National Institute, BARC Training School Complex, Mumbai, India
e-mail: kbose@actrec.gov.in

K. Bose (ed.), *Textbook on Cloning, Expression and Purification of Recombinant Proteins*, https://doi.org/10.1007/978-981-16-4987-5_9

9.1 Introduction

Purification of the protein is an indispensable prerequisite for its structural and functional characterization [1]. This process involves a series of steps that begin from cloning in a vector and transformation in a suitable host system as discussed in Chap. 2. With the increase in number of recombinantly expressed proteins for basic research for clinical and industrial applications, the complexity pertaining to their purifications increase manifold. Post bacterial cell lysis, usually, a simple two or sometimes three step protocol is followed for purification of recombinant proteins that include affinity chromatography (His_6, GST, or MBP tags), ion exchange (cation/anion) or IEX chromatography, and the final polishing by gel filtration chromatography (GFC) that have been elaborated in Chaps. 6, 7, and 8. Although the caveat of obtaining >95% pure recombinant protein for structural and functional studies in a laboratory setup is generally achieved through these steps, certain proteins do require special attention. Separation of highly hydrophobic proteins [2], using the abovementioned tools, often does not produce expected results. Such proteins have to be separated by modifying certain parameters such as using a stationary and mobile phase that binds and elutes proteins with large hydrophobic patches.

With rapid advancement in chromatography, various properties of proteins have been continuously harnessed so as to devise distinct purification strategies for different genre of proteins. Two of the recently developed chromatographic methods for such proteins include RPC and HIC that utilize surface hydrophobicity of the proteins for their separation [3].

In this chapter, we will therefore focus on using the principle of surface hydrophobicity of the protein as the defining factor for its separation and purification [4]. The chromatographic methods discussed in this chapter include reversed phase chromatography (RPC) and hydrophobic interaction chromatography (HIC). However, it is important to note that these are not stand-alone techniques and mostly give desired level of purity when used in combination with affinity, IEC, and/or GFC.

The term reversed phased stems from the fact that there is a reversal in the stationary phase, which is hydrophobic unlike the other conventional chromatographic techniques. The hydrophilic mobile phase enables the apolar protein molecules to get adsorbed onto the stationary phase while allowing the other impurities to elute out of the column [5]. This phenomenon is termed as surface adsorption chromatography which will be described in detail in this chapter, along with protocols pertaining to purification of recombinant proteins and troubleshooting tips.

9.2 Principle of Surface Adsorption Chromatography

Adsorption generally refers to a surface phenomenon in which molecules from gas, liquid, or solid adhere to a solid surface. This adhesion creates a film of the molecules often called the adsorbate onto the solid surface or adsorbent. However, in chromatographic separation, the adsorbent is the stationary phase and the adsorbate is the mixture of macromolecules or analytes that adhere to the stationary phase and get separated. Since this chapter focuses on recombinant protein purification, the analytes here are protein molecules. The separation is based on the principle of competitive replacement of analytes by the mobile phase (that is elution buffer) [6]. Few important factors on which adsorption of protein molecules depend are:

1. The binding strength between the protein and the stationary phase
2. The surface area of the stationary phase resin
3. Displacement of the mobile phase by the bound protein
4. The binding strength between stationary phase (matrix) and the mobile phase

The ability or the strength with which the mobile phase can elute the bound protein sample is often termed as the eluotropic strength [7]. For example, a mobile phase with a high eluotropic strength will elute the protein more quickly. Thus, depending on the stationary phase used, a suitable mobile phase with a good eluotropic strength must be chosen.

The different types of stationary phase support generally used in adsorption chromatography include:

1. Polar acidic support (silica)
2. Polar basic support (alumina)
3. Nonpolar support (polystyrene)

The major distinguishing factor between normal phase and reversed phase chromatography is that the former uses polar support as stationary phase while the latter uses nonpolar supports. Thus, solutes (proteins) are retained based on their polarity.

This chapter would elaborate more on RPC and HIC that are based on the principle of adsorption chromatography using nonpolar matrix as the stationary phase to separate hydrophobic (nonpolar) proteins.

9.3 Reversed Phase Chromatography

As mentioned above, the stationary phase or the ligands on the matrix are hydrophobic or nonpolar in reversed phase chromatography (RPC) [8]. This technique is specifically important for certain hydrophobic proteins that cannot be efficiently purified using the other three chromatographic techniques described in the previous chapters [9, 10]. This technique uses very stringent hydrophobic medium for stronger interactions and hence can separate macromolecules or polypeptides of very small size difference [11, 12]. However, this high resolution is often achieved at the expense of loss of native conformation and activity of proteins, since organic solvents denature proteins and hence limits its use in studies where a functional protein or enzyme is not required [13].

9.3.1 Principle of Protein Separation in RPC

The separation in this type of chromatography takes place based on the hydrophobic interaction between the nonpolar solute molecules in the mobile phase and the hydrophobic ligands on the stationary phase matrix as illustrated in Fig. 9.1. This technique is based on the partitioning principle to separate the components of the mixture. The partitioning of the solute is dependent on many factors like binding properties of the stationary matrix, hydrophobic character of the solute, and the overall composition of the mobile phase. The initial conditions for chromatography are such that they favor adsorption of the hydrophobic components onto the matrix followed by a change in conditions that favor desorption of the solute molecules back into the mobile phase [14, 15].

For eluting the proteins adhered to the matrix, normal phase chromatography uses the isocratic elution principle, where the elution buffer is consistent for a particular elution profile [16]. However, generally in case of reversed phase chromatography, gradient elution is used. Under aqueous conditions, strong adsorption of proteins on the reversed phase matrix takes place, thus attaching the proteins to the matrix. At the time of desorption, a very narrow range of organic solvent is required to elute the proteins adhered to the matrix. Since any biological mixture would contain protein molecules with a diverse affinity for the matrix, a range of different concentrations of the organic elution buffer is used for eluting the bound protein in a gradient manner.

Some of the powerful attributes of RPC such as high resolving capacity, reproducibility, and recovery of the proteins has made it a popular tool in separation of proteins and peptides for certain specific studies. Moreover, RPC allows the separation conditions to be adjusted such that the researcher can either choose to bind the biomolecules that are of interest and allow other nonspecific molecules to pass through or vice versa. However, the former method is preferred for better yield. Given its efficiency in separation of peptides [17], the major use of this technique lies in its analytical applications rather than preparative [18]. Nevertheless, it is useful in

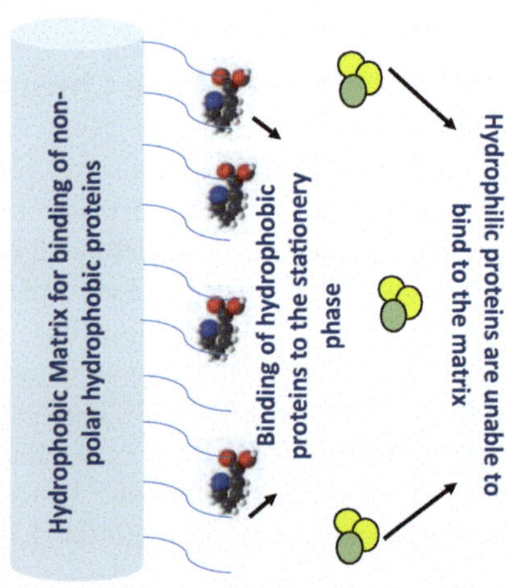

Fig. 9.1 Diagrammatic representation of reversed phase chromatography matrix. The figure represents the binding of nonpolar hydrophobic protein molecules to the reversed phased matrix. As shown in the figure, the hydrophilic or polar protein molecules that cannot bind to the matrix are washed out of the column

obtaining highly purified proteins in preparative scale for studies as well as offers separation of chemically and structurally similar proteins.

9.3.2 Overview of Steps in RPC

The first and foremost step in purification involves overexpressing the protein of interest in a suitable host system. The protocol standardization for RPC depends on the nature of protein and how the biomolecule would be used after purification.

9.3.2.1 Column Equilibration

As observed and followed for every other chromatographic technique, the column equilibration constitutes the first step in RPC as depicted in Fig. 9.2. The packed column is equilibrated with a reversed phase medium under suitable conditions of pH, ionic strength, and polarity (hydrophobicity). Certain organic solvents can be used to refine the polarity of the mobile phase, ensuring that it is low enough so as to dissolve partially hydrophobic molecules but at the same time high enough for binding of the hydrophobic molecules to the hydrophobic matrix [19].

9.3.2.2 Binding of the Extract on Column (Capture)

In the next step, the mixture of proteins or the sample solute that needs to be separated is allowed to bind on to the equilibrated matrix (Fig. 9.3). This is mainly done to isolate, concentrate, and stabilize the protein to be separated. The solvent used to equilibrate the packed column and the one used for dissolving the sample to be loaded is usually the same. The sample is allowed to bind on to the column by adjusting the flow rate such that optimum binding occurs. Washes with the column equilibration buffer are given for removing the unbound solute molecules. Sometimes, addition of ion-pairing agents such as mild acids (0.1% Trifluoroacetic acid or TFA) that enhance the hydrophobic interactions between the protein sample and the matrix are used in the buffer. They do so by binding to the charged groups on proteins and thus suppressing their effect on hydrophobic interactions.

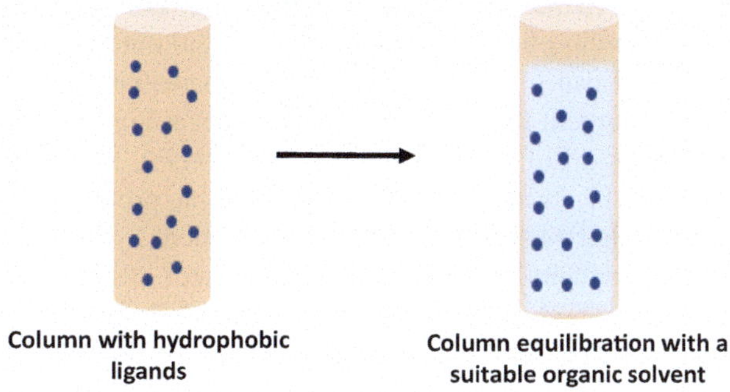

Fig. 9.2 The figure represents the column equilibration step in RPC. The matrix is allowed to equilibrate in a buffer of suitable pH, ionic strength, and hydrophobicity in which the protein will be purified

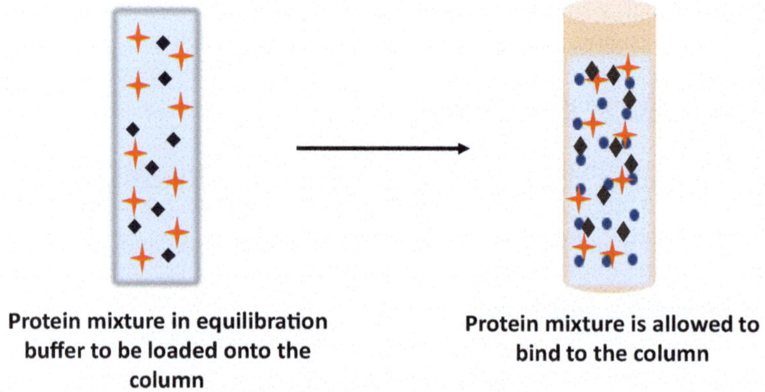

Fig. 9.3 The figure represents binding step in RPC in which the protein of interest is allowed to bind onto the resin

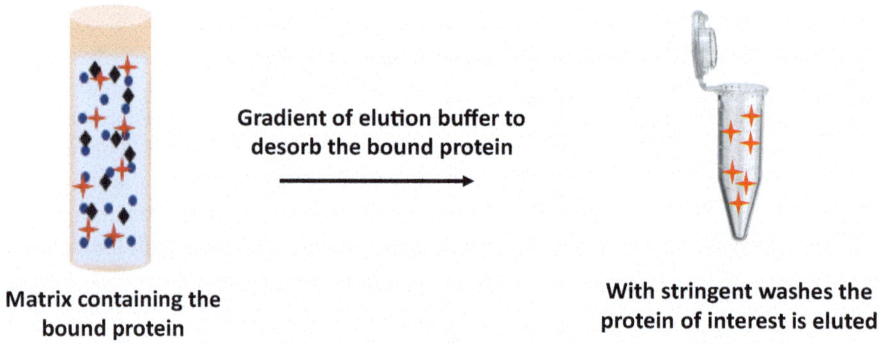

Fig. 9.4 The figure represents hydrophobic protein elution step in RPC where the bound protein is eluted using an elution buffer of increasing organic solvent content

9.3.2.3 Elution of Protein Molecules (Desorption)

After allowing the solute molecules to bind onto the matrix, they are then desorbed (Fig. 9.4) using a gradient of mobile phase with increasing percentage of organic solvent content that gradually allows increase in the stringency of washes and helps in desorption of bound solutes (here protein), from the matrix depending on their hydrophobic content. The less hydrophobic proteins would elute out of the column in the earlier washes as compared to the more hydrophobic proteins. Usually, the pH of the gradient of elution buffers is kept constant. The eluants from different gradients of elution buffers are collected in separate tubes and later run on an SDS-PAGE to check for purity of proteins of interest in the eluted fractions.

9.3.2.4 Cleaning of Column and Storage

The next step involves the cleaning of the matrix for reusing in the next round of separation. This is achieved by washing the column with 100% organic solvent so that any tightly bound hydrophobic molecules will be eluted, thus removing all bound biomolecules before its reuse. This is followed by re-equilibrating the column with the initially used solvent for storage until next set of purifications.

A reversed phase chromatography medium usually comprises hydrophobic ligands attached to a chemically and mechanically stable matrix. The commercially available base matrix is either made of silica or polystyrene (Table 9.1) [11], the latter being more robust and stable at higher pH conditions [20]. The diameter of the bead determines its particle size, which is important for its ability towards efficient separation. The larger particle size media are more apt for large scale preparative scale purification.

9.3.3 Protocol for Protein Purification Using RPC

Step 1: Preparation for Purification

This step is required to decide the buffer system and ion-pairing agents that would be used so that the protein binds onto the column.

Step 2: Column Equilibration

The most widely used columns for RPC are C8 and C18, where the numbers denote the length of alkyl chains attached to the stationary phase. The longer the alkyl chain, more is the hydrophobicity. Thus, C18 is the most preferred column for RPC where long retention times are needed as it has the highest level of hydrophobicity. The protocol for RPC column equilibration mentioned here is for C18 [21, 22].

Table 9.1 A detailed summary of the variety commercially available media (matrix) for reversed phase chromatography with its specifications and applications in protein purification

Column name	Resin	Use
1. Sephasil C4 (for protein) Sephasil C8 and C18 (for peptides)	Silica (5 μm bead size)	a. Purification of recombinant proteins and synthetic peptides b. High resolution analysis
2. Sephasil C4 (for protein) Sephasil C8 and C18 (for peptides) *(Numbers denote length of alkyl chains attached to the resin)*	Silica (12 μm bead size)	a. Purification of proteins, nucleotides and peptides
3. SOURCE 5, SOURCE 15, SOURCE 30 (*Numbers denote the bead size of the resin*)	Polystyrene/ divinylbenzene	a. These resins are very stable at high pH b. Used for purification of synthetic peptides, oligonucleotides and recombinant proteins c. Resin stays stable to some extent even with fluctuations in flow rate and pressure

Conditioning or equilibration of the column is an important step prior to loading the protein to enable optimal protein binding. The equilibration buffer is the same as the one used for dissolving protein sample.

Composition of equilibration buffer:

Buffer A: 0.1% TFA in water, pH 2.0 (the pH of the buffer and the concentration of organic modifier used is usually kept low to enhance the hydrophobic interactions between the protein and the matrix since the charged groups on the protein are masked in acidic conditions so that they don't interfere in the hydrophobic interactions).

Buffer B: NaOH 10 mM in water, pH 12 (higher pH ensures that all the hydrophobic and tightly bound impurities in the form of proteins or peptides elute before loading the column for next round of protein purification).

Steps for equilibration are as follows:

- Give three column volumes (CV) of washes with buffer B.
- Run a gradient from 100% buffer B to 100% buffer A.
- Equilibrate the column in buffer A until all signals on the monitor are stable.

Note: The solvents and the additives that constitute the mobile phase should be of highest purity often referred to as HPLC (high pressure liquid chromatography) grade. Before adding the mobile phase onto the column, it should be degassed under vacuum to prevent gas formation. If any particulate solid additive is to be added to the mobile phase, it must first be filtered to prevent clogging of the column.

Step 3: Loading Protein onto the Matrix

Before loading the sample onto the RPC column, it should be free of any particulate matter to avoid column blockage. Dissolve the protein sample in abovementioned equilibration buffer A. Centrifuge the samples at 10,000 × *g* for 10 min or filter through 0.22 or 0.45 μm filter. The filter should be resistant to the solvents used in buffers. After filtering the sample, load it onto the column immediately to prevent any side reaction such as oxidation.

Step 4: Elution

Most of the RPC protocols contain two elution buffer systems eluent A and eluent B. Eluent A consists of a lesser concentration of the organic modifier about 5% and eluent B consists of higher organic modifier concentration about 80%. Both the eluents should contain at least 0.1% of an acid which would act as ion-pairing agent [23].

Elution buffer for silica-based matrix: (mostly used for proteins and peptides):

Eluent A: 0.1% TFA in water, 5% acetonitrile.

Eluent B: 0.1% TFA in acetonitrile (maximum 80%).

Eluent A is added to the column first followed by eluent B in which the protein is eluted. The buffers need to be filtered through a 0.22 μm filter to remove particulates and degassed under vacuum before use. Composition of elution buffers used is dependent on the choice of base matrix [24].

Eluted protein is then run on SDS-PAGE for determination of purity. Clean fractions are either used immediately for studies that are previously designed or are flash-frozen for long-term storage.

Step 5: Column Regeneration and Storage

Common protocol followed for cleaning a reversed phase column is as follows:

- At a low flow rate equilibrate the column with mobile phase A, which is 0.1% TFA in water.
- Run a gradient through the column from 100% mobile phase A to 100% mobile phase B (mobile phase B is 0.1% TFA in 2-propanol).
- Equilibrate the column with 100% mobile phase B and then bring it back to 100% mobile phase A.
- Equilibrate the column with the mobile phase buffer used for next round of purification. If the mobile phase used for purification has a composition which differs significantly from mobile phase A in which the column is present (**step 3**), the new mobile phase must be introduced into the column slowly in a linear gradient.

The columns are generally stored in 20% ethanol at 4 °C. Seal the column well to prevent drying of the column.

9.3.4 Some Important Factors that Govern Optimum Separation and Resolution in RPC

9.3.4.1 Length of the Separation Column

Unlike smaller peptides, the separation of larger proteins using reversed phase chromatography is not much dependent on the length of the column. Since desorption of large proteins takes place in a very narrow concentration range of organic solvent used for elution, the column length does not make any difference in improving the resolution of the reversed phase chromatography [25]. Moreover, use of gradients of the organic solvent as mobile phase further reduces the dependency on column length for higher resolution.

9.3.4.2 Flow Rate

Flow rate is often thought to have an effect on resolution of separated molecules in column chromatography. For reversed phase separation, lower flow rates decrease the resolution for larger molecules, although opposite effect is observed for separation of smaller molecules. This might be due to occurrence of longitudinal diffusion of larger molecules if they are retained for a larger amount of time due to lower flow rates [26].

9.3.4.3 Temperature

Temperature has an effect on the viscosity of the mobile phase used for separation; the latter has more profound effect on smaller peptides compared to protein molecules.

9.3.4.4 Mobile Phase

The mobile phase or buffer in reversed phase chromatography comprises commonly used organic solvents/modifiers such as acetonitrile and methanol. Isopropanol is used at times to enhance the elution strength. Tetrahydrofuran is another type of organic modifier used for elution. In addition, the mobile phase usually contains low levels of trifluoroacetic or phosphoric acid. The role of the acids is to protonate the matrix and form ion pairs with the eluates. Ion-pairing agents such as perchlorate can be used at neutral pH.

Any organic solvent used should not absorb in the UV range since elution of sample is majorly detected by UV absorbance. Generally, acetonitrile or ethanol in water serves as the best options since they do not interfere with the absorption measurements of proteins and peptides [27].

9.3.4.5 Gradient Elution

Gradient refers to a range of different concentrations of the organic solvent used as the mobile phase. The gradient of the organic modifier is such that at the start of the gradient, the organic modifier is highly polar (more aqueous content) and proceeds towards a less polar (less aqueous content) state. A broad gradient is usually preferred when the parameters for reversed phase chromatography are to be standardized [28].

9.3.4.6 Retention of Proteins

The retention strength increases roughly with the size and the hydrophobicity of a given substance. In case of proteins and larger polypeptides, the prediction of the retention behavior becomes difficult due to the complexity of their three-dimensional structures. Moreover, under the conditions of RPC with its acidic and hydro-organic mobile phases, many proteins tend to denature and unfold either partially or completely when adsorbed on the resin. Oxidation, deamidation, aggregation, and fractionation are also possible. The use of RPC in preparative work commonly requires refolding of the protein into its native configuration after separation. Although challenging, this has been shown to be possible for a number of peptides and smaller proteins, which are of interest to the pharmaceutical industry such as human insulin and human growth hormone (hGH). Moreover, studies have shown that use of salts in the elution buffer stabilize the native conformation of the proteins of interest [13]. Nevertheless, purifying larger proteins in their native forms by RPC is still a big challenge and needs further technical advancements in the field.

9.3.5 Different Uses of RPC

9.3.5.1 Desalting and Protein Concentration

The major drawback of chromatographic techniques like gel filtration is sample dilution. However, with RPC that relies on the principle of adsorption, the protein gets concentrated while being eluted. Furthermore, sample concentration gets coupled with buffer exchange or desalting and the concentrated macromolecules or elutes are collected in organic solvents of low polarity [29].

9.3.5.2 High Resolution Separations

It is often possible to obtain high resolution separation with RPC due to its robustness. Such applications include separation of individual peptide fragments from an enzymatically digested protein sample. Certain parameters need special attention such as column flow rate, column length, and the mobile phase composition to achieve such high resolution [18].

9.3.5.3 Large Scale Preparative Purification

High resolution separation combined with scaling up is used for large scale protein purifications. The purification is first standardized on a smaller bead size matrix and then scaled up retaining the selectivity of the matrix by using larger bead size [30].

9.3.6 Applications

RPC has found applications in many analytical and preparative biochemical separations of hydrophobic biomolecules with good quality resolution and recovery. Thus, it is has become one of the most routinely applied separation techniques for specific needs in protein laboratories today. It is mostly used as a polishing step when all the conventional chromatographic techniques such as affinity, IEC, and GFC do not produce desired level of purity for certain proteins with high hydrophobic contents.

9.4 Hydrophobic Interaction Chromatography

As discussed above, RPC helps in the separation and purification of hydrophobic (nonpolar) moieties based on hydrophobic interactions between the matrix and the analyte. One more technique that makes use of a similar principle for separation of hydrophobic protein molecules is hydrophobic interaction chromatography (HIC). The main difference between HIC and RPC is that, the latter uses less stringent conditions and binding affinity between the stationary phase and the macromolecules is less.

9.4.1 Principle of Hydrophobic Interactions

Protein folding is a very organized and ordered process in which they fold by minimizing the exposure of nonpolar (hydrophobic) residues to water [31, 32]. In this process of minimizing the exposure to water, the protein molecules tend to associate with each other and aggregate in order to achieve thermodynamic stability that is a state of lower energy [33]. The three-dimensional structure of a protein is the result of intra- and intermolecular interactions between their well-formed secondary structures. The sum of hydrophobic residues in the buried and exposed patches determines the degree of hydrophobicity of protein. Since the number of hydrophobic amino acids varies, their distribution and strength of hydrophobicity are the characteristic features of each protein, and hence specific separation is performed with hydrophobic supports or matrices [34].

9.4.2 Difference between RPC and HIC

Unlike RPC, where solutes are adsorbed and separated in the nonpolar stationary liquid phase, the solutes (proteins) are adsorbed and separated on a stationary solid phase in HIC [35]. Furthermore, the surface of an RPC medium is more hydrophobic compared to HIC. HIC works better in more polar and less denaturing environment (organic solvents) as compared to RPC. The consequence is that, eluted proteins are mostly denatured in RPC. However, in HIC, care is taken to maintain native conformations of proteins with minimum structural damage and retention of activity of enzymes, thus making it more popular in research laboratories compared to RPC.

9.4.3 Principle of HIC

The term Hydrophobic Interaction Chromatography was coined by Hjertén et al. in 1973 [36]. He also described this technique as a salt-mediated separation of proteins. The separation is largely based on the hydrophobic interactions between the sample protein and the adsorbent matrix. Addition of salts leads to engagement of water molecules in solvating predominantly the salt ions, thus significantly reducing the number of water molecules available for interacting with the hydrophilic amino acids as shown in Fig. 9.5.

The principle behind interaction of hydrophobic ligands adsorbed onto the stationary phase and the sample hydrophobic protein is similar to the interaction between two hydrophobic molecules (proteins) leading to an increase in the overall entropy as shown in Fig. 9.6 [37].

Thus, addition of salts would either favor or disrupt hydrophobic interactions depending on their influence on the polarity of the solvent. The influence of these interactions often follows the Hofmeister series [5]. Hydrophobic interaction promoting anions are listed in the descending order starting from the anion with maximum strength to increase hydrophobic interactions. Ions that promote

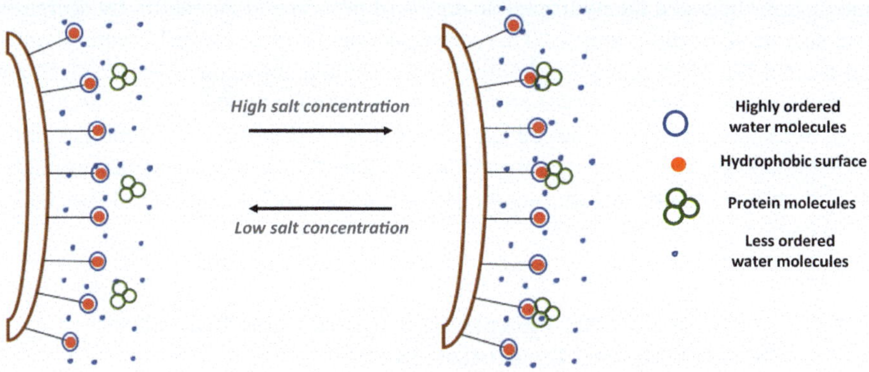

Fig. 9.5 Effect of salts on hydrophobic interactions: Water molecules shield the hydrophobic surfaces of protein molecules as well as the matrix ligands. Thus, hydrophobic molecules merge together so as to minimize the area exposed to water, thereby maximizing the entropy. Addition of salts enhances the hydrophobic interactions as explained above and control the equilibrium of the hydrophobic interaction

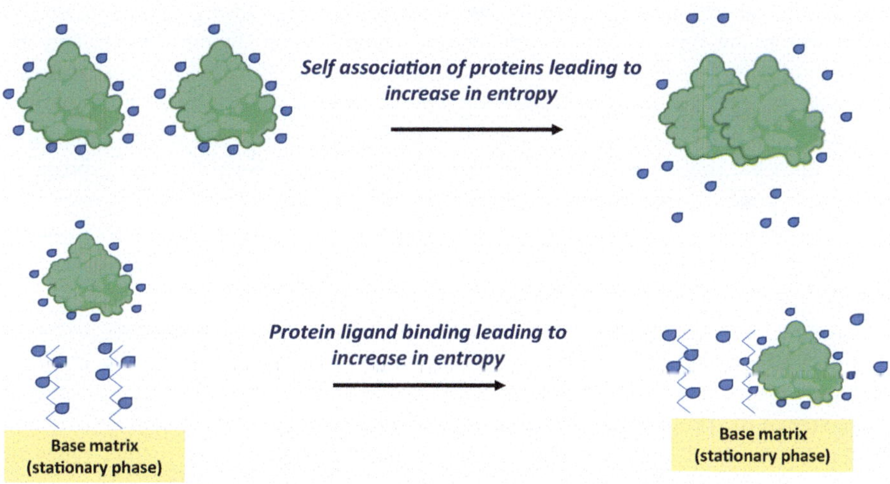

Fig. 9.6 Diagrammatic representation showing that the association of a hydrophobic ligand with hydrophobic matrix is similar to the association between two hydrophobic proteins driven by an increase in the entropy

hydrophobic interactions are termed as lyotropes and those that disrupt these interactions are termed as chaotropes [38]. The best examples of lyotropic salts that facilitate hydrophobic interaction in aqueous solution are ammonium sulfate and sodium chloride. These salts are commonly added when protein purification is done using HIC [5].

A protein sample is applied onto the HIC column matrix. The sample application takes place under high salt conditions. High salt conditions aid in promoting interactions between the hydrophobic region of protein and the matrix by decreasing the solvation of the sample. This phenomenon is often called "salting out of proteins" [39]. More the hydrophobicity of the sample, lesser the salt needed. Elution buffer contains a lower salt concentration to elute the sample. Organic modifiers or mild detergents are sometimes added to the elution buffer to help in efficient elution of the bound protein. Organic solvents help in the elution of proteins that are tightly bound onto the matrix by altering the polarity of the mobile phase and thus help in weakening the interaction [40].

9.4.4 Some Important Factors that Govern the Optimum Separation and Resolution in HIC

9.4.4.1 Stationary Phase

In HIC, hydrophobic ligands are immobilized on stationary phase that is known as the *base matrix*. Common examples of base matrices for stationary phase include agarose, polystyrene-divinyl benzene, methacrylate, and silica. Ligands with varying degrees of hydrophobicity are commercially available that are coupled to the base matrix for adsorbing the protein. Hydrophobicity in stationary phases differs according to the type of ligand used, ligand chain length, ligand density, and the type of solid support/matrix [5]. For example, short n-alkyl hydrocarbons as shown in Table 9.2 are the most preferred ligands as they possess a pure hydrophobic character that helps in protein separation at a high resolution. Aryls (phenyl groups added onto the matrix as ligands) are sometimes used as ligands as they promote both hydrophobic and aromatic interactions. They lack overall charge, do not exhibit pure hydrophobic character, and are a mixture of hydrophobic and aromatic characters. The hydrophobicity of stationary phase and the strength of interaction increases with the increase in number of n-alkyl residues [5].

Another factor affecting the strength of hydrophobic interactions between immobilized ligand and protein is ligand density, also known as *ligand loading*. Although protein surface hydrophobicity is considered the most influencing factor, it is the size of the protein that determines the relationship between ligand density and protein retention. Ligand density can be manipulated to obtain distinct types of stationary phases with different selectivity and binding capacities [41, 42].

Table 9.2 List of popular commercially used matrices for HIC along with the ligands that can be conjugated to the matrices for enhancing hydrophobic interactions [5]

Matrix name	Base matrix material	Ligands used for attachment
SOURCE 15	Polystyrene/ divinylbenzene	Phenyl group, Butyl groups, Octyl group, Ether group, Isopropyl group
Sepharose high performance	6% Agarose	
Sepharose fast flow	6% Agarose	
Sepharose fast flow	4% Agarose	

9.4.4.2 Matrix

Matrix of the HIC column contributes significantly towards the degree of hydrophobic interaction [43]. An inert matrix with high binding capacity ensures minimum nonspecific interactions with the sample. Furthermore, a matrix with high physical and chemical stability stays unaltered under extreme conditions such as high salt concentrations. The most widely used supports are hydrophilic carbohydrates (e.g., cross-linked agarose), silica, or synthetic copolymer materials (polystyrene/divinylbenzene). Using the same type of ligand, the selectivity of the stationary phases can bring about a change in function and use of the different types of supports. The initial HIC matrix was made of polysaccharides such as agarose, cellulose, and dextran. These matrices were extensively used for conventional HIC but the drawback of using these matrices was that they could withstand only low pressure. Nowadays, inorganic supports like silica- or polymer-based HIC columns are generally used. Some commercially available solid supports as well as the latest silica-based porous particles are available with alkyl-amide or butyl phases [43]. State-of-the-art columns can now withstand pressure drop of up to 100–400 bar. Table 9.2 below summarizes the commercially available matrices for HIC.

9.4.4.3 Mobile Phase

Characteristics of mobile phase such as type and concentration of salt used, pH, temperature, and presence of additives greatly influence the selectivity and retention in HIC. Different types of salts have the ability to increase hydrophobic interactions depending on their type and concentration. The addition of these salts increases the protein-ligand interaction in HIC. The effect of different salt concentration on protein precipitation follows the order in the lyotropic series, also known as Hofmeister series (Table 9.3) for protein precipitation from aqueous solution [5]. In this series, salts are arranged based on their decreasing ability of protein precipitation. In general, sodium, potassium, and ammonium sulfates produce high precipitation effect by increasing ligand protein interactions as well as stabilize protein structure. Apart from salt type, another important parameter that influences the HIC retention is salt concentration. The binding of protein onto HIC matrix will increase linearly at a specific salt concentration and will continue to increase in an exponential manner at higher concentrations. On the basis of their position in lyotropic series, different salt concentration is required to produce same precipitation effect. Stronger salts are required in low concentration (1–1.5 M), whereas weaker salts are required at higher concentrations (3–5 M) to maintain same retention. The salts used in HIC must have high solubility in order to avoid salt precipitation when used at higher concentrations.

Table 9.3 List of anions and cations in the Hofmeister series that act as lyotropes and chaotropes [5]

Anions	$PO_4^{3-} > SO_4^{2-} > CH3COO^- > Cl^- > Br^- > NO_3^- > CLO_4^- > I^- > SCN^-$
Cations	$NH_4^+ > Rb^+ > K^+ > Na^+ > Cs^+ > li^+ > Mg^{2+} > Ca^{2+} > Ba^{2+}$

In the table above, the anions and cations listed are in the order of decreasing strength of hydrophobic interactions (decreasing lyotropic strength). Thus, phosphate and ammonium ions have a strong tendency to promote hydrophobic interactions and are strong lyotropes. The lyotropic strength decreases from left to right for both cations and anions and thus the ions towards extreme right do not promote hydrophobic interactions effectively.

9.4.4.4 pH

The role of pH on protein retention in HIC is complex. Generally, the hydrophobic interactions between the ligand and proteins decrease with increase in pH (up to 9–10), which is due to increase in hydrophilicity resulting from altered protein charge. On the other hand, there is increase in hydrophobic interactions with decrease in pH. However, it is usually observed that retention of proteins in HIC changes drastically at pH range of 5–8.5, and this shift is protein dependent as the overall charge on every protein is distinct [44]. This suggests that this characteristic could be optimally harnessed for efficient protein purification by HIC.

9.4.4.5 Temperature

The binding of protein on the adsorbent in HIC is an entropy-driven process. By increasing the temperature, there is an increase in the interactions between the protein and the adsorbent. The strength of van der Waals forces that contribute to hydrophobic interactions increases with increasing temperature. However, an opposing effect can be seen because of the effect of temperature on the conformational state of different proteins and their solubility. So, the purification process that is developed at room temperature might not replicate in the cold room or vice versa. This parameter can also be modulated during elution step to achieve weaker interactions and separate protein under milder conditions without denaturation [45].

9.4.4.6 Additives

Addition of additives has various effects on the purification process. They not only improve protein solubility or modify its conformation but also aid in the elution process [46]. For example, adding a low concentration of water-miscible alcohol (ethanol), detergents (Triton-X-100), and chaotropic salts to the mobile phase results in weakening of protein-ligand interaction. The nonpolar region of alcohol and salts competes for the ligands, resulting in displacement of bound protein from the ligand. Chaotropic salts assist in decrease in retention of protein via disruption of the structure of protein. At higher concentration, these salts can denature protein or make them nonfunctional. Therefore, use of the chaotropes in the elution buffer is only done when there is quite strong hydrophobic interaction between protein and the ligand or during column cleaning.

9.4.5 Sample Preparation for HIC

For the target protein to bind onto the HIC adsorbent, the buffer that contains the protein sample must have a high salt concentration. The type of salt and its precise concentration should be standardized to prevent precipitation and obtain optimum yield [47]. Salt concentrations within the range of 0.5–2.0 M are widely used in HIC. The concentration of salt may be greater than 2.0 M only when a stronger binding is needed and the protein is stable in the presence of such high salt concentration [48].

Note: Use of extreme buffer pH should be avoided to maintain a stable interaction between protein and the adsorbent.

9.4.6 Overview of Important Steps in HIC

Basic steps of protein separation remain the same for RPC and HIC. Given below is a brief description of some important purification steps for HIC.

9.4.6.1 Equilibration of HIC Column

The buffer used for equilibration is the same in which the protein is kept for binding. The salt concentration and the pH of the buffer should be such that the protein binds to the column. The flow rate is maintained in a way that the adsorbent remains stable. The range of flow rates that can be used are often mentioned by the manufacturer's guidelines. The velocity of the buffer flowing through the column is usually maintained between 50 and 300 cm/h [5, 49].

9.4.6.2 Protein Loading

In this step, the protein is allowed to bind onto the adsorbent under optimal pH and salt concentrations that would promote hydrophobic interactions. The nonspecifically bound proteins are washed away in the buffer washes that are given before the protein is eluted due to their lesser affinity towards resin. 3–10 CVs of washes can be given to get rid of the impurities. The salt concentration used for washing steps is intermediate to that of the loading and the elution steps [5].

9.4.6.3 Elution of the Bound Protein

The routinely used method of elution comprises a buffer with a decreasing gradient of salt concentration for separation of the protein of interest from the impurities and subsequent elution from the resin (adsorbent). The salt concentration gradient helps in decreasing the affinity between the bound protein and the adsorbent, thus aiding in the elution step. Alternatively, in another approach, the impurities are allowed to wash away in the elution steps, while the protein of interest stays bound to the column [5]. Apart from this, use of organic solvents, chaotropic agents, and detergents is also possible as described below.

There are four ways of collecting the elutes from the column [5]:

1. Using a decreasing salt concentration gradient—such conditions help in decreasing the hydrophobic interactions between the bound protein and adsorbent and facilitates desorption of the protein.
2. Use of organic solvents—such solvents often termed as the organic modifiers (e.g., acetonitrile and isopropanol) change the polarity of the solvent, thus helping in the elution of the bound protein.
3. Use of chaotropic salt is another way to disrupt hydrophobic interactions.
4. Use of detergents—these have been widely used specifically to purify membrane proteins using HIC. Detergents are known as protein displacers and help in eluting the protein from the column.

It is important to note that proteins can also be eluted under denaturing conditions (presence of high amounts of organic solvents) where the native conformation is not retained. However, this approach is advisable either when the protein's native structure and/or its function is not of prime importance or there is a possibility of refolding the protein back to its native form (reversible change in conformation) post purification. However, if the biological activity of the protein is not of concern, then strong alcoholic eluents can be used.

9.4.6.4 Column Regeneration

The adsorbents in HIC can be used for multiple rounds of purification. Therefore, to increase their durability, the resins need to be regenerated after few rounds of protein purification. The cleaning methods employed and the reagents used for cleaning should be suitable for the base matrix and the ligands used, as their stability is to be maintained. 6 M guanidine hydrochloride is often used and preferred for removing the tightly bound proteins from the matrix. If detergents have been used in the purification buffers, they are often removed by adding ethanol or methanol in the regeneration step. Caustic solution of 1 M NaOH is used for sanitization of the matrix. Post regeneration, the column is stored in the storage solution as mentioned in the guidelines given by the manufacturer [5]. The caveat to be followed is that the storage solution should not impair the stability of the base matrix and the hydrophobic ligands used.

9.4.7 Standard Protocol for HIC

Step 1: Preparation for Purification

This step is required to decide the buffer system and the salt that would be used for the protein to bind onto the column. Salt concentration used should be such that the protein binds effectively onto the column and at the same time should not precipitate. The range of salt concentration used is from 0.5 to 2.0 M and the pH between 5 and 7.

Step 2: Column Equilibration

Any of the columns mentioned in Table 9.2 can be used for HIC since these columns are stable enough to withstand harsh conditions like high salt concentration and pH.

Equilibrate the column in the same buffer in which the protein is present. Give 3–5 column washes with the buffer maintaining the flow rate between 50 and 300 cm/h. The pH and the salt concentration should be optimum for binding the protein.

The duration of equilibration step is 2 h.

Following options for equilibration buffers can be explored for a protein purification on HIC column.

1. 50-mM Phosphate (Na_2HPO_4 and NaH_2PO_4) + 3-M Ammonium Sulfate, pH 7.0
   Ammonium sulfate is the preferred salt when the protein to be separated does not bind effectively to the resin. It is a precipitant that favors hydrophobic interactions [5].
2. 50-mM Phosphate (Na_2HPO_4 and NaH_2PO_4) + 3-M sodium chloride, pH 5.0
   Sodium chloride can be used when the protein has a strong hydrophobic character by itself. Moreover, if the pH is kept low, it further promotes hydrophobic interactions since the acidic groups interact with charged amino acids that may interfere with the hydrophobic interactions [5].

Step 3: Loading Protein onto the HIC Matrix

The protein sample is loaded onto the HIC column and allowed to bind. After an interval of 1–2 h, the column is washed with 3–10 CVs of equilibration buffer used in previous step to get rid of impurities.

Wash buffer composition used:

50 mM Phosphate buffer (Na_2HPO_4 and NaH_2PO_4), pH 7.0.

Make up the volume to 1 l with purified water and filter through 0.2-μm filter. Set the pH to 7.0

The wash buffers may or may not contain salt. If it contains salt, its concentration should be less than the protein loading buffer.

Step 4: Elution

A decreasing linear salt gradient of about 10 CVs is used for protein elution from column. The fractions are collected and analyzed for the presence of protein either by measuring the absorbance at 254 nm or by running the elutes on an SDS gel.

Elution buffers used are:

1. 50-mM Phosphate (Na_2HPO_4 and NaH_2PO_4) + 0–1.0 M Ammonium Sulfate, pH 7.0
2. 50-mM Phosphate (Na_2HPO_4 and NaH_2PO_4) + 0–1.0 M Sodium Chloride, pH 7.0

The salt concentration here is lowered for the protein to elute from the resin. Rest of the buffer composition remains the same as that used for column equilibration.

Step 5: Column Regeneration and Storage
The reusable column is regenerated after few rounds of purification as described above and stored in 20% ethanol for next round of purification.

The important steps in HIC are mentioned in Fig. 9.7 below.

9.4.8 Applications

HIC that exploits the hydrophobic interaction of proteins has been widely used in combination with other purification strategies that separates according to parameters that include size (Gel filtration), charge (Ion exchange chromatography), and ligand affinity (Affinity chromatography). The order in which different purification techniques are combined determines the outcome of the separation process.

9.4.8.1 HIC in Combination with Ion Exchange Chromatography

HIC could be used along with Ion exchange chromatography. Since both of these techniques work on the principle of hydrophobic interactions and charge, respectively, the eluted material from IE column can be directly loaded with minimal sample treatment onto HIC columns. For example, recombinant HIV reverse transcriptase was purified using multistep purification strategy involving ion exchange, ammonium sulfate precipitation, and HIC [50]. Here, the protein of interest is first purified using IEX by applying the lysate directly to the column. After this first step of purification based on charge, the hydrophobic protein was concentrated using ammonium sulfate precipitation and loaded onto the HIC column for further purification.

9.4.8.2 HIC in Combination with Gel Filtration Chromatography

HIC provides an efficient way of concentrating a sample using a strong salt solution like ammonium sulfate and this serves as a major advantage of using this technique prior to gel filtration where sample volume is limited and a high concentration is required. Also, a hydrophobic protein is not effectively purified on GFC resins; hence an HIC step before GFC becomes important. Here, the protein purified from the HIC resin is concentrated using ammonium sulfate and applied onto the GFC resin. This purification strategy was successfully employed for purification of human pituitary prolactin protein [51].

9.4.8.3 HIC for Studying Changes in Protein Conformation

HIC can also be used to identify any change in the protein conformation since it may alter the physicochemical properties including hydrophobicity of protein. This alteration with respect to hydrophobicity parameter can be explored by HIC. This is because any change in the protein conformation may result in the rearrangement of hydrophobic residues affecting the binding of the protein onto the HIC resin. For

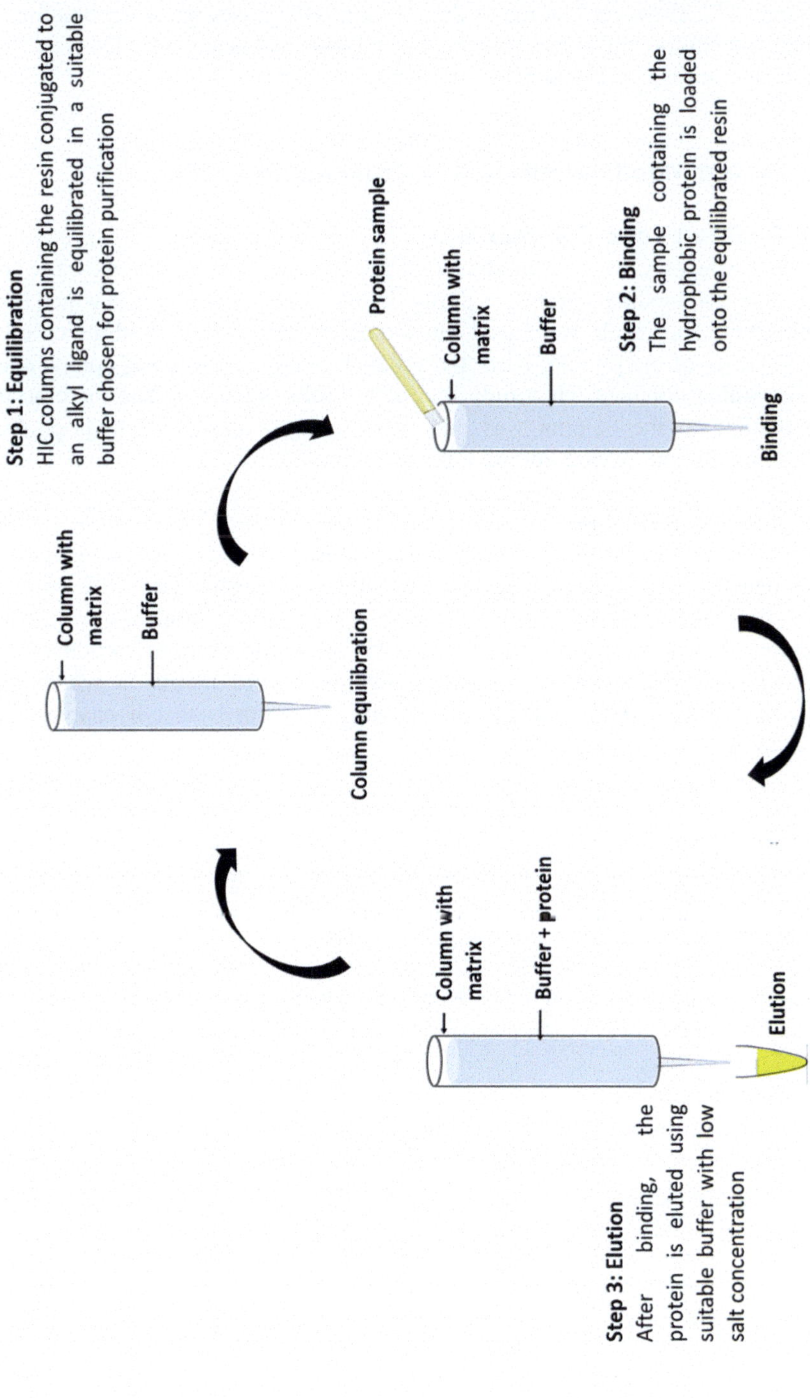

Fig. 9.7 Schematic representation of the general steps followed for purifying a hydrophobic protein using HIC

example, conformational isomers of α_2_macroglobin were separated by HIC using this principle [5]. However, the major drawback of HIC arises when the binding is too strong and elution buffer requires organic solvents and detergents. This leads to protein denaturation as previously discussed [5].

9.4.9 Recent Modifications and Improvements in HIC

9.4.9.1 Dual Salt Load Conditioning

As the name suggests, this is a modification of the process in which two salts are used for increased binding and protein purity. In one of the applications in which two salts, phosphate and citrate were used showed nearly two-to-three-fold increase in binding capacity of the protein. Also, the product quality (purity) was comparable and same as that obtained when only one of the salts was used. The increase in protein binding to the column was due to increased exposure of hydrophobic surfaces of the protein when two different salts are used [5].

9.4.9.2 Improved Resins

High quality resins developed recently have been shown to have a very good binding capacity due to optimal resin pore size, which aids in more than 90% protein recovery from the column. The most recent type of resin termed *monolith* is considered to be the fourth-generation matrix [5]. *Monolith* resins are produced by polymerizing a liquid precursor into a continuous porous matrix in which the particles are coalesced. The advantage of using monolith is that it enables the scientists to control two parameters together:

- Advantage of optimizing the nature of material and its porosity that would in turn affect the separation.
- Allows scientists to optimize and control the pore size of the resins that in turn affects the permeability of the material. Controlling the pore size also allows regulation of pressure fluctuations in the column.

Though these columns allow a faster rate of separation, one of the drawbacks of using *monolith* is its lower binding capacity due to reduced surface area [5].

9.4.9.3 Membrane HIC

Another recent advancement in HIC is the use of membranes instead of resins as they are found to provide an increased flow rate [52]. Membranes also have the advantage of lower processing times and lesser buffer volume as compared to the resin. Membranes have often been used to get rid of impurities that interfere with the purity of the desired protein. Cellulose membranes have been widely used for protein and enzyme purification with some common hydrophobic ligands like octyl, butyl, phenyl groups and polyethylene glycol. Use of membranes for HIC made the process significantly faster.

9.4.9.4 Flow Through Mode

The flow through mode has been applied to HIC to make it a high throughput technique [5]. In this mode, the impurities and the aggregates are retained onto the membrane and the product of interest flows through the column. The aggregates and the impurities have more affinity towards the resin than the product; hence they are retained and eventually are separated from the desired product. This modified technique has many advantages over conventional HIC including requirement of less salt in elution buffer and elimination of gradient elution step.

9.4.10 Advantages of Using HIC

Some of the proteins have the tendency to aggregate and hence these aggregates need to be separated from their desirable monomeric form. HIC is the best tool that can be used to separate such aggregates as it is more efficient in providing superior selectivity in separation of misfolded proteins [5].

One of the disadvantages of HIC is the use of high salt concentration for enhancing the hydrophobic interactions. The high salt concentration might interfere with the downstream processing and further biophysical and functional characterizations.

9.5 Conclusion

Reversed Phase Chromatography (RPC) and Hydrophobic Interaction Chromatography (HIC) have found wide applications at the lab research level. Purification of hydrophobic proteins has been a challenge for researchers, which was majorly overcome with the help of RPC and HIC. To cater to the unmet needs of recombinant hydrophobic protein purifications, several research endeavors have been undertaken for improvement in the RPC and HIC techniques. As a consequence, newly designed columns have become more robust in operations and protein recovery. Despite few shortcomings, RPC and HIC, combined with other conventional chromatographic techniques, have helped researchers to purify many hydrophobic proteins with very high purity.

9.6 Troubleshooting for RPC and HIC

Problem	Cause	Solution
1. Clogged column	Presence of protein aggregates and particulates	Replace column, use 10% dextran sulfate or 3% Polyvinylpyrrolidone (PVP) to remove particulates before chromatography

(continued)

Problem	Cause	Solution
	Protein precipitates formed due to removal of stabilizing agents from the buffer system	Clean/regenerate the column, use different eluent that maintains the protein in stable state
2. Protein does not elute in the organic solvent gradient	pH of the buffer is not maintained	Adjust the pH so that the protein remains stable
3. Low resolution	Selectivity of the matrix is poor Column is overloaded Flow rate is too high	Change the ion-pairing agent (e.g., TFA) Clean/regenerate the column; load less sample Decrease the flow rate to get well-resolved peaks
4. Cracks in packed bed	Air leak in the column creating excess pressure	Check for leaks in the column, repack the column

Acknowledgments The authors thank Advanced Centre for Treatment, Research and Education in Cancer (ACTREC) for providing necessary infrastructure and resources for successful completion of the chapter. The authors acknowledge Ms. Chanda Baisane, Bose Lab, ACTREC for formatting the manuscript.

Problems

Multiple Choice Questions

1. A mobile phase with high eluotropic strength would lead to:
   (a) Tighter binding between protein and matrix
   (b) Faster elution of the protein from matrix
   (c) No effect on elution
   (d) Lowering the resolving capacity
2. Ion-pairing agents are used for:
   (a) Enhancing elution
   (b) Increasing resolving capacity
   (c) Enhancing hydrophobic interactions between matrix and protein
   (d) Equilibrating the matrix
3. One of the advantages of HIC over RPC is:
   (a) It helps in the separation of hydrophobic proteins
   (b) It has high resolving capacity
   (c) It uses nonpolar mobile phase for elution
   (d) It does not denature protein

Subjective Questions

1. A moderately hydrophobic protein was separated using HIC in 50 mM phosphate buffer (Na_2HPO_4 and NaH_2PO_4), 100 mM NaCl at pH 7.0. It was found that the protein yield was very less after performing HIC. What parameters should be changed or worked upon to increase the yield of the protein?
2. A hydrophobic protein was purified by RPC for which 0.1% TFA (trifluoroacetic acid) in water, pH 2.0 (buffer A) and 10 mM NaOH in water, pH 12 (buffer B) were used as equilibration buffers. A 10–60% gradient of acetonitrile was used as the organic modifier for elution. The eluates contained negligible amount of the target protein. Also, there were a few impurities along with the desired protein. What could be the probable solution for these issues?

References

1. Berg JM, Tymoczko JL, Stryer L. Biochemistry. 5th ed. New York: W H Freeman; 2002.
2. Crabb JW, Heilmeyer LMG. Micropreparative protein purification by reversed-phase high-performance liquid chromatography. J Chromatogr. 1984;296:129–41.
3. Fausnaugh JL, Kennedy LA, Regnier FE. Comparison of hydrophobic-interaction and reversed-phase chromatography of proteins. J Chromatogr. 1984;317:141–55.
4. Malmsten M. Ellipsometry studies of the effects of surface hydrophobicity on protein adsorption. Colloids Surf B: Biointerfaces. 1995;3(5):297–308.
5. McCue JT. Chapter 25: Theory and use of hydrophobic interaction chromatography in protein purification applications. In: Burgess RR, Deutscher MP, editors. Methods in enzymology. New York: Academic Press; 2009. p. 405–14.
6. Rifai N, Wittwer ARHCT, Hoofnagle A. Principles and applications of clinical mass spectrometry, small molecules, peptides, and pathogens. 1st ed. Netherland: Elsevier; 2018.
7. Patel HB, Jefferies TM. Eluotropic strength of solvents: prediction and use in reversed-phase high-performance liquid chromatography. J Chromatogr A. 1987;389:21–32.
8. Aguilar M-I. Reversed-phase high-performance liquid chromatography. In: Aguilar M-I, editor. HPLC of peptides and proteins: methods and protocols. Totowa, NJ: Springer New York; 2004. p. 9–22.
9. Josic D, Kovac S. Reversed-phase high performance liquid chromatography of proteins. Curr Protoc Protein Sci. 2010;61:8.7.1–8.7.22.
10. Mehta A. Principle of reversed-phase chromatography HPLC/UPLC. 2012. https://pharmaxchange.info/2012/12/principle-of-reversed-phase-chromatography-hplcuplc-with-animation/.
11. Reversed phase chromatography, principles and methods. Buckinghamshire: Amersham Biosciences.
12. Shabram P, Scandella C. Adenoviral vectors for gene therapy. New York: Academic Press; 2016.
13. McNay JL, O'Connell JP, Fernandez EJ. Protein unfolding during reversed-phase chromatography: II. Role of salt type and ionic strength. Biotechnol Bioeng. 2001;76(3):233–40.

14. Horvath C, Melander W. Liquid chromatography with Hydrocarbonaceous bonded phases; theory and practice of reversed phase chromatography. J Chromatogr Sci. 1977;15(9):393–404.
15. Janson J-C. Protein purification: principles, high resolution methods, and applications, Methods of biochemical analysis, vol. 54. New York: John Wiley & Sons; 2011.
16. Schellinger AP, Carr PW. Isocratic and gradient elution chromatography: a comparison in terms of speed, retention reproducibility and quantitation. J Chromatogr A. 2006;1109(2):253–66.
17. Field JK, Euerby MR, Petersson P. Investigation into reversed phase chromatography peptide separation systems part II: an evaluation of the robustness of a protocol for column characterisation. J Chromatogr A. 2019;1603:102–12.
18. Dorsey JG, Dill KA. The molecular mechanism of retention in reversed-phase liquid chromatography. Chem Rev. 1989;89(2):331–46.
19. Valkó K, Snyder LR, Glajch JL. Retention in reversed-phase liquid chromatography as a function of mobile-phase composition. J Chromatogr A. 1993;656(1):501–20.
20. Davankov VA, et al. Hypercrosslinked polystyrene as a novel type of high-performance liquid chromatography column packing material: mechanisms of retention. J Chromatogr A. 2003;987 (1):67–75.
21. Shen C-H. Quantification and analysis of proteins. In: Diagnostic Molecular Biology. New York: Academic Press; 2019.
22. Mant CT, Zhou NE, Hodges RS. Correlation of protein retention times in reversed-phase chromatography with polypeptide chain length and hydrophobicity. J Chromatogr A. 1989;476:363–75.
23. Guo D, Mant CT, Hodges RS. Effects of ion-pairing reagents on the prediction of peptide retention in reversed-phase high-resolution liquid chromatography. J Chromatogr A. 1987;386:205–22.
24. Jennissen HP, Heilmeyer LMG. General aspects of hydrophobic chromatography. Adsorption and elution characteristics of some skeletal muscle enzymes. Biochemistry. 1975;14(4):754–60.
25. Bietz JA. Separation of cereal proteins by reversed-phase high-performance liquid chromatography. J Chromatogr A. 1983;255:219–38.
26. McCalley DV. Effect of temperature and flow-rate on analysis of basic compounds in high-performance liquid chromatography using a reversed-phase column. J Chromatogr A. 2000;902 (2):311–21.
27. Brettschneider F, et al. Replacement of acetonitrile by ethanol as solvent in reversed phase chromatography of biomolecules. J Chromatogr B. 2010;878(9):763–8.
28. Snyder LR, Dolan JW, Gant JR. Gradient elution in high-performance liquid chromatography: I. Theoretical basis for reversed-phase systems. J Chromatogr A. 1979;165(1):3–30.
29. Pohl T, Kamp RM. Desalting and concentration of proteins in dilute solution using reversed-phase high-performance liquid chromatography. Anal Biochem. 1987;160(2):388–91.
30. DeAmicis C, et al. Comparison of preparative reversed phase liquid chromatography and countercurrent chromatography for the kilogram scale purification of crude spinetoram insecticide. J Chromatogr A. 2011;1218(36):6122–7.
31. Chandler D. Interfaces and the driving force of hydrophobic assembly. Nature. 2005;437 (7059):640–7.
32. Pace CN, et al. Forces contributing to the conformational stability of proteins. FASEB J. 1996;10(1):75–83.
33. Schauperl M, et al. enthalpic and entropic contributions to hydrophobicity. J Chem Theory Comput. 2016;12(9):4600–10.
34. Queiroz JA, Tomaz CT, Cabral JMS. Hydrophobic interaction chromatography of proteins. J Biotechnol. 2001;87(2):143–59.
35. Fausnaugh JL, Kennedy LA, Regnier FE. Comparison of hydrophobic-interaction and reversed-phase chromatography of proteins. J Chromatogr A. 1984;317:141–55.
36. Eriksson KO, et al. Application of high-performance chromatographic and electrophoretic methods to the purification and characterization of glucose oxidase and catalase from *Penicillium chrysogenum*. J Chromatogr. 1987;397:239–49.

37. Breiten B, et al. Water networks contribute to enthalpy/entropy compensation in protein-ligand binding. J Am Chem Soc. 2013;135(41):15579–84. (1520–5126 (Electronic))
38. Collins KD. Ions from the Hofmeister series and osmolytes: effects on proteins in solution and in the crystallization process. Methods. 2004;34(3):300–11.
39. Curtis RA, et al. Protein-protein interactions in concentrated electrolyte solutions. Biotechnol Bioeng. 2002;79(4):367–80. (0006–3592 (Print))
40. Arakawa T, et al. Solvent modulation of column chromatography. Protein Pept Lett. 2008;15 (6):544–55. (0929–8665 (Print))
41. Lin F-Y, et al. Microcalorimetric studies of interactions between proteins and hydrophobic ligands in hydrophobic interaction chromatography: effects of ligand chain length, density and the amount of bound protein. J Chromatogr A. 2000;872(1):37–47.
42. Deitcher RW, et al. A new thermodynamic model describes the effects of ligand density and type, salt concentration and protein species in hydrophobic interaction chromatography. J Chromatogr A. 2010;1217(2):199–208.
43. Ochoa JL. Hydrophobic (interaction) chromatography. Methods Mol Biol. 2011;681:431–7. (0300–9084 (Print))
44. Xia F, Nagrath D, Cramer SM. Effect of pH changes on water release values in hydrophobic interaction chromatographic systems. J Chromatogr A. 2005;1079(1):229–35.
45. Haidacher D, Vailaya A, Horváth C. Temperature effects in hydrophobic interaction chromatography. Proc Natl Acad Sci. 1996;93(6):2290.
46. Shukla AA, et al. Preparative purification of a recombinant protein by hydrophobic interaction chromatography: modulation of selectivity by the use of chaotropic additives. Biotechnol Prog. 2002;18(3):556–64.
47. Senczuk AM, et al. Hydrophobic interaction chromatography in dual salt system increases protein binding capacity. Biotechnol Bioeng. 2009;103(5):930–5.
48. Szepesy L, Rippel G. Comparison and evaluation of HIC columns of different hydrophobicity. Chromatographia. 1992;34(5):391–7.
49. McCalley DV. A study of column equilibration time in hydrophilic interaction chromatography. J Chromatogr A. 2018;1554:61–70.
50. Unge T, et al. Expression, purification, and crystallization of the HIV-1 reverse transcriptase (RT). AIDS Res Hum Retrovir. 1990;6(11):1297–303.
51. Department of Molecular Biology & Biotechnology. Protocols and tips in protein purification. Sheffield: The University of Sheffield; 2002.
52. Ghosh R. Separation of proteins using hydrophobic interaction membrane chromatography. J Chromatogr A. 2001;923(1–2):59–64. (0021–9673 (Print))

Purification of Difficult Proteins

10

Saujanya Acharya, Roshnee Bose, and Kakoli Bose

Abstract

Protein expression and purification have been reviewed comprehensively over the years by researchers. Despite the advances in technology made in this field of research, a huge number of proteins still baffle researchers by the difficulty posed in producing sufficient amounts of protein amenable for downstream processes such as structural and biochemical studies. Membrane proteins are an important case in point. This chapter compiles the most popular resources available for the production of a few categories of challenging proteins. While a number of strategies have been described to tackle the problems of expression and purification, an optimal protocol well suited for a specific protein is usually a result of extensive screening.

Keywords

Purification · Membrane protein · Toxic protein · Inclusion bodies

S. Acharya · K. Bose (✉)
Integrated Biophysics and Structural Biology Lab, Advanced Centre for Treatment, Research and Education in Cancer (ACTREC), Tata Memorial Centre, Mumbai, India

Homi Bhabha National Institute, BARC Training School Complex, Mumbai, India
e-mail: kbose@actrec.gov.in

R. Bose
Integrated Biophysics and Structural Biology Lab, Advanced Centre for Treatment, Research and Education in Cancer (ACTREC), Tata Memorial Centre, Mumbai, India
e-mail: kbose@actrec.gov.in

K. Bose (ed.), *Textbook on Cloning, Expression and Purification of Recombinant Proteins*, https://doi.org/10.1007/978-981-16-4987-5_10

10.1 Introduction

Recombinant protein production technology has revolutionized the field of biological and biomedical science. It has not only made it possible to produce large amounts of proteins that might be limited in their natural sources, but also provide material for use in functional, structural, and biochemical studies [1]. Moreover, recombinant proteins are being increasingly used for a number of medical conditions (such as insulin) and even produced at an industrial scale [2]. It is indeed intriguing that de novo proteins not found in nature are also being produced with the long-term goal of developing multifunctional drug vehicles at the nanoscale range [3, 4]. Despite the advances in technology, protein production has often hit roadblocks. While facing the task of purification of proteins one might come across proteins that "misbehave." Problems might include negligible expression, poor yield, aggregation, denaturation, improper folding, toxicity, or nonfunctionality to mention a few [5, 6].

These "difficult-to-purify" proteins can be classified into different groups and the most prominent categories include membrane proteins, toxic proteins, and inclusion bodies. Membrane proteins in this regard need special mention as they comprise 30% of the cell's protein pool and more than 50% of currently available drugs in the market [7, 8]. Nonetheless, this category of proteins is notorious for the challenges it presents during production and as a result a great deal of information is available from investigators who have optimized ideal conditions for expression and purification. Due to the diverse nature of proteins belonging to a particular group, developing a generalized protocol suited for purification is almost an impossible feat to achieve and every protocol needs to be customized for the protein of interest. However, this chapter attempts to highlight key clues to approach the purification of difficult-to-purify proteins with a special focus on membrane proteins. In addition, two other categories, namely toxic proteins and inclusion bodies, will be briefly discussed. Note should be taken that the procedures employed for the latter two categories often overlap with that of membrane proteins and have been described in common unless otherwise indicated.

10.2 Types of Challenging Proteins and their Purification

10.2.1 Membrane Proteins

10.2.1.1 Overview

Membrane proteins (MPs) constitute around one third of a cell's proteins in living organisms and are important targets for over half of all drugs [7]. They participate in important biological processes such as cell signaling, molecular transport, photosynthesis, bioenergetic processes, and catalysis [9, 10]. Structurally, they can be broadly classified into four main categories: Peripheral, Amphitropic, Lipid-linked, and Integral (Fig. 10.1) [9–11].

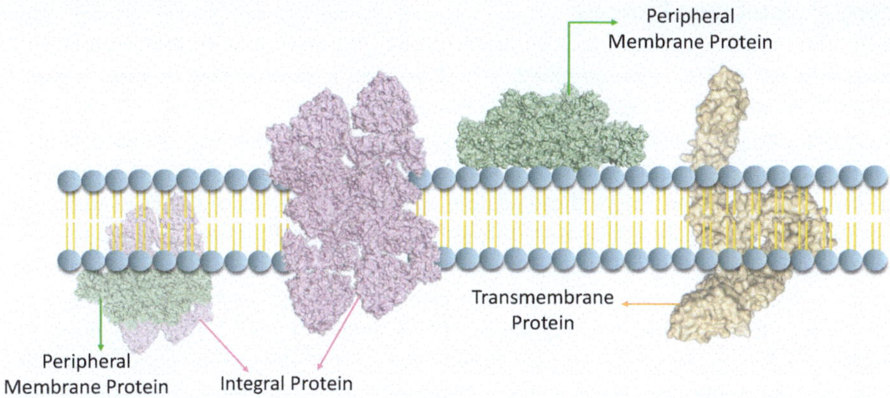

Fig. 10.1 Types of membrane proteins. The cartoon represents a biological lipid bilayer in which membrane proteins are embedded. The peripheral proteins are loosely attached to one side of the membrane (outer or inner), while the integral proteins remain deeply embedded. The integral proteins that span the entire lipid layer are known as transmembrane proteins

Peripheral Proteins

Peripheral proteins do not penetrate the width of the lipid membrane and are loosely attached on the surface through electrostatic interactions or hydrogen bonds [10]. Therefore, these proteins can be easily removed by changes made in pH or salt concentration and rendered water soluble. Cytochrome c and MBP are two common examples of peripheral proteins [12, 13]. Cytochrome c readily dissociates from the mitochondrial membrane by treatment with 0.15 M KCl, thereby demonstrating the ease with which peripheral proteins can be separated.

Amphitropic Proteins

Amphitropic proteins are a relatively new class of proteins, which exhibit dual nature, i.e., they can shuttle between a globular water-soluble and a loosely membrane-bound form through electrostatic and hydrophobic forces [9, 14]. The shift in form happens through a conformation change triggered by modifications such as phosphorylation, acylation, or ligand binding, which in turn exposes a previously hidden membrane binding site on the protein. Subsequently, contact with the membrane is made by inserting a hydrophobic stretch of amino acids, such as alpha-helix, into the lipid bilayer. Colicin A, Alpha-hemolysin, and globular pyruvate oxidase from *Escherichia coli* are some classic examples of amphoteric proteins.

Lipid-linked Proteins

Lipid-linked proteins are proteins that are tethered to the membrane through one or more lipid groups that are covalently attached to them [9, 10]. These proteins are usually covalently modified by three different types of lipids, isoprenoid groups, fatty acyl groups, and glycoinositol phospholipids (GPIs), and are referred to as "prenylated," "fatty acylated," and "GPI-linked," respectively.

Integral Membrane Proteins

Integral membrane proteins are characterized by regions that help permanently lock them into the plasma membrane [9, 10]. They can exist in different forms: type I integral membrane protein, type II integral membrane protein, C-terminally anchored integral membrane protein, type II signal anchored protein and a multi-spanning membrane protein based on the orientation and mode of integration of the hydrophobic region of these proteins. Depending on the number of times the membrane-spanning domain passes through the plasma membrane, they are referred to as bitopic (single pass) or polytopic (multi pass). On the other hand, integral proteins that remain associated with just one side of the membrane are known as monotopic, for example, prostaglandin H2 syntheses 1 and 2 [9]. The nonpolar interaction forces between the membrane spanning hydrophobic regions and the lipid bilayer counter the forces trying to push this region out into a water-filled area ensuring that the integral proteins stay in place and never leave the membrane. These proteins can be removed only by detergents, nonpolar solvents, and denaturants that disrupt the hydrophobic interactions in the plasma membrane and are therefore relatively difficult to isolate (Fig. 10.2). Functionally, they act as channels, transporters, and receptors and are often involved in cell adhesion [9, 11]. Examples of such proteins include Integrins, Insulin receptors, and Cadherins [15, 16].

10.2.1.2 Problems Encountered

A number of factors make the expression, purification, crystallization, and structure resolution of membrane proteins difficult such as their relatively hydrophobic surfaces, flexibility, low levels of expression, and instability [6]. Naturally, membrane proteins exist in very low abundance that makes purification from their natural sources challenging. Moreover, overexpression in other model systems often leads to low expression or production of nonfunctional proteins due to inadequate membrane insertion and protein folding apart from the possibility of leading to cell toxicity (Fig. 10.3) [17]. Choosing an appropriate expression system is therefore a crucial step in this process. Purification of membrane proteins is challenging since they are embedded in the lipid bilayer. Furthermore, protein purification handling and analysis involves procedures that have been designed for aqueous environments [6, 17]. Solubilization may be accompanied with drastic protein loss due to aggregation/denaturation and/or loss of activity which present further challenges in characterizing these molecules.

10.2.1.3 Conventional Strategies Employed for Purification

A successful strategy includes the selection of appropriate host strain, vector, promoter, selection markers, optimal gene sequence and methods employed for expression and purification of such proteins [18]. The following sections succinctly describe all of the necessary steps involved.

Source

Natural: In case the membrane proteins are produced abundantly in their natural source, a highly specific initial purification method that does not require a tag is

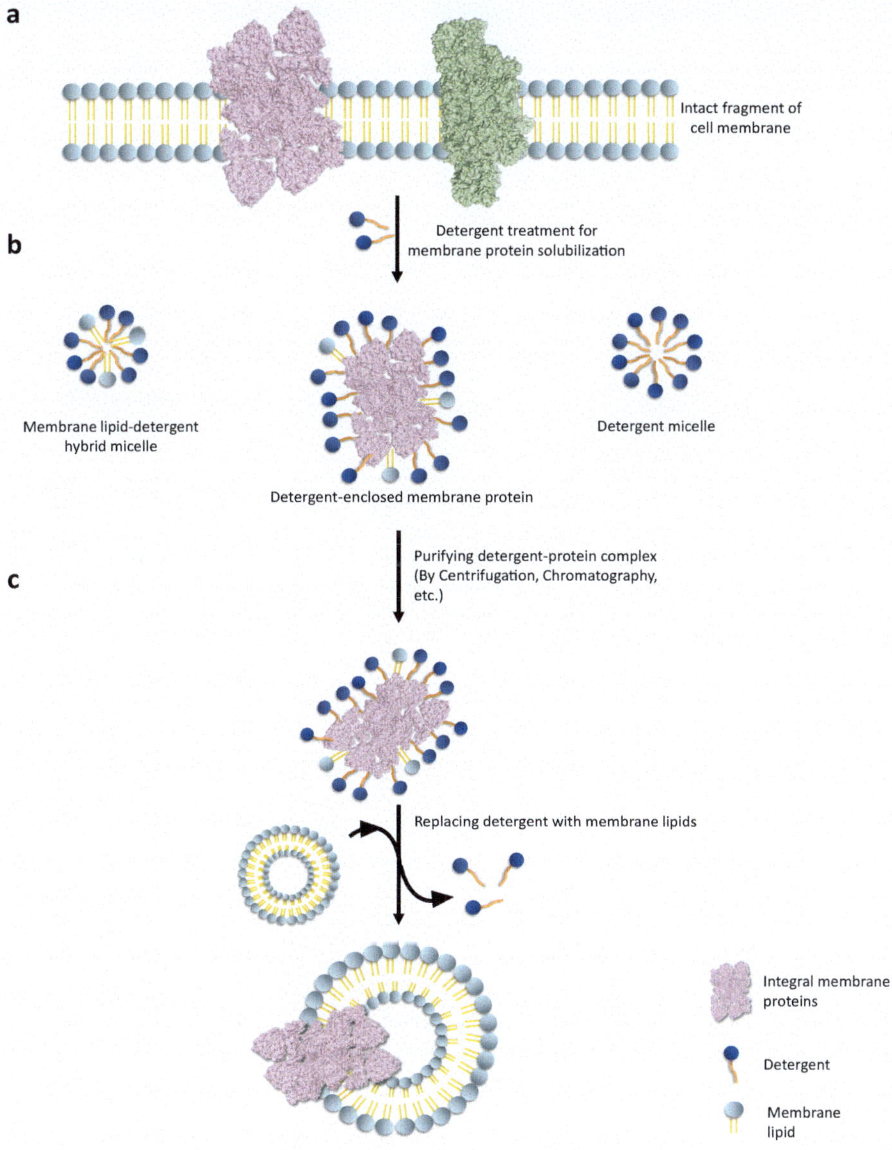

Fig. 10.2 Solubilization of integral proteins by detergents. The schematic diagram represents the steps involved in integral protein solubilization from the membrane using detergents. (**a**) Low concentrations of detergent are used to perturb the membrane generating micelles containing lipid, protein, and detergent. (**b**) These complexes are subsequently purified by centrifugation, chromatography, etc. (**c**) The excess detergents are then removed to prevent complications in downstream purification processes by transferring the membrane protein into a liposome or into a different detergent

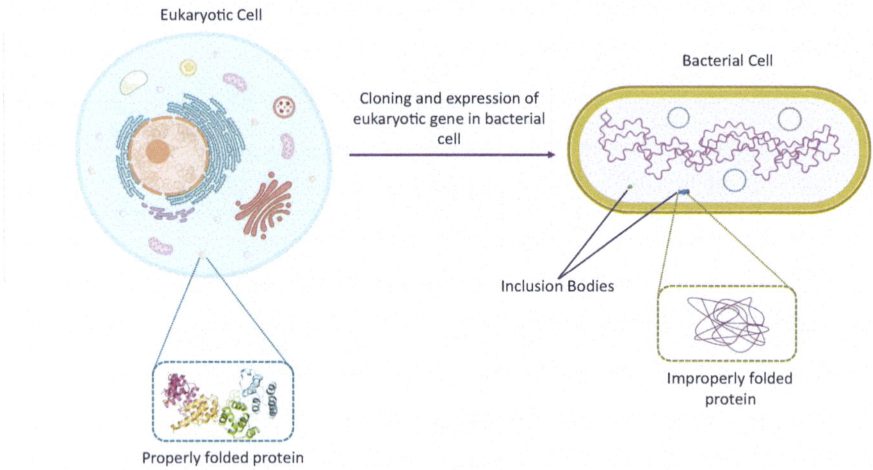

Fig. 10.3 Misfolding of proteins in heterologous protein expression. The figure highlights the problem of protein misfolding encountered during heterologous expression of eukaryotic genes in prokaryotic hosts (bacterial cell). There are multiple reasons for misfolding and may happen due to inadequate membrane insertion, lack of appropriate posttranslational modifications, or absence of folding aids such as chaperones. These may lead to formation of insoluble inclusion bodies in the bacterial host or lead to cell toxicity

employed. Ion exchange chromatography is usually used in combination with other classical chromatography methods with gel filtration as the final purification step. The final step is usually beneficial in removing aggregates and impurities while simultaneously allowing for buffer exchange. For example, Bovine Rhodopsin, a G-coupled protein receptor (GPCR), is purified from bovine retina where the protein is expressed in high levels [19]. However, the use of natural sources can be limiting if the protein is expressed in low amounts naturally. Furthermore, it does not provide the possibility of genetically modifying these proteins either for easy purification or labeling for nuclear magnetic resonance (NMR)/crystallographic studies.

Heterologous hosts: Recombinant membrane proteins can alternately be expressed in a variety of hosts that can be either prokaryotic or eukaryotic. *E. coli, Rhodobacter* spp., and *Lactococcus lactis* are common examples of prokaryotic systems whereas yeast, insect, and mammalian cells serve as eukaryotic hosts [20–22].

Expression Systems

Maximizing membrane protein expression requires choosing the most appropriate combination of key elements such as the vector and expression host. Conventional cloning methods have evolved to allow parallel expression screening through rapid cloning of genes into multiple expression vectors simultaneously. For example, *Gateway technology* is based on site-specific recombination by bacteriophage lambda to integrate DNA into bacterial chromosome. The integration is catalyzed by Int/IHF enzymes between att recombination sites. The process is reversible and excision is catalyzed by the Xis enzyme. This method is widely used to clone membrane protein (MP) ORFs into vectors [23]. Creator and the fragment exchange (FX) cloning are some of the other widely used techniques for high-throughput cloning of MPs [24]. The Creator system uses the Cre-loxP recombinase method to rapidly transfer gene of interest into multiple expression vectors [25]. The FX cloning system on the other hand uses Type IIS restriction enzyme and negative selection markers to clone into "initial cloning vectors" with different antibiotic resistances. The ORF can be subcloned from these initial vectors to several different expression host vectors readily [26]. However, it is impossible to predict if a protein of interest will express sufficiently and be easily purified or characterized. Therefore, once the gene of interest is cloned in expression vectors, it is ideal to test different expression systems for production of MPs. Moreover, it is also important to keep in mind that each of these systems has its *pros* and *cons*.

Prokaryotic Cells

Prokaryotic cells such as *E. coli, R. sphaeroides,* and *Lactococcus lactis* are preferred choices for expression of membrane proteins (both prokaryotic and eukaryotic MPs) due to the range of advantages they offer [20–22, 27]. It is quick, cheap and easy to use. It also allows for rapid screening of constructs for high yield. However, these systems have certain limitations when it involves heterologous production of MPs, which include the absence of posttranslational modifications, toxicity, and the requirement of codon-optimized genes [28]. Although, the high costs of the eukaryotic expression systems for eukaryotic MP production have paved the way for cost-effective bacterial-based alternatives.

Over time, the *E. coli* system still remains a favorite since it is easy to grow and innumerable genetic/molecular/biochemical methods have been optimized for MP production in *E. coli*. The BL21 (DE3) strain is driven by the lacUV5 promoter and the bacteriophage T7 RNA polymerase (T7 RNAP) [29]. This leads to the synthesis of more mRNA and therefore more protein. However, overproduction of MPs can often lead to cell toxicity or integration into inclusion bodies which may not yield functional proteins after extraction and refolding. Hence a number of strains were developed to minimize this effect and increase MP yields. The C41 (DE3) and C43 (DE3) strains, which are derivatives of BL21 (DE3), were developed to produce "toxic" membrane proteins by lowering the expression of T7 RNAP [30]. Similarly, the Lemo21 (DE3) strain fine tunes the expression of T7 RNAP by encoding the T7 Lys gene under control of a rhamnose promoter. The T7 RNAP expression in turn provides a broad range of expression for the protein of interest. GFP-fusions of MPs have been successfully produced in Lemo21 systems, which also ensure that the proteins are well folded for further structural/functional assays. BL21 (DE3) pLysS, BL21 (DE3) Codon Plus, and Rosetta™ 2(DE3) cells have also been used successfully in the homologous and heterologous production of membrane proteins [30].

L. lactis, a gram-positive lactic acid bacterium, compatible with a wide range of expression vectors and inducible promoters, is also used for MP production such as the human KDEL receptor [20]. *L. lactis* has multiple features that are advantageous for the production of MPs [20]. Its slower growth rate helps difficult proteins to fold properly, allows co-production of chaperones along with MPs, expression of multiple plasmids, and aids incorporation of amino acid derivatives (such as Seleno-methionine) in MPs produced for further structural characterization. Another gram-positive bacterium, *B. subtilis*, which is widely used for production of secretory proteins, has now been modified to eliminate its stress-responsive systems to make this system amenable to large amounts of MP production [30]. With a large variety of hosts available for MP production, it is always preferable to choose a host system that closely resembles the natural host of the target MP.

Eukaryotic Cells

Yields of eukaryotic membrane proteins in prokaryotic hosts are usually not sufficient for functional and structural studies. Therefore, eukaryotic hosts are increasingly used for the production of eukaryotic MPs. *S. frugiperda*, HEK 293, and yeast cells including *S. cerevisiae* and *P. pastoris* have been used for the production of MPs, which constitute around 26% of all eukaryotic membrane proteins [31, 32].

MP production in yeast is fast (however, needs longer time than bacteria), economical, easy to cultivate, amenable to genetic modifications, and also provides for posttranslational modifications that are absent in bacteria such as disulfide-bond formation, prenylation, phosphorylation, acylation, and O- and N-linked glycosylation [33]. Adenosine A2A and the histamine H1 GPCRs from *P. pastoris* have been successfully characterized [34, 35]. *P. pastoris* is able to produce large amounts of protein due to its ability of reducing production of toxic ethanol [32]. Additionally, it can be grown in bioreactors to maximize yield. Protein production in this system is also governed by methanol-inducible promoters, which result in high yield without the need for a high copy number gene. *S. cerevisiae*, on the other hand, has been extensively studied as a host strain and therefore a number of genetic and biochemical methods have been standardized for optimal protein production [31, 33]. Rat TRPV2 channel and human GLUT1 glucose transporter in complex with cytochalasin are MPs that have been characterized from *S. cerevisiae* [36, 37] . The ability of *S. cerevisiae* for in vivo homologous recombination also makes it well suited for high-throughput production of MPs.

Higher eukaryotic hosts such as baculovirus/insect cells and various mammalian cells are frequently used for the production of eukaryotic MPs especially when PTMs and native folding are important [38]. Moreover, they provide a more native lipid environment for proper protein function. Insect *Spodoptera frugiperda* (Sf9, Sf21) and *Trichoplusia ni* (Hi5) infected by the baculovirus *Autographa californica* multinucleopolyhedrovirus (AcMNPV) are the most commonly used systems for membrane protein expression [39]. Baculovirus systems are also safe since they do not infect mammals. The lytic, dsDNA virus is noninfectious and its promoters are inactive in mammalian cells. However, it is relatively expensive to use this system due to its complex metabolism mechanism and inability to grow in minimal medium. It is also expensive to produce isotope-enriched proteins for advanced structural characterization like NMR [30]. Chinese hamster ovary cells (CHO), human embryonic kidney cells (HEK293), baby hamster kidney cells (BHK-21), and monkey kidney fibroblast cells (COS-7) are some commonly used mammalian cell lines

[40]. Yield is usually low and varies widely depending on the MP of interest. Also, they are quite expensive in producing sufficient amounts of target protein for structural analysis. These systems are used for MPs that require specific PTMs and/or subcellular environments for optimal folding and activity, which is not possible in prokaryotes or lower eukaryotes.

Cell-free (CF) Expression System

This system is rapidly becoming a fast, efficient, open, and inexpensive method for production of membrane proteins directly into detergent micelles without the limitations of conventional expression systems described earlier [41]. It also solves the problems of protein translocation into the lipid bilayer, yield (>1–2 mg/mL), background expression, and detergent compatibility. Moreover, this system operates over short reaction times and small volumes. The ease with which CF allows the labeling of proteins (fluorescent amino acids, seleno-methionine, $^{15}N/^{13}C$) or the addition of protein-specific supplements such as substrates, chaperones, inhibitors, and coenzymes makes it an attractive system as well [41].

CF systems can be differentiated on the basis of a variety of factors such as kinds of extracts used, method of expression, and type of reaction setup [42]. The most commonly used extracts are *Escherichia coli*, wheat germs, insect, yeast, CHO, *HeLa* cells, and rabbit reticulocytes. However, only a specific type of posttranslational modification/(s) may be allowed in each of these systems, which could be limiting, e.g., di-sulfide bridges (*E. coli*, wheat germ) and glycosylations (reticulocytes). MPs such as epidermal growth factor receptor proteins, transporters, and KcsA potassium channel have been successfully expressed in CHO, *E. coli*, and insect CF systems, respectively. MPs can be produced in CFs as precipitates in a non-hydrophobic environment, called precipitate mode. In detergent-based mode, they are produced directly in detergent micelles where they remain soluble. Lastly, in lipid-based mode, they are produced within liposomes and may be integrated into them. Reaction setups consist of either the one- or the two-compartment system (*aka* continuous exchange CF method) where the reaction mix is separated by a semi-permeable membrane from a feeding mixture containing low molecular weight compounds that are continually exchanged between the two compartments.

In CF systems, the cell line of choice is cultured and lysed keeping the ribosomal activity in the lysate intact. The cell extract is subsequently prepared from the lysates using different clarifying methods and then used to produce the protein of interest [43]. During extract preparation, the presence of exogenous (supplemented) or endogenous microsomes (consisting of endoplasmic reticulum fragments) assists correct folding of membrane proteins, which is crucial to the process. For example, rabbit reticulocyte, wheat germ, and *E. coli* require exogenous addition of microsomal structures for membrane protein production during extract preparation, while *HeLa*, CHO, and insect platforms are equipped with endogenous microsomes resulting from rupturing of their endoplasmic reticulum. While the CF platforms may vary depending on the organism used, the basic steps of the process remain the same [44].

CF expression kits provide an easy method of production of MPs and contain individually purified components required for transcription and translation. They complete protein synthesis in just 2–4 h and can be set up easily. The most integral components of this method are the genetic material and the reaction mixture. The genetic material containing the protein of interest can be linear, circular, or a PCR fragment. The essential components of this template include in-frame start-stop codons, a suitable promoter sequence, upstream ribosome binding site (RBS), downstream spacer, and a terminator. mRNA can serve as template too; however pre-processing is required in eukaryotic systems to yield a mature mRNA. In CF expression, the commercially available vectors already have the required features for optimal transcription and translation incorporated in them that minimizes the hassles of complex processing steps. The reaction mix contains the transcriptional and translational machinery including RNA polymerase, ribosomes, tRNA, amino acids, cofactors, energy source, and cellular components for adequate protein folding. These are mostly derived from cell extracts. The general methodology involves thawing the expression components on ice: cell extract, RNase inhibitor, RNA polymerase, protein synthesis reaction buffer (nucleotides, amino acids and energy source). This is followed by adding the components in the required ratio to make a working reaction mix, and incubation at 37 °C with shaking for 2–4 h. The proteins can thereafter be analyzed by SDS-PAGE or western blot. Usually, a 50 μL reaction yields 0.5 mg/mL of protein. The protein of interest can be subsequently purified using affinity chromatography such as Ni-NTA column.

Summary of the Expression Systems

A brief summary of the popular expression systems along with their advantages and disadvantages have been tabulated below (Table 10.1) [28, 30, 33, 42, 44–47].

Techniques for Purification

Construct Design

One of the most important criteria in membrane protein production is the design of the expression construct with focus on promoter strength and fusion tags. The construct should help in optimal transcription, translation, and translocation of protein to the membrane followed by easy detection and isolation. It is also important that the design should not alter the structure–function of the protein grossly. An ideal expression vector includes a selection marker, an origin for replication, and a suitable promoter-terminator flanking the sequence.

Sequences, Tags, and Cleavage Sites

Signal sequences, which are short N-terminal amino acid stretches targeting the protein to the membrane, are often added or modified since they have a huge effect on protein production and help increase expression [30]. For example, a signal sequence named "α-mating factor pre-pro-sequence" is used for expression in *P. pastoris* and *S. cerevisiae* to increase secretion [48]. Similarly, protein coding sequences are often modified to facilitate higher yield of protein for structural studies. Two commonly used approaches include the introduction of single or

Table 10.1 A brief summary of expression systems

Expression system	Cell lines and popular strains	Advantages	Disadvantages
Prokaryotic	*E. coli* C41 (DE3), C43 (DE3): express MPs and toxic proteins from various organisms. BL21 (DE3) pLysS: has p15A origin, pLysS plasmid produces T7 lysozyme to control range of expression, suitable for MPs and toxic proteins. BL21 (DE3) CodonPlus: carries extra copies of the argU, ileY, and leuW tRNA genes. The tRNAs encoded by these genes recognize the AGA/AGG (arginine), AUA (isoleucine), and CUA (leucine) codons. Rosetta2 (DE3): universal translation due to 7 additional tRNAs for rare codons. Lemo21 (DE3): Rhamnose-dependent tunable T7 RNAP expression, reduces inclusion body formation, suitable for MPs and toxic proteins, when grown without rhamnose it is same as pLysS containing strain.	1. Fast growth 2. Inexpensive 3. High yield 4. Easy to culture 5. Suitable for bioreactors 6. Amenable to genetic and biochemical modifications to optimize yield 7. Allows cheap NMR active isotope labeling	1. Minimal PTMs available 2. Overexpression may lead to inclusion bodies 3. Codon usage different from eukaryotes
	L. lactis NZ9700, NZ9700: nisin-controlled gene expression system (NICE) for production of MPs.		
	B. subtilis B. subtilis168 with overexpression of SecDF, B. subtilis WB600 with co-expression of GroES, DnaK, PrsA are some engineered strains for high yield of heterologous proteins.		
Lower Eukaryotic	*Saccharomyces cerevisiae* BJ2168, DSY-5: protease-deficient strains used in structural characterization	1. Relatively fast growth 2. Inexpensive 3. High yield 4. Suitable for bioreactors 5. Eukaryotic PTMs available.	1. Glycosylation not same as mammalian cells 2. Culture time more than bacteria
	Pichia pastoris SMD1163: codon optimized, N-linked glycosylation site		

(continued)

Table 10.1 (continued)

Expression system	Cell lines and popular strains	Advantages	Disadvantages
	removed KM71: no modifications	6. Suitable for secreted or intracellular proteins 7. Allows cheap NMR active isotope labeling 8. Endotoxins eliminated 9. Easy to culture	
Higher eukaryotic	Insects *Spodoptera frugiperda* *Sf*21: derived from pupal ovarian tissue of the fall army worm *Sf*9: clonal isolate of Sf21 Ideal for easy transfection, purification of recombinant proteins	1. High expression 2. Relatively fast growth 3. Optimal protein folding 4. Moderately scalable 5. Extensive PTMs 6. Endotoxin-free	1. Expensive 2. Glycosylation not same as mammalian cells 3. Viral infection may lyse cell and lead to protein degradation
	Mammalian HEK293: well suited for both transient and stable transfection CHO: good growth in suspension systems and bioreactors	1. Good expression 2. Moderately scalable 3. Optimal protein folding 4. Extensive PTMs 5. Endotoxin-free	1. Expensive 2. Complex factors needed for growth
Cell free	Cell extracts used *E. coli*, wheat germ, reticulocytes: minimal posttranslational modifications available Insect, yeast, CHO, HeLa cells: adequate posttranslational modifications available	1. Systems are commercially available and easy to set up 2. Scalable 3. Conditions can be easily optimized 4. Use of non-natural amino acids/isotope labeling for NMR 5. PCR products also serve as template 6. Endotoxin-free	1. Limited PTMs 2. Expensive 3. Good yield

multiple point mutations and replacement of unstructured loops with stably folding domains [30, 45, 49]. These methods are usually aimed at increasing the stabilization of the proteins and the counter heterogeneity generated by posttranslational PTMs. Both these methods have been used to crystallize GPCRs from insect and bacterial cell lines [30].

Epitope tags and fusion partners make the task of detection and purification simple and quick. Carboxy-terminal (C-terminal) tags along with signal peptides at the N-terminus are preferred while designing constructs [45, 49]. Multiple tags have also been advantageous as is the case with GPCRs. It is advisable to insert a cleavage

Table 10.2 Tags used for protein purification

Tag	Features
Poly-histidine tag (His)	1. 3–12 residues 2. Small size does not affect protein structure or function 3. Low toxicity 4. Easy purification using IMAC 5. Well suited for protein-protein interaction studies
FLAG	Well suited for protein-protein interaction studies
GFP	1. Easy to monitor protein folding and yield 2. Well suited for protein-protein interaction studies
AviTag	1. Tighter binding 2. Allows biotinylation. Can be used for biotin-streptavidin interactions
Others: poly-Arg, FLAG, c-myc, streptavidin-based, GST	1. Easy purification 2. Well-suited for protein-protein interaction studies

site between the sequences for easy removal of the tag after purification, if required. The table below (Table 10.2) summarizes the different tags that are routinely used as well as their key features.

Cleavage sites are incorporated in constructs to allow the option of removing tags using site-specific proteolysis by an enzyme [45, 49]. Tobacco etch virus (TEV) protease, enterokinase, thrombin, and factor Xa have been successfully used for this purpose [30]. TEV particularly is popular due to its easy production, viability, and specificity. Chemical reagents for tag removal such as cyanogen bromide (CNBr) make for a cheap option; however, they are harsh and may affect the protein. These chemicals have been used for purification of inclusion bodies nonetheless. Inducible self-cleaving tags have been developed to avoid limitations of the conventional methods. For example, in the IMPACT™ system, tags rely on an internal sequence (inteins) that self-excise upon activation and join the remaining portions (exteins) to produce two independent products [50]. Controlled intracellular processing (CIP) involves proteolytic cleavage of the tag intracellularly from an independently expressed tagged protein upon induction

Promoters and Plasmids

In *E. coli*, most MPs are expressed using the T7 RNA polymerase (T7RNAP), T5-, ara-, and tet-promoters based system [29, 30]. T7 based expression systems are compatible with the DE3 gene, whereas the T5 promoter can be paired with any *E. coli* strain. Overproduction of MPs can be toxic for the host cells and therefore repressor systems in *E. coli* help tune the protein production by different concentrations of inducers. In the *lacI* repressor system, the *lacI* tetramer binds T7RNAP attachment site (T7 promoter, pT7) and blocks transcription of the gene [29, 30]. Isopropyl β-D-1-thiogalactopyranoside (IPTG) displaces *lacI* and triggers transcription. Additionally, the amount of protein can be titrated using different concentrations of IPTG. However, basal expression or "leaky" expression of protein is an issue in the absence of the inducer. In this regard, the *ara* and *tet* expression

Table 10.3 Promoters and plasmids used for membrane protein production in *E. coli*

Promoter	Plasmid	Inducer
T7/lac (T7 bacteriophage promoter)	pET, pRSET	IPTG
T5/lac (T7 bacteriophage promoter)	pQE	IPTG
araBAD (arabinose metabolic operon promoter)	pBAD	Arabinose
tet (tetracycline promoter)	pASK-IBA	Anhydrotetracycline (AHT)
tac (*E. coli* tryptophan operon and lac operon promoter derived)	pMAL, pGEX	IPTG

systems provide much stronger repressors [30]. In the *ara* expression system, *araC* dimer (repressor) transcribed from the *araC* promoter, creates a DNA loop and blocks both *araC* and *ara*BAD (pBAD) promoters. Induction by arabinose causes the displacement of *araC* and start of transcription from pBAD promoters. This system allows protein expression strictly when it is induced; however, there is no scope for tuning the protein expression by varying inducer concentration. In the *tet* system, *tetR* repressor is displaced by tetracycline and *tetA* codes for antibiotic resistance to tetracycline. It is also compatible with any *E. coli* strain. The table below (Table 10.3) shows some of the popular promoter-plasmid combinations suitable for MP production.

In yeast, strong inducible (e.g., pAOX1) promoters are used, which trigger expression only after the cell density has reached an optimal level [45]. This ensures that cell toxicity is minimal due to leaky expression of toxic MPs. Some commonly used inducible promoters in yeast are GAL1 (Galactokinase), GAL10 (α-D-galactose-1-phosphateuridyltransferase), and MET25 (O-acetylhomoserine sulfhydrylase) in *S. cerevisiae* and AOX1, AOX2, and FLD1 in *P. pastoris* [31, 45]. Constitutive promoters (e.g., pGAP: Glyceraldehyde-3-phosphate dehydrogenase and pTEF: Translation elongation factor 1) are also not uncommon. *S. cerevisiae* make use of high copy number plasmids that make up for promoters that are weaker than that of *P. pastoris*

In baculovirus system, pFastBac vector under the control of the strong coat protein forming promoter (AcMNPV polyhedron promoter) is used [45]. Other commonly used promoters include p6.9 and p10. While, pFastBac-Dual, pOET, pVL1392-3, and pBAC-1 are the plasmids of choice compatible with the aforementioned vectors [30]. For mammalian cells, both transient and stable transfection methods are compatible. Strong constitutive promoters in mammalian systems include the human cytomegalovirus (CMV from herpes virus) immediate early promoter and the simian virus 40 (SV40) early promoter [30]. Inducible systems make use of the Tet operator sites (TetO) downstream of the CMV promoter. When induced with tetracycline, TetR repressor is not allowed to bind TetO site allowing transcription. A variety of plasmids are compatible with the promoters as shown below (Table 10.4).

Table 10.4 Promoters and plasmids used for membrane protein production in yeast

Promoter	Type and source of promoter	Compatible plasmids
CMV	Constitutive, human cytomegalovirus	pcDNA3.1, pCMV, pEG BacMam
EF1a	Constitutive, human elongation factor 1 alpha	pEF
SV40	Constitutive, simian vacuolating virus 40	PSF-SV40
CMV/TO	Inducible, hybrid of CMV promoter and tetracycline operator 2 (TetO2) sites designed for high-level tetracycline-regulated expression	pcDNA4/TO, pACMVtetO CMV

Antibiotic and Drug-based Screening

Assessing plasmid stability in a particular bacterial host and medium is a beneficial step to consider before large-scale production of proteins [28, 51]. The procedure includes screening of a few independent colonies plated on agar with and without antibiotic in varying dilutions. Stability is determined by the number of colonies obtained under both conditions. If number of colonies is the same, it is advisable to proceed with large-scale preparations. Otherwise reviewing appropriate hosts such as C41(DE3)/C43(DE3)/Lemo21(DE3)/BL21(DE3) pLysS, which are amenable for stably expressing genes, is considered prudent before moving on to the next step [28]

Another strategy to screen *E. coli* strains with higher protein production for MPs include expressing the gene of interest from two separate plasmids with different cytoplasmic selection markers, such as one plasmid containing a kanamycin resistance marker and the other plasmid containing a mouse dihydrofolate reductase (trimethoprim resistant) marker [28]. Colonies screened on appropriate medium with the two selection drugs are further selected considering that it indicated clones with increased capability of target protein production. Similarly, MPs with β-lactamase drug resistance marker have also made it possible to select strains with improved protein production by screening for increased ampicillin resistance [30]

Culture Growth Conditions

Poor and unmonitored culture conditions can result in reduced cell growth, accumulation of proteins in inclusion bodies, and/or cell toxicity. Some factors that are taken into consideration for improved yield include lowering the culture temperature, optimizing the inducer concentration, and choosing a suitable induction time [30, 47]. Shorter induction time periods work well for MPs since they stand the risk of autoproteolysis due to cell toxicity. For example, to increase the stability of plasmids coding MPs when propagated in *E. coli*, the bacteria may be cultured at 30 °C instead of the usual 37 °C. A dose-dependent screening of inducer concentration (for example 0.01–2 mM of IPTG) is also beneficial to identify an optimum expression level of the protein of interest. For large-scale cultures, protocols involving cell growth to an OD of 0.6–1.0 followed by induction for 1–3 h before harvesting have also been employed [30, 47]

Insect cells grow ideally at 27–28 °C without the need of CO_2. Both serum-containing and serum-free media work well; however, it is advisable not to abruptly change the type of media, which might lead to loss of cells as they adapt to a specific medium [52]. Cultures can be in suspension or adherent. Adherent cultures are loose, so EDTA or harsh trypsinization is not required to detach them. Several different factors can be optimized for maximum protein production such as cell density, the ratio of cells to virus, time length of infection, use of protease inhibitors, and changing growth temperature [52]. These conditions need to be optimized depending on the protein of interest. In higher eukaryotes, complex media with nutrients that satisfy the requirements of growth, maintenance, as well as protein expression is an absolute necessity [52]. Additionally, growth factors such as insulin or insulin-like growth factor (IGF) are also required. Parameters such as pH, osmolarity, and redox potential should also be monitored [52]

Detergents and Buffers

Detergents are an important consideration in protein purification since they coat hydrophobic regions of the protein to allow solubilization by forming micelles. Chaotropic agents (such as I^-, ClO_4^-, and SCN^-) that disrupt hydrogen bonds and decrease the hydrophobic effect are also essential in favoring protein disaggregation [53]. It is ideal to use agents that do not disrupt protein structure since the process of refolding a denatured protein is extremely challenging. A suitable detergent/buffer for purification needs to be determined based on the nature of the protein and involves a lot of trial and error; more so since the choice of agent may also affect downstream processes. Buffers containing high salts are also often used as they aid solubilization by decreasing the electrostatic interactions between proteins and charged lipids of the membrane. Peripheral proteins are easily dissociated and agents commonly used for the process include acidic/alkaline buffers, metal chelators (EDTA, EGTA), chaotropic ions, denaturing agents (8 M urea or 6 M guanidine hydrochloride), and salt solutions/high ionic strength buffers (1 M NaCl or KCl) [53, 54]. Integral proteins on the other hand are separated using organic solvents or detergents (ionic, nonionic, or zwitterionic) [54, 55]. It is imperative that detergent quality should not be compromised upon and care must be taken to use compatible additives in the buffer that do not affect detergent solubility, critical micelle concentration (CMC), cloud point, and aggregation number [53–55]. CMC is the cutoff concentration above which micelles form, whereas cloud point is the temperature at which micelles aggregate to form a separate phase that is immiscible with the aqueous phase. A detergent screening step is usually helpful in such cases. Detergent mixtures can also be tried. Ideally, detergent concentrations should be above the CMC during solubilization; however, during purification it should be ten times lower. The following table (Table 10.5) lists some of the detergents routinely used in purifications.

It is also worthwhile to keep in mind that detergents used for solubilization may not work well with downstream processes such as biophysical characterization or crystallization. Therefore, after initial extraction, excess detergent should be removed or exchanged for an alternative detergent so that it does not affect the

Table 10.5 Different types of detergents used for purification of membrane proteins [53]

Type of detergent	Examples
Ionic	Cetyltrimethylammonium bromide (CTAB), Sodium cholate, Sodium deoxycholate
Nonionic	*N*,*N*-bis(3-D-gluconamidopropyl)cholamide (Big Chap), Octaethylene glycol monododecyl ether ($C_{12}E_8$), Triton X-100, Triton X-114
Zwitterionic	3-[(3-cholamidopropyl) dimethylammonio]-1-propanesulfonate (CHAPS), 3-[(3-cholamidopropyl) dimethylammonio]-2-hydroxypropane-1-sulfonate (CHAPSO), Dodecyldimethylamine oxide (LDAO)

stability of the proteins. Techniques include chromatographic supports, thorough washing with buffer, precipitation, ultrafiltration, and dialysis for detergent exchange or removal [53–55]

10.2.2 Toxic Proteins

10.2.2.1 Overview

Proteins which cause death of host cells on expression or problems in cultivation and maintenance during the growth phase are termed as toxic proteins [56, 57]. This is a problem one often encounters with heterologous expression of proteins in *E. coli*, even though thousands of proteins have been successfully expressed to very high levels in BL21 (DE3). *E. coli* system provides an economical way of mass producing important/commercial proteins. Therefore, cell toxicity can prove to be a major impediment to this process. Literature suggests that about 80% recombinant proteins expressed are toxic to some degree for their hosts, and about 10% of these cause significant damage [57]. While insoluble and dysfunctional proteins do not belong to this category, they do increase the metabolic burden of the host cell. However, some partially soluble or functional proteins may still be toxic to the host cell. A classic example of such a toxic protein is the mammalian gene bax, a member of the bcl-2 family and a positive modulator of apoptosis [58].

10.2.2.2 Problems Encountered

Protein toxicity can manifest itself in different ways. Obtaining few or no colonies compared to regular transformation experiments is most likely a sign of toxicity. The transformation efficiency of vectors coding toxic proteins is significantly lower than that of control vectors or vectors of nontoxic proteins. Observing a smaller percentage of positive clones is also a common phenotype. In *E. coli*, leaky promoters such as *lac* allow some expression of protein in the absence of inducer [30]. This might lead to plasmid instability and/or loss of plasmid, a common problem with ampicillin markers. Such cultures grow slowly and cannot reach normal cell density or do not grow at all. Upon induction, some proteins are so toxic that they inhibit growth and also kill host cells [57].

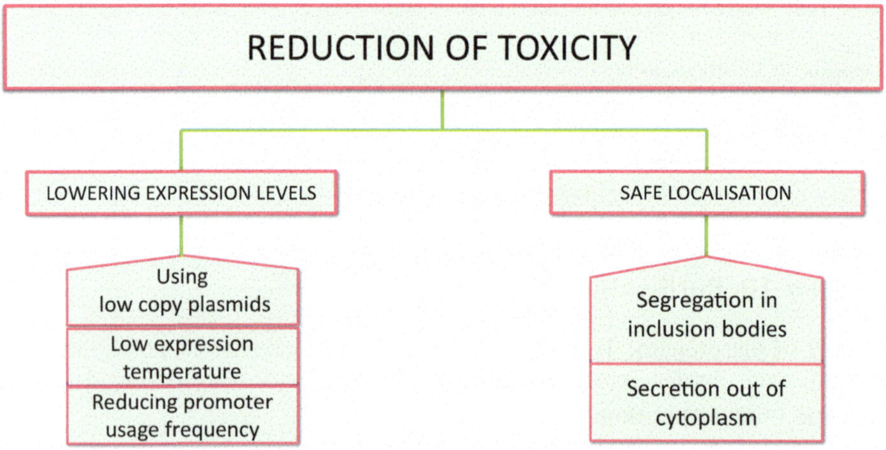

Fig. 10.4 Overcoming protein toxicity. The figure underscores the multipronged approach involved in tackling protein toxicity. The left section of the diagram shows how leaky or overall expression can be lowered until induction by a number of ways such as using low-copy plasmids, reducing promoter usage frequency, or lowering temperature. Additionally, the right section of the diagram shows how reduced cell growth or death can be prevented by sequestering toxic proteins in inclusion bodies or secreting them directly into the cytoplasm

10.2.2.3 Conventional Strategies Employed for Purification

Solutions to tackling the problem of cell toxicity include a multipronged strategy. This includes ways in which the toxic gene can be tolerated by host cells during growth, efficient expression, and high production of the protein on induction without killing off the cells (Fig. 10.4).

Plasmid Stability/Toxicity Test

At the very outset, it is advisable to test the culture to determine the fraction of cells that still carry the target plasmid (plasmid stability). The general method has been discussed in detail in Sect. 10.2.1.3.7 under “Antibiotic and Drug-Based Screening.” Similarly, the plasmid toxicity test is a comparison of the number of colonies obtained on agar plates with and without the inducer. Toxicity is indicated by the absence of colonies on plates containing antibiotic and inducer despite successful transformation and growth in the presence of antibiotic. This preliminary test is a good point to start and decide on adopting an appropriate strategy subsequently.

Expression Hosts

A tight control of toxic gene expression or preventing “leakiness” is one of the key factors responsible for successful production of toxic proteins in *E. coli* [57]. In this regard, some host strains work a great deal better with toxic proteins than the parent BL21 (DE3) cells. Host strains that include C41 (DE3), C43 (DE3), and Lemo21 (DE3) are well suited for the production of toxic proteins by allowing “tunable”

Table 10.6 *E. coli* strains used for production of toxic proteins

E. coli strains	Features
BL21-AI	Combination of T7 RNAP and P_{BAD} promoter
PB4144	This strain has a temperature-sensitive allele of the supF suppressor that converts stop codon UAG to tyrosine
CopyCutter™ and EPI400™	Can regulate the copy number of ColE1-type plasmids (containing colicin E1 gene)
ABLE C and ABLE K	Can reduce the copy number of ColE1-derived plasmids by 4- and 10-fold, respectively
BL21(DE3)NH	Ability to tolerate the expression of unstable genes
ECF529, ECF530	Have copies of plasmid replication regulator π controlled by P_{BAD} promoter, which allows a tight control of the plasmid copy number

Table 10.7 Vectors commonly used for cloning toxic proteins

Vectors	Features
pETcoco-1	Allows arabinose-dependent control of copy number
pOU61	Allows thermal control of copy number
pETKmS1, 2, 3	Includes stabilizing sequences
pLAC11, 22, 33	Use of the highly repressible full-length *lac* promoter-operator region

expression whose features have been covered in great depth in Sect. 10.2.1.3.2 under "Prokaryotic Cells." C41(DE3) or C43(DE3) strains particularly can enhance the plasmid stability for toxic proteins [30]. Some specialized strains not covered in the aforementioned section have been highlighted in the table below (Table 10.6) [46, 57].

Construct Design

Promoters, Plasmids, and Tags

Toxic protein production can be regulated by multiple ways. Some of these include: the use of T7 RNAP with tightly regulated promoters, use of low copy number plasmids, constitutive expression of phage T7 lysozyme from the pLysS/pLysE plasmids to reduce basal expression, and the use a strong terminator upstream of inducible promoter [30, 57]. In case of T7 based promoters and the co-expression of lysozyme from a compatible pLysS or pLysE, the amount of T7RNAP expressed far overcomes the effect of the repressor upon subsequent induction by IPTG. The "leakiness" of *lac* promoters and how this condition can be salvaged to benefit toxic protein production has been highlighted in Sect. 10.2.1.3.7 under **Construct Design—Promoters and Plasmids**. Moreover, suitable promoters such as the arabinose promoter and the benefits of pLysS/pLysE plasmids have also been discussed briefly in the previous sections. Some other vectors not covered earlier have been highlighted in the table below (Table 10.7) [46, 57].

Large tags such as GST or Trx (thioredoxin) are known to alleviate toxicity. However, in some cases tags are also known to do otherwise, so choice of tags is extremely important. Expressing only individual/hydrophilic/less toxic domains instead of the full length protein may solve the problem completely [30, 45, 49]. Alternatively, direct secretion of the protein in the periplasmic space or in the medium may circumvent the problem of toxic protein accumulation inside the cell and allow for easy purification. While it might not be a very convenient method, expressing toxic proteins in inclusion bodies as aggregates is also a strategy that is sometimes adopted. The aggregated proteins are not toxic and can be recovered by denaturation and refolding. However, this is a difficult purification strategy since refolding and obtaining proteins in fully functional form might be quite challenging, which is discussed in detail in Sect. 10.2.3 under *Inclusion bodies*.

Culture Conditions

Optimizing culture conditions can help efficient toxic protein expression immensely. Induction conditions such as induction time, concentration, and temperature can be manipulated [30, 47, 52]. Induction time should be monitored so as to not exceed optimum time and cell density to prevent cell death/protein degradation. A range of inducer concentrations (IPTG 0.1–1.0 mM) should be tested and the lowest concentration allowing optimal expression should be chosen. Very high concentrations burden the cell's machinery. Lower concentrations are usually preferred for toxic proteins. Lower temperatures eliminate protein toxicity and increase protein solubility. Longer induction time goes well with lower temperatures because of reduced protein synthesis. A commonly used temperature range includes 25–15 °C.

Antibiotic screening using high concentrations such as up to 200 mg/mL helps in retaining the expression vector as described in Sect. 10.2.1.3.7 under "Antibiotic and Drug-Based Screening." Alternately, use of antibiotic timentin (~75 μg/mL) instead of ampicillin is also recommended [30]. Growing cultures by directly inoculating colonies from an agar plate is one of the unconventional methods of growing toxic proteins. Similarly, addition of 1% glucose to lactose-rich culture medium to repress induction of the *lac* promoter by lactose is also used [6, 57].

10.2.3 Inclusion Bodies

10.2.3.1 Overview

Expression of huge amounts of recombinant proteins in *E. coli* (>2% of total cellular proteins) sometimes leads to the formation of aggregates referred to as "inclusion bodies" [59]. They usually gather in the cytoplasm or are secreted in the periplasmic space if signal sequences are used in the vector coding the proteins. Culture conditions that might induce formation of inclusion bodies include high temperature, high inducer of concentration, strong promoters, high copy number of target gene, high hydrophobicity, lack of eukaryotic chaperones or posttranslational machinery and reduced environment of bacterial cytosol [60]. Under such conditions, protein translation rate is unusually high that puts the quality control machinery of the cell under pressure resulting in the formation of partially folded/misfolded protein aggregates. However, some advantages of inclusion body production of the target protein are high amount and purity. These proteins remain protected in inclusion bodies away from action of proteolytic enzymes. It also offers a way to produce toxic proteins without killing the cell. Inclusion bodies can form in yeast, insect, and mammalian cells as well. *E. coli* inclusion bodies appear as dense, refractile spherical/cylindrical particles of 0.2–1.5 μm in diameter with rough or smooth edges. Found on either one or both poles of the cell, on cell division, they are inherited by only one of the daughter cells. Interleukin-1β, β-lactamase and Human granulocyte-colony stimulating factor (GCSF) form inclusion bodies while retaining biological activity. They also possess secondary structures and are referred to as "non-classical inclusion bodies" [59].

10.2.3.2 Problems Encountered

Inclusion bodies impose a major hurdle in purifying recombinant proteins from the host. Purification involves isolation, solubilization, and refolding to yield functional proteins [61]. Despite technological advances, extraction and refolding yields very poor yield of final product. Extraction of protein from washed pellets using chaotropes such as urea and guanidine hydrochloride (GdnHCl) often leads to complete loss of existing secondary structures and results in aggregation of protein molecules even during the refolding process [60].

10.2.3.3 Conventional Strategies Employed for Purification

When faced with a problem of inclusion bodies, one can either optimize to make the target protein soluble or purify protein from the pellets and refold (Fig. 10.5). Briefly, changes in culture conditions such as reduced temperature, induction time, inducer concentration, cell density in addition to solubility-increasing tags, co-expression of chaperonins and use of appropriate expression hosts help in increasing solubility [60, 62]. Some of these factors have been discussed comprehensively in the previous sections. This section focuses specifically on purification of proteins from solubilized pellets.

Cell Disruption Isolation and Solubilization

Cell rupture methods can be mechanical such as sonication/homogenization, or chemical (preferred) by use of cell lysis agents such as lysozyme [59, 60]. Inclusion bodies are separated by centrifugation due to higher density; however, these extracts are still contaminated with host proteins, cellular debris, and RNA and need to be washed with detergents such as deoxycholic acid and Triton X-100 [59, 60]. Addition of DNase at this stage also removes contaminating DNA. The washed pellets are then solubilized using denaturants such as urea and guanidine hydrochloride (GdnHCl). β-mercaptoethanol (BME) or dithiothreitol (DTT) are added to prevent incorrect formation of disulfide bonds in case the protein contains multiple cysteine residues [59]. Since high concentration of these reagents often leads to total loss of protein structure with problems in refolding, a "mild" method of solubilization is recommended [59]. Mild solubilization using agents such as Tris–HCl buffer, low concentration of Dimethyl Sulfoxide (DMSO), n-propanol, and sarcosyl is beneficial for extraction of "non-classical inclusion bodies" with no need for refolding. For

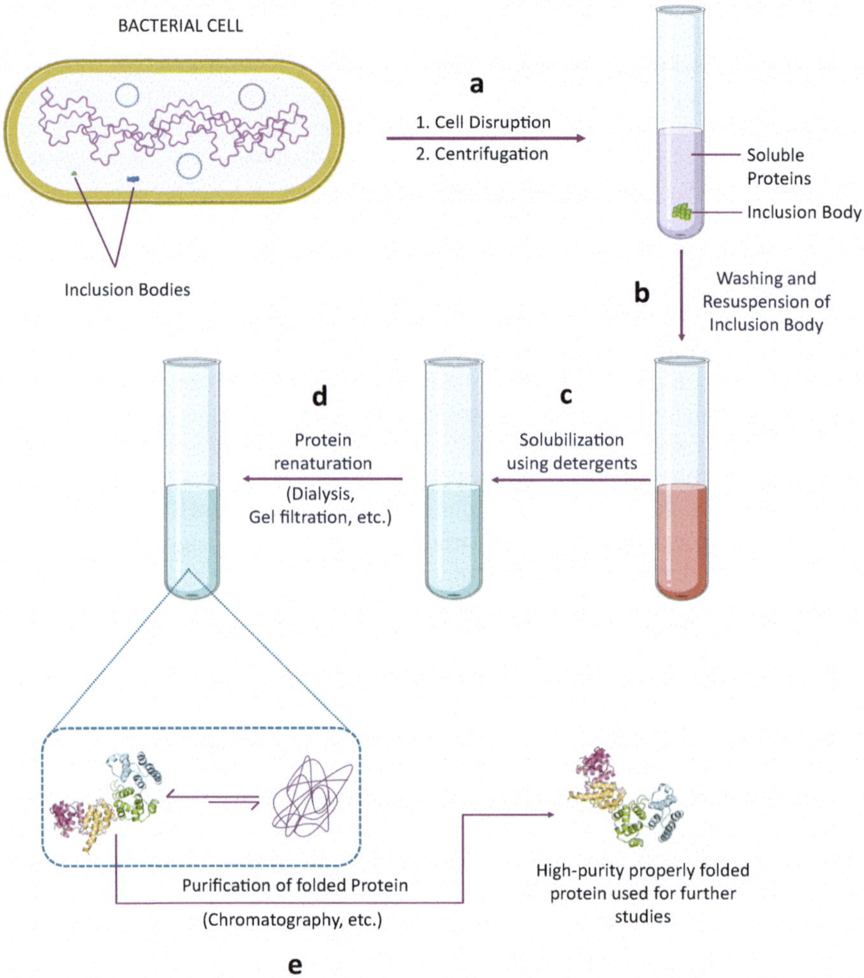

Fig. 10.5 Purification of proteins from inclusion bodies. The figure enumerates the basic steps in purification involving solubilization, extraction, and refolding to yield functional proteins. (**a**) Cells are disrupted through mechanical or chemical methods and inclusion bodies are separated by centrifugation. (**b**) These are subsequently washed to remove unwanted chemicals, (**c**) solubilized using detergents, and (**d**) renatured or refolded to yield well-folded functional proteins. The renaturation process usually involves dialysis and/or gel filtration. (**e**) The renatured proteins are finally purified using a suitable technique such as chromatography

classical inclusion bodies as well reagents such as lower concentrations of urea, organic solvents (ethanol, propanol to enhance folding), and high pH buffers in combination with 2 M urea as well as detergents (SDS, CTAB, N-Lauroylsarcosine, Lauroyl-L-glutamate) are recommended for mild and efficient solubilization. High hydrostatic pressure (2–4 kbar) is also an efficient method that works by disrupting aggregates and molecular interactions of inclusion bodies [59]. Subsequent removal of pressure spontaneously aids in refolding of the proteins. Temperature is rarely used as a primary denaturing agent but reagents such as Gdn-HCl or urea are commonly used and are more effective at elevated temperatures (37–60 °C) [59]. However, it is to be noted that no single method works for all target proteins and choice of reagent needs to be optimized.

Refolding

Refolding of the extracted proteins is achieved by removal of the denaturants and is a critical step in purification of proteins from inclusion bodies. Denaturant is generally removed by dilution and dialysis in a refolding buffer [63]. Conventional methods of dilution and dialysis have many disadvantages such as requirement of huge amounts of buffer, low yield, aggregation, and unsuitability for large-scale protein production [64]. These bottlenecks were overcome with recent developments of various chromatographic methods allowing on-column refolding. Size exclusion, affinity chromatography, and ion exchange are used most extensively, while hydrophobic interaction chromatography is sparingly used [60, 64]. Size exclusion chromatography separates the folded form from the misfolded aggregated proteins on elution, with the refolding buffer. In other methods, proteins are on immobilized solid support before refolding [61]. More recent methods make the use of micro-fluidic chips and urease enzyme. In micro-fluidic chips, the denaturant is passed in a controlled manner through laminar flow in micro-channels. This allows control over denaturant concentration and has been used successfully for refolding difficult proteins like citrate synthase [60, 61, 63]. The urease method involves gradual removal of urea from the solubilized protein solution mediated by a urease enzyme-catalyzed reaction, thus reducing the requirement of large volumes of refolding buffer [60].

Another important criterion is the composition of the refolding buffer. One strategy that mimics in vivo refolding makes use of pro-peptides in the refolding buffer to increase the refolding yield [59, 64]. A pro-peptide is generally found between the signal peptide and the mature part of a polypeptide and helps proteins fold in vivo. Other strategies use chaperones and other folding catalysts like peptidylprolyl *cis-trans* isomerase or protein disulfide isomerase. However, there is no universal buffer for refolding and the constituents need to be screened for each protein of interest. One can choose from a variety of additives, which act as refolding enhancers as listed below (Table 10.8) [59–61].

Table 10.8 List of additives generally used as constituents of refolding buffers for purification of insoluble proteins

Category	Additives
Chaotropes	Urea, Guanidine hydrochloride
Amino acids	Glycine, Arginine, Proline
Sugars and polyhydric alcohols	Sucrose, Polyethylene glycol, Sorbitol, Glycerol
Others	Sulfobetaines, substituted pyridines and pyrolles, Acid substituted aminocyclohexanes

10.3 Case Studies of Challenging Proteins

This section describes simplified protocols collected from literature that have been optimized for a specific sample. These would provide simple steps for designing and implementing one's own protocols for handling challenging proteins.

10.3.1 Purification of the Recombinantly Expressed Membrane Protein Ammonium Transporter (AmtB) from E. coli

This protocol is an example of a simple and cost-efficient method of efficiently purifying AmtB for crystallization, by screening the most efficient detergents [65]. *A total of 26 detergents, 4 types of chromatography columns, and various buffer conditions were screened using a 96-well plate format.*

10.3.1.1 Materials

1. Expression construct of AmtB gene with N-terminal FLAG and C-terminal 6-His.
2. *E. coli* C41(DE3) host cells.
3. Detergents: FOS-CHOLINE-10 (FC10), FC11, FC12, HEGA-10, Nonyl maltoside, Decyl maltoside, Undecyl maltoside, Dodecyl maltoside, CHAPS, CHAPSO, Nonyl thiomaltoside, Decyl thiomaltoside, Undecyl thiomaltoside, Dodecyl thiomaltoside, Cymal-6, Cymal-7, LDAO, TDAO, C8E4, C8E6, C10E5, C10E5, C12E8, Octyl glucoside, Nonyl glucoside, Triton X-100, Triton X-114.
4. Buffer A (pH 8.0): 20 mM Tris–HCl, 50 mM NaCl, 0.5 mM Tris(2-carboxyethyl) phosphine hydrochloride (TCEP), 1 mg/mL cell wall disrupting enzyme lysozyme, complete protease inhibitor cocktail EDTA-free and 10 units/mL endonuclease benzonase.
5. Buffer P (pH 8.0): 20 mM Tris–HCl, 300 mM NaCl, 0.5 mM reducing agent TCEP.

10.3.1.2 Method

1. Transformed *E. coli* C41 (DE3) cells were cultured in a shake flask at 37 °C till an OD_{600} of 0.8 was reached.
2. The culture was induced with 0.1 mM IPTG at 20 °C overnight.
3. 1 mL aliquots of culture were transferred to a 96-well plate and cells were harvested by centrifugation at 3000 × *g* for 10 min.
4. In each well, pellets were resuspended in 50 μL of Buffer A and a detergent (1–2% concentration) from those listed in materials.
5. Cells were lysed at 4 °C for an hour and the suspension passed through a 96-well filter plate to remove inclusion bodies and precipitates.
6. Filtrate was added to 25 μL Ni-NTA agarose resin in a 96-well plate format containing Buffer P and an appropriate detergent.
7. Binding was allowed by 15 min agitation at 4 °C. Unbound material was removed by centrifugation at 100 × *g* for 30 s.

8. The resin was washed with 20 column volumes (CV) of 40 mM imidazole in the Buffer P containing the appropriate detergent at 100 × *g* for 30 s. The bound recombinant membrane proteins were eluted in 30 μL Buffer P containing 500 mM imidazole and the respective detergent, by centrifugation at 100 × *g* for 1 min.
9. These samples were then analyzed by dot blot and SDS PAGE. 1 μL of sample was transferred to nitrocellulose, dried, and the target 6-His-tagged proteins were detected using western blotting probe [65]. The purity of the samples was analyzed by SDS–PAGE.
10. The effect of different imidazole concentrations and four different types of metal affinity columns was also investigated and optimal conditions were chosen for large-scale purification and crystallization.
11. FC12 at 0.1% and TALON resin were chosen for large-scale purifications. Gel filtration was also used to improve purity of AmtB. Some other detergents showing low efficiency were also included in the large-scale purification to study their effect in purification and crystallization.
12. AmtB was extracted with FC12 and subsequently purified with FC12, UDM, DDM, Cy6, Cy7, OG, and LDAO.

AmtB crystals appeared in the first screen after 3 days in the presence of DDM and Cymal-6. The crystals could diffract to ~11 Å.

10.3.2 Extraction of Proteins from Inclusion Bodies in *E. coli*

This protocol describes the solubilization, extraction, and refolding of recombinant protein from inclusion bodies [66, 67].

10.3.2.1 Materials

1. Reagents: 40% glucose (protein stabilizer), Ampicillin (100 mg/mL), IPTG (100 mg/mL).
   Stock solutions: lysozyme (50 mg/mL), DNase (1 mg/mL in 50% glycerol), 75 mM NaCl, $MgCl_2$ (0.5 M in dH_2O to remove impurities).
2. Buffers.
   Solution buffer (pH 8.0): 50 mM Tris–Cl, 25% sucrose (protein stabilizer), 1 mM NaEDTA (chelating agent), 0.1% NaAzide, 10 mM DTT (reducing agent).
   Lysis buffer (pH 8.0): 50 mM Tris–Cl, 1% Triton X-100, 1% Na deoxycholate (ionic detergent), 100 mM NaCl, 0.1% NaAzide (bacteriostatic agent), 10 mM DTT.
   Washing buffer with Triton (pH 8.0): 50 mM Tris–Cl, 0.5% Triton X-100, 100 mM NaCl, 1 mM NaEDTA, 0.1% NaAzide, 1 mM DTT.
   Washing buffer without Triton (pH 8.0).
   Refolding buffer: 100 mM Tris–Cl, 400 mM L-Arginine, 2 mM NaEDTA, 0.5 mM oxidisedglutathione, 5 mM reduced Glutathione, protease inhibitors, pH 8.0.

Elution buffer for FPLC (pH 8.0): 150 mM NaCl, 20 mM Tris–Cl pH 8.0.

10.3.2.2 Method

1. Transformed *E. coli* cells were cultured in a shake flask with glucose and amp till an OD_{600} of 0.7 was reached.
2. Culture was induced with IPTG for 3 h and pelleted by centrifugation.
3. Pellet was resuspended in 13 mL solution buffer and sonicated at 50% level 4–5, 30 pulses on ice.
4. To this suspension 100 μL lysozyme, 250 μL DNase I, and 50 μL $MgCl_2$ were added, vortexed, followed by addition of 12.5 mL lysis buffer.
5. After 30 min at room temperature, 350 μL NaEDTA was added and dropped in liquid N_2. Thereafter, 200 μL $MgCl_2$ was added after thawing sample for 30 min at 37 °C.
6. When viscosity decreased, 350 μL NaEDTA was added and centrifuged at 4 °C to retrieve pellet.
7. Pellet was then resuspended in 10 mL washing buffer with Triton, sonicated on ice, and centrifuged to discard the supernatant.
8. Pellet was again resuspended in 10 mL washing buffer without Triton, sonicated on ice, and centrifuged to discard the supernatant.
9. Thereafter, pellet was dissolved in 9 mL of 8 M GdnHCL (pH 8.0) with 4 mM DTT till it dissolved.
10. 1 mL of dissolved protein was added dropwise with a syringe to the refolding buffer while vigorously stirring.
11. This solution was then stirred slowly for 8 h at 4 °C. **Steps 10** and **11** were repeated twice.
12. Finally sample was filtered and concentrated till 12 mL in a 200 mL concentrator to get a good concentration of the protein. OD_{280} was measured.
13. For higher purity, sample was then passed through size exclusion column and eluted in elution buffer.
14. Samples were then analyzed on SDS-PAGE for purity.

10.4 Conclusion

Recombinant protein production technology has made huge advances over time. An integral part of this process involves overcoming challenges faced by researchers in expressing and purifying "difficult" proteins. These types of proteins belong to a number of categories such as membrane proteins, toxic proteins, inclusion bodies, monoclonal antibodies, and intrinsically disordered proteins. The amount of literature available in each category is exhaustive and beyond the scope of a single chapter. However, the most salient categories of challenging proteins have been dealt with in this chapter with an aim to summarize the common problems encountered and the general methods one can use to work around them. These strategies can be employed at different stages, from as early on as designing a suitable expression construct to optimizing conditions of purification much later on. While the three categories of proteins discussed in the chapter differ in nature, the techniques used for optimal production of these proteins often overlap. Overall, the conventional ways to tackle these proteins include an appropriate construct design, optimizing expression and culture conditions as well utilizing efficient methods and suitable reagents for purification. Having said that, each protein is unique and these general strategies need to be modified and customized for each individual candidate.

Acknowledgments The authors thank Advanced Centre for Treatment, Research and Education in Cancer (ACTREC) for providing necessary infrastructure and resources for successful completion of the chapter. The authors acknowledge Ms. Chanda Baisane, Bose Lab for formatting the manuscript.

Problems

Multiple Choice Questions

1. When designing a construct to adequately express a toxic protein in a host system which of the following do you need to take into consideration:
   (a) A tightly regulated promoter
   (b) A low copy number plasmid
   (c) Use a tag that alleviates toxicity
   (d) All of the above
2. If you were tasked with quickly purifying a large amount of a membrane protein to study the posttranslational modifications, the system you think would be best suited for the job would be:
   (a) E. coli
   (b) Saccharaomyces cerevisiae
   (c) Spodoptera frugiperda
   (d) Cell-free system
3. Isolation of inclusion bodies can be achieved by:
   (a) Centrifugation
   (b) Use of detergents
   (c) Both A and B
   (d) Neither A nor B

Subjective Questions

1. A 30 kDa protein with a PI of 7.0 was expressed in BL21 (DE3) host cells using a pET28a vector with N- and C-terminal Histidine tags. On performing an expression check using centrifuged bacterial pellets, the protein was found to be abundantly produced. However, post-purification the protein was hardly visible on the SDS-PAGE gel. An expression check

(continued)

using the supernatant and pellets of the centrifuged cells separately indicated that the protein was insoluble.

Lysis buffer: 20 mM Tris–Cl (pH 8), 300 mM NaCl, 10 mM Immidazole, 5% Glycerol, pH 8.

What are some of the steps that could be taken to salvage the protein from the inclusion bodies?

2. The same protein 30 kDa protein (Q3a) could not be made to express in the soluble fraction despite trying out multiple strategies. What should be the best bet at this point?

References

1. Wingfield PT. Overview of the purification of recombinant proteins. Curr Protoc Protein Sci. 2015;80:6.1.1–6.1.35.
2. Pham PV. Chapter 19 - Medical biotechnology: techniques and applications. In: Barh D, Azevedo V, editors. Omics technologies and bio-engineering. New York: Academic Press; 2018. p. 449–69.
3. Grayson KJ, Anderson JLR. Designed for life: biocompatible de novo designed proteins and components. J R Soc Interface. 2018;15(145):20180472.
4. Quijano-Rubio A, Ulge UY, Walkey CD, Silva D-A. The advent of de novo proteins for cancer immunotherapy. Curr Opin Chem Biol. 2020;56:119–28.
5. HjertÉN S, Pan H, Yao K. Some general aspects of the difficulties to purify membrane proteins. In: Peeters H, editor. Protides of the biological fluids, vol. 29. Netherland: Elsevier; 1982. p. 15–25.
6. Carpenter EP, Beis K, Cameron AD, Iwata S. Overcoming the challenges of membrane protein crystallography. Curr Opin Struct Biol. 2008;18(5):581–6.
7. Ott CM, Lingappa VR. Integral membrane protein biosynthesis: why topology is hard to predict. J Cell Sci. 2002;115(10):2003.
8. Newport TD, Sansom MSP, Stansfeld PJ. The MemProtMD database: a resource for membrane- embedded protein structures and their lipid interactions. Nucleic Acids Res. 2019;47 (D1):D390–D7.
9. Hedin LE, Illergård K, Elofsson A. An introduction to membrane proteins. J Proteome Res. 2011;10(8):3324–31.
10. Yeagle PL. Chapter 10 - Membrane proteins. In: Yeagle PL, editor. The membranes of cells. 3rd ed. Boston: Academic Press; 2016. p. 219–68.
11. Guo L, Wang S, Li M, Cao Z. Accurate classification of membrane protein types based on sequence and evolutionary information using deep learning. BMC Bioinform. 2019;20(25):700.
12. Spaar A, Flöck D, Helms V. Association of cytochrome c with membrane-bound cytochrome c oxidase proceeds parallel to the membrane rather than in bulk solution. Biophys J. 2009;96 (5):1721–32.
13. Hu J, Qin H, Gao FP, Cross TA. A systematic assessment of mature MBP in membrane protein production: overexpression, membrane targeting and purification. Protein Expr Purif. 2011;80 (1):34–40.
14. Quagliariello E, Palmieri P. Structure and function of membrane proteins. Netherland: Elsevier; 1983. p. II.
15. Maître J-L, Heisenberg C-P. Three functions of cadherins in cell adhesion. Curr Biol. 2013;23 (14):R626–R33.

16. Clemmons DR, Maile LA. Minireview: integral membrane proteins that function coordinately with the insulin-like growth factor I receptor to regulate intracellular signaling. Endocrinology. 2003;144(5):1664–70.
17. Zorman S, Botte M, Jiang Q, Collinson I, Schaffitzel C. Advances and challenges of membrane-protein complex production. Curr Opin Struct Biol. 2015;32:123–30. (1879-033X (Electronic))
18. Mancia F, Love J. High-throughput expression and purification of membrane proteins. J Struct Biol. 2010;172(1):85–93.
19. Blankenship E, Lodowski DT. Rhodopsin purification from dark-adapted bovine retina. Methods Mol Biol. 2015;1271:21–38. (1940–6029 (Electronic))
20. King MS, Boes C, Kunji ERS. Chapter 4 – Membrane protein expression in Lactococcus lactis. In: Shukla AK, editor. Methods in enzymology, vol. 556. New York: Academic Press; 2015. p. 77–97.
21. Schlegel S, Hjelm A, Baumgarten T, Vikström D, de Gier J-W. Bacterial-based membrane protein production. Biochim Biophys. 2014;1843(8):1739–49.
22. Zoonens M, Miroux B. Expression of membrane proteins at the *Escherichia coli* membrane for structural studies. Methods Mol Biol. 2010;601:49–66. (1940–6029 (Electronic))
23. Reece-Hoyes JS, Walhout AJM. Gateway recombinational cloning. Cold Spring Harb Protoc. 2018;2018(1):pdb.top094912-pdb.top.
24. Bruni R, Kloss B. High-throughput cloning and expression of integral membrane proteins in *Escherichia coli*. Curr Protoc Protein Sci. 2013;74:29.6.1–29.6.34. (1934–3663 (Electronic))
25. Festa F, Steel J, Bian X, Labaer J. High-throughput cloning and expression library creation for functional proteomics. Proteomics. 2013;13(9):1381–99.
26. Geertsma ER. FX cloning: a versatile high-throughput cloning system for characterization of enzyme variants. Methods Mol Biol. 2013;978:133–48. (1940–6029 (Electronic))
27. Erbakan M, Shen YX, Grzelakowski M, Butler PJ, Kumar M, Curtis WR. Molecular cloning, overexpression and characterization of a novel water channel protein from *Rhodobacter sphaeroides*. PLoS One. 2014;9(1):e86830.
28. Rosano GL, Ceccarelli EA. Recombinant protein expression in *Escherichia coli*: advances and challenges. Front Microbiol. 2014;5:172. <bacterial-strains-for-protein-expression.pdf>
29. Kesidis A, Depping P, Lode A, Vaitsopoulou A, Bill RM, Goddard AD, et al. Expression of eukaryotic membrane proteins in eukaryotic and prokaryotic hosts. Methods. 2020;180:3–18.
30. Routledge SJ, Mikaliunaite L, Patel A, Clare M, Cartwright SP, Bawa Z, et al. The synthesis of recombinant membrane proteins in yeast for structural studies. Methods. 2016;95:26–37.
31. Byrne B. Pichia pastoris as an expression host for membrane protein structural biology. Curr Opin Struct Biol. 2015;32:9–17.
32. Baghban R, Farajnia S, Rajabibazl M, Ghasemi Y, Mafi A, Hoseinpoor R, et al. Yeast expression systems: overview and recent advances. Mol Biotechnol. 2019;61(5):365–84. (1559–0305 (Electronic))
33. Fraser NJ. Expression and functional purification of a glycosylation deficient version of the human adenosine 2a receptor for structural studies. Protein Expr Purif. 2006;49(1):129–37. (1046–5928 (Print))
34. Shiroishi M, Kobayashi T, Ogasawara S, Tsujimoto H, Ikeda-Suno C, Iwata S, Shimamura T. Production of the stable human histamine H_1 receptor in Pichia pastoris for structural determination. Methods. 2011;55(4):281–6. (1095-9130 (Electronic))
35. Moiseenkova-Bell V, Wensel TG. Functional and structural studies of TRP channels heterologously expressed in budding yeast. Adv Exp Med Biol. 2011;704:25–40.
36. Kapoor K, Finer-Moore JS, Pedersen BP, Caboni L, Waight A, Hillig RC, et al. Mechanism of inhibition of human glucose transporter GLUT1 is conserved between cytochalasin B and phenylalanine amides. Proc Natl Acad Sci. 2016;113(17):4711. <CPPS-10-591.pdf>
37. Marheineke K, Grünewald S, Christie W, Reiländer H. Lipid composition of Spodoptera frugiperda (Sf9) and *Trichoplusia ni* (Tn) insect cells used for baculovirus infection. FEBS Lett. 1998;441(1):49–52. (0014–5793 (Print))
38. Ooi A, Wong A, Esau L, Lemtiri-Chlieh F, Gehring C. A guide to transient expression of membrane proteins in HEK-293 cells for functional characterization. Front Physiol. 2016;7:300.

39. Khambhati K, Bhattacharjee G, Gohil N, Braddick D, Kulkarni V, Singh V. Exploring the potential of cell-free protein synthesis for extending the abilities of biological systems. Front Bioeng Biotechnol. 2019;7:248.
40. Schneider B, Junge F, Shirokov VA, Durst F, Schwarz D, Dotsch V, et al. Membrane protein expression in cell-free systems. Methods Mol Biol. 2010;601:165–86.
41. Krinsky N, Kaduri M, Shainsky-Roitman J, Goldfeder M, Ivanir E, Benhar I, et al. A simple and rapid method for preparing a cell-free bacterial lysate for protein synthesis. PLoS One. 2016;11 (10):e0165137.
42. Gregorio NE, Levine MZ, Oza JP. A user's guide to cell-free protein synthesis. Methods Protoc. 2019;2:1.
43. He Y, Wang K, Yan N. The recombinant expression systems for structure determination of eukaryotic membrane proteins. Protein Cell. 2014;5(9):658–72.
44. Bowers LM, LaPoint K, Anthony L, Pluciennik A, Filutowicz M. Bacterial expression system with tightly regulated gene expression and plasmid copy number. Gene. 2004;340(1):11–8.
45. Kubicek J, Block H, Maertens B, Spriestersbach A, Labahn J. Expression and purification of membrane proteins. Methods Enzymol. 2014;541:117–40.
46. Lin-Cereghino GP, Stark CM, Kim D, Chang J, Shaheen N, Poerwanto H, et al. The effect of α-mating factor secretion signal mutations on recombinant protein expression in Pichia pastoris. Gene. 2013;519(2):311–7.
47. Karyolaimos A, Ampah-Korsah H, Zhang Z, de Gier JW. Shaping *Escherichia coli* for recombinant membrane protein production. FEMS Microbiol Lett. 2018;365:15. <NEB Impact system.pdf>.<Protein Expression - E.coli - Decreasing Protein Toxicity - EMBL.pdf>
48. Pedro AQ, Queiroz JA, Passarinha LA. Smoothing membrane protein structure determination by initial upstream stage improvements. Appl Microbiol Biotechnol. 2019;103(14):5483–500.
49. Arnold T, Linke D. The use of detergents to purify membrane proteins. Curr Protoc Protein Sci. 2008;Chapter 4:Unit 4.8.1–4.8.30.
50. Smith SM. Strategies for the purification of membrane proteins. Methods Mol Biol. 2017;1485:389–400.
51. Lin S-H, Guidotti G. Chapter 35 Purification of membrane proteins. Methods Enzymol. 2009;463:619–29. <toxic proteins.pdf>
52. Saïda F, Uzan M, Odaert B, Bontems F. Expression of highly toxic genes in *E. coli*: special strategies and genetic tools. Curr Protein Pept Sci. 2006;7(1):47–56. (1389–2037 (Print))
53. Donnelly MI, Stevens PW, Stols L, Su SX, Tollaksen S, Giometti C, et al. Expression of a highly toxic protein, Bax, in *Escherichia coli* by attachment of a leader peptide derived from the GroES cochaperone. Protein Expr Purif. 2001;22(3):422–9.
54. Singh A, Upadhyay V, Upadhyay AK, Singh SM, Panda AK. Protein recovery from inclusion bodies of *Escherichia coli* using mild solubilization process. Microb Cell Factories. 2015;14:41.
55. Palmer I, Wingfield PT. Preparation and extraction of insoluble (inclusion-body) proteins from *Escherichia coli*. Curr Protoc Protein Sci. 2004;Chapter 6:Unit 6.3.
56. Kumar D. Chromatographic protein refolding/renaturation. In: Reference module in life sciences. Netherland: Elsevier; 2018. <Applied and Environmental Microbiology-1991-Strandberg-1669.full.pdf>.
57. Singh SM, Panda AK. Solubilization and refolding of bacterial inclusion body proteins. J Biosci Bioeng. 2005;99(4):303–10. (1389–1723 (Print))
58. Yamaguchi H, Miyazaki M. Refolding techniques for recovering biologically active recombinant proteins from inclusion bodies. Biomol Ther. 2014;4(1):235–51.
59. Niegowski D, Hedrén M, Nordlund P, Eshaghi S. A simple strategy towards membrane protein purification and crystallization. Int J Biol Macromol. 2006;39(1):83–7. <Isolation of proteins from.pdf>
60. Palmer I, Wingfield PT. Preparation and extraction of insoluble (inclusion-body) proteins from *Escherichia coli*. Curr Protoc Protein Sci. 2012;70(1):6.3.1–6.3.20.

Protein Quantitation and Detection

11

Raghupathi Kummari, Rashmi Puja, and Kakoli Bose

Abstract

The precise measurement of protein concentration is a prerequisite for proceeding towards biochemical, biophysical and structural characterizations of a purified recombinant protein. Accuracy in measurement of protein concentration is essential towards quantitative assays including enzyme kinetics, protein-protein interaction studies using biophysical probes, protein folding and structural analyses where reproducibility of the experimental parameters within a narrow error range becomes essential both in basic and industrial research. With advancement in recombinant DNA technology and its wide application in obtaining pure protein of interest, requirement of reliable protein quantitation tools became imperative. This growing need urged researchers to develop various assays for determining concentrations of proteins with precision and high sensitivity. Although no tool is universally applicable, investigators can choose the most appropriate quantitation assay based on sample amount, amino acid contents, etc. This chapter describes fast and accurate protein quantitation tools that are routinely used in research laboratories and can be adopted in high-throughput format as well. Furthermore, it provides a detailed protocol of protein detection and purity determination method using sodium dodecyl sulphate poly acrylamide gel electrophoresis (SDS PAGE).

R. Kummari · R. Puja · K. Bose (✉)
Integrated Biophysics and Structural Biology Lab, Advanced Centre for Treatment, Research and Education in Cancer (ACTREC), Tata Memorial Centre, Mumbai, India

Homi Bhabha National Institute, BARC Training School Complex, Mumbai, India
e-mail: kbose@actrec.gov.in

K. Bose (ed.), *Textbook on Cloning, Expression and Purification of Recombinant Proteins*, https://doi.org/10.1007/978-981-16-4987-5_11

Keywords

Quantitation · Lowry · Bradford · BCA assay · SDS PAGE · Beer Lamber Law · UV spectroscopy

11.1 Introduction

Recombinant proteins need to be quantified prior to their structural and functional analyses. Biophysical studies demand proteins of high purity and of a precise concentration. Therefore, methods to accurately quantitate proteins are very important for their subsequent characterizations. This chapter therefore focuses on protein quantification using spectroscopic methods in solution. These methods are fast, do not require any treatment (chemical and enzymatic), and can be done with a crude mixture or purified protein. These spectroscopic probes involve UV absorption, colorimetric dye-binding as well as fluorescent based methods [1]. They have the flexibility of either being used for measuring concentrations of individual proteins or in high-throughput formats to cater to the requirements of specific experiments. Furthermore, these are routine laboratory experiments with minimal instrumentation and inexpensive reagents that are readily available in any laboratory. The selection method can vary depending upon the sample type (peptides, proteins, chemical modifications or post-translational modifications, etc.). However, we will focus the discussion primarily on quantitation techniques pertaining to recombinant proteins obtained from bacterial expression systems.

With progress in recombinant protein technology, parallel development of protein quantitation tools has also taken place. Different conventional spectroscopic techniques including UV absorption measurement, bicinchoninic assay (BCA), Bradford assay, Lowry assay and so forth have been fine-tuned over the past decade enabling their applications in high-throughput set-ups as well. However, it is important to note that every tool has its own set of advantages and limitations and has to be chosen based on its suitability in the particular case. For example, UV absorption measurements at 280 nm might not be the best choice for proteins having none or minimal number of aromatic acids, mainly tryptophan [2]. In that case, if the analyte is pure and consists of one particular protein in greater proportion (for example, a protein purified using affinity chromatography), Bradford and Lowry assays can be used. However, if the protein is expressed in eukaryotic system and assumed to be highly glycosylated, then Bradford and Lowry methods should be avoided [1]. Furthermore, since recombinant protein obtained at the end of purification process is often of limited quantity and sometimes might be of lower concentration, it is important that the assay methods should be highly sensitive and require minimal amount of protein samples. Hence, there are distinct criteria for different proteins (as they differ in amino acid side chain sequences, isoelectric pI, secondary structure and prosthetic groups) that need to be taken into consideration while choosing the best quantitation technique, which have been discussed later in this chapter in detail. This chapter concludes by describing protein detection method using SDS PAGE that is routinely used in all laboratories.

11.2 Types of Protein Quantitation Assays

There are several types of spectroscopic assays available for protein quantitation and each has its own *pros* and *cons* as described in Table 11.1 (adapted and modified from [3, 4]). No single assay is universally suitable and acceptable for all the samples. It is the researcher to decide the type of assay to be used based on the sample's major content and volume, required accuracy for the experiment to be performed, incubation time, reproducibility and presence of additional interfering substances [1]. Some of the assays are currently available in microplate format making them faster, easier, more sensitive and high-throughput. Comparison of different types of spectroscopic techniques with their *pros* and *cons* is provided in Table 11.1 below.

Table 11.1 Different protein quantitation techniques

Name of the method, detection limit and wavelength range	Mechanism	Advantages	Limitations
Bradford Assay 20–2000 μg/ml 470 nm	• Protein-Coomassie blue dye complex formation	• Very sensitive • Fast and simple • Wide range buffer compatibility • Inexpensive	• Requirement of standards • Wide variation in response to different proteins • Nonlinear standard curve
Lowry Assay 10–1000 μg/ml 750 nm	• Proteins reduce cupric to cuprous followed by Folin–Ciocalteu reduction	• Simple procedure • Sensitive • Microplate format is possible	• Easily influenced by variations in tyrosine and tryptophan residues • Time-consuming • Can be interfered by small compounds
BCA Assay 20–2000 μg/ml 562 nm	• Reaction occurs with reduction of Cu^{2+} to Cu^{+}	• Very sensitive • Rapid • Has a wide range of buffer compatibility • Response with little changes between different types of proteins	• At room temperature, the reaction doesn't complete with ease
UV Absorption 0.1–100 μg/ml 280 nm	• Absorption by aromatic amino acids (tyrosine, tryptophan and phenylalanine)	• Sample can be easily recovered • Requirement of very small volume • Fast and simple • Inexpensive	• Least sensitive method Easily influenced by presence of nucleic acids and other additives such as detergents and denaturants

Adapted from Refs. [3–5]

General Considerations and Necessary Precautions

1. Micro well plates versus cuvette-based studies:

   For any colorimetric study using spectroscopic tools, there is a choice between high-throughput analyses with multiple samples (using microplate) versus cuvette-based assays for single proteins. If there are multiple samples, and each sample needs to be precisely measured in duplicates or triplicates, micro plates are the best option for such situations. Furthermore, they are fast and require very little protein sample. However, it is to be noted that 96-well quartz plates are expensive and might go through scratches from multiple handling and hence care should be taken while using them. In addition, this method does not comply very well with UV absorption method of protein quantitation, where individual protein sample needs to be blank corrected. On the other hand, if there is a single or very less number of samples, cuvette-based assays using absorption spectrophotometer is most appropriate. For cuvette-based assays, it is recommended to use a two-beam instrument where a reference cell is present that can be used for buffer measurement for blank correction. This prevents loss of accuracy due to instrumental drift as well as saves time for blanking each sample separately [1]. However, for any spectroscopic measurement, extreme care must be taken to clean and dry the cuvette as fingerprints and dust particles pose impediment to experimental precision. An appropriate washing solution should be used for cleaning the cuvettes and dried with a stream of Nitrogen gas before proceeding with the experiment [1].

2. Interfering substances:

   It is difficult to obtain a sample devoid of all interfering substances, which might be a by-product of certain steps in protein purification or stabilization. For example, His-tagged protein solutions elute from the Ni-NTA column with a high concentration of imidazole, and similarly GST-tagged proteins contain a considerable amount of reducing agent (e.g. Dithiothreitol or DTT) to prevent aggregation of the tag [6]. Some proteins require additional stabilizing agents like glycerol or detergents for maintaining their stability post purification.

   Therefore, prior to selecting the best protein quantitation method, it is imperative to understand the tolerance level of each assay for such reagents or additives. A list of threshold levels of additives for each type of assay can be considered prior from the tables provided [1, 7–10]. Nonetheless, it is recommended to remove the additives in an additional purification step if possible. For example, imidazole can be removed through desalting using an additional gel filtration chromatographic step as described in earlier sections of this book (Chap. 8). However, it is important to note that these additional steps might often cause dilution, incomplete recovery, precipitation, and aggregation of the protein sample, which can be minimized by following certain steps as discussed in [11]. For biophysical and structural studies, where precision in protein concentration is of foremost importance, standard graphs can be plotted in two different ways to rule out the possibility of interference of additives in the buffer. Comparison of the slopes of plots with the protein in water, and the other in the buffer solution would confirm buffer interference if any [12].

3. Selecting the standards and interpreting the results:
   The total number of side-chains of amino acids in each protein is different and they behave differently in each type of assay. A standard is a highly purified homogenous sample protein with known concentration and easy availability. The standard proteins are important for assay calibration and hence their selection needs to be done judiciously. The most widely used standard is BSA, which is inexpensive, stable and its pure form can be obtained commercially in a large quantity [13]. Immunoglobulin g (IgG) and bovine gamma globulin (BGG) are also some of the few standards used [14]. However, care should be taken to see whether the standard is similar in amino acid composition to the protein of interest; otherwise there will be either over- or underestimation of the protein concentration as discussed later in the chapter.
   It is important to measure the concentration of standards in triplicates to reduce the error rate in the assay. Although a standard curve can be obtained with fewer points, it is advisable to choose at least three to four points below and above the test samples' concentration range. The number of standards mostly depends upon the non-linearity of the curve and degree of accuracy. Point-to-point interpolation method can be used for calculations, using equation for linear regression line obtained from minimum two points (just below and above the sample). The quadratic curve fit for a nonlinear standard curve can be used for calculating concentration using computer software [15].
4. Preparation of the sample:
   It is advisable to prepare fresh sample every time; if it is not possible, addition of antimicrobial agents and protease inhibitors are recommended to prevent the microorganism growth and the degradation of the sample. The addition of stabilizers and detergents must be within the assay's permissible range [1, 7–10]. It is recommended to remove the large particles using a 0.2-micron filter or centrifugation at high speed to separate the aggregates and other contaminants. In microplate format, lower volume sample (<5 μl) might cause pipetting errors for viscous protein solutions; therefore, it is recommended to dilute the samples appropriately. Regular change of pipette tips is recommended to prevent cross-contamination from well to well.

11.2.1 Different Types of Protein Quantitation Assays

11.2.1.1 Dye-Based Assays

These assays that include BCA, Lowry, and Bradford are the routine laboratory methods to estimate total amount or concentration of proteins of interest. All these techniques depend on chemical reactions that lead to changes in colour, the intensity of which is read as the output spectroscopically at a particular wavelength. For example, the BCA assay utilizes the property of certain amino acids to reduce Cu (II) to Cu (I) in an alkaline milieu to produce a purple coloured bicinchoninic acid that shows an emission maximum at 562 nm [9]. Similarly, the Lowry assay shows an emission peak at 750 nm, and in Bradford assay, the wavelength shifts from 465 to 595 nm [8, 16].

Although there are several advantages of these assays that include inexpensive and quick methods (within 3 h), certain limitations restrict their universal applicability. Since the dye-based assays depend on the chemical reactions, the protein environment and amino acid composition play pivotal role in determining the efficacy of the assays on a case-to-case basis. Furthermore, the accuracy of these colorimetric assays depends on proper calibration using protein standards and hence the appropriateness of those chosen standards becomes crucial for any particular experiment.

The efficiency of the assays depends upon certain important caveats as mentioned below:

- Ensuring accessibility of the reaction site in the protein to the dye used in the assay.
- Both the reference protein (for standardization) and protein under examination should have similar amino acid composition.
- Since certain materials (salts and other additives) might hamper proper measurements through interference. For example, certain additives such as polyethylene glycol (PEG) that is used for protein stabilization might act as a deterrent through steric hindrance.
- Certain proteins with predominance of small non-polar amino acids (e.g. collagen) are less sensitive to colour development due to lesser dye-binding capacity. Moreover, the quantitation becomes underestimated if it is calibrated with a dissimilar protein like BSA that has high dye-binding ability [17].

To circumvent the problem with accuracy as mentioned above, especially when the proteins used for calibration are quite often dissimilar, a technique known as amino acid analysis (AAA) becomes important [18]. Briefly, this method relies on the concentration of free amino acids in the protein after they are subjected to hydrolysis [19]. Although initially it was a complex and labour-intensive process, recent developments have made the procedure more precise and simple [20].

The different types of dye-based assays are described in the subsequent sections.

11.2.1.2 Bradford (Coomassie Blue) Assay

Principle

First described by Bradford [7], at acidic pH, the anionic form of the dye, Coomassie Brilliant Blue G-250, binds preferentially to certain amino acids present in the protein (arginine, histidine, phenylalanine, tryptophan and tyrosine) and gets stabilized [21, 22]. This leads to shift in its absorbance maximum (termed as metachromatic shift) from 465 to 595 nanometres. A pictorial representation of the assay is provided in Fig. 11.1. The limitation of the assay is that the majority of the

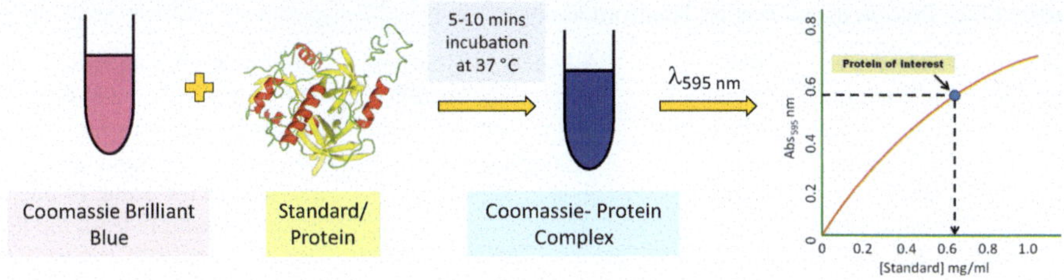

Fig. 11.1 Schematic representation of Bradford assay. The figure depicts determination of concentration of unknown protein from the fit to the standard curve plotted by measuring absorbance at 595 nm wavelength (λ_{595}) using standard protein solutions of known concentration

signal is due to its interaction with arginine residues and hence the assay is very much sequence-specific and varies from protein to protein.

Metachromasia is a phenomenon of characteristic alteration in the colour of staining process that is carried out in biological tissues shown by certain stains after their binding to particular substances, which are present in tissues, called chromotropes.

This concept of metachromasia requires the presence of polyanions within the tissue. When a concentrated basic dye solution such as toluidine blue is used for staining such tissues, the bound dye molecules result in the formation of dimeric and polymeric aggregates. The spectrum of light absorption of these dye aggregates varies from the individual monomeric dye molecules. Cell and tissue structures, which contain high concentrations of ionized sulphate and phosphate groups such as the ground substance of cartilage, mast cell granules, and rough endoplasmic reticulum of plasma cells, show the phenomenon of metachromasia.

Protocol

Reagents required:

Coomassie Brilliant Blue G-250 (Bio-Rad Laboratories), ethanol, phosphoric acid, standard protein (e.g. BSA).

Preparation of the dye solution:

To prepare 1 L of dye solution, 100 mg of Coomassie Brilliant Blue −250 is dissolved in 50 ml of 95% ethanol followed by addition of 85% phosphoric acid (H_3PO_4) with continuous stirring. Upon complete dissolution of the dye, adjust the volume to 1 L with distilled water and filter to remove precipitates if any. For long-term storage, it is advisable to store the solution at 4 °C [1].

Experimental Procedure

- Prepare BSA standards (1, 0.5, 0.25 and 0.125 mg/ml from 1 mg/ml of BSA stock in a buffer that is suitable for Bradford assay).
- Add standard solutions as well as proteins of interest into cuvettes or microwell plates (as per the protocol). Blank solutions (buffer only) will also be added in separate cuvettes or in wells of the 96-well plate for background corrections.

Table 11.2 Protein estimation by Bradford assay

BSA standard conc. (µg/µl)	Add 5 µl of each of the serially diluted BSA standard to each well	Total amount of protein (µg) in 5 µl solution	Add 100 µl Bradford reagent and read at 595 nm on a microplate reader	Plot graph with known concentrations of standard solutions versus their observed absorbance and extrapolate the value of the unknown sample from the fit
0		0		
0.0625		0.3125		
0.125		0.625		
0.25		1.25		
0.5		2.5		
1		5		

- Bradford reagent (as prepared above) is added to the proteins as well as blank solutions in the ratio of circa 40:1.
- After 5–10 min of incubation at room temperature, measure the absorbance of each sample (cuvette-based) or the microplate at 450 and 595 nm.
- Plot a graph with concentrations of the standard protein (known) on the X-axis versus either the absorbance values at 595 nm or the ratio of 595/450 nm on the Y-axis.
- The standard curve obtained can be fitted to a polynomial equation and the concentration of the unknown protein can be obtained from the fit [23] as shown in Table 11.2.

Note: Ensure that the temperature of the Bradford reagent attains room temperature before adding it to the protein solutions. The experiment has a threshold size-limit for proteins and peptides as minimum number of desired amino acids needs to be present to get a proper spectroscopic signal [21].

11.2.1.3 Lowry Assay

Principle

The Lowry *aka* alkaline copper reduction assay [24] has been modified to a two-step procedure that is based on combination of two simple steps that include reduction of copper [25] and Folin-phenol method [26]. Since the second step enhances the reaction obtained from **step 1**, it is one of the most accurate and sensitive methods of protein quantitation [24]. Briefly, it involves reduction of $Cu2^+$ to Cu^+ by proteins in an alkaline environment (biuret reaction) followed by a subsequent enhancement step where Folin–Ciocalteu reagent (active component is mixture of phosphomolybdic and tungstic acid) is reduced to emit an intense blue colour with an emission maximum of ~750 nm [27] as shown in Fig. 11.2. This method also very much depends on the presence of certain amino acids including tyrosine, tryptophan, cysteine, and histidine and hence colour variation might occur for different protein samples [8].

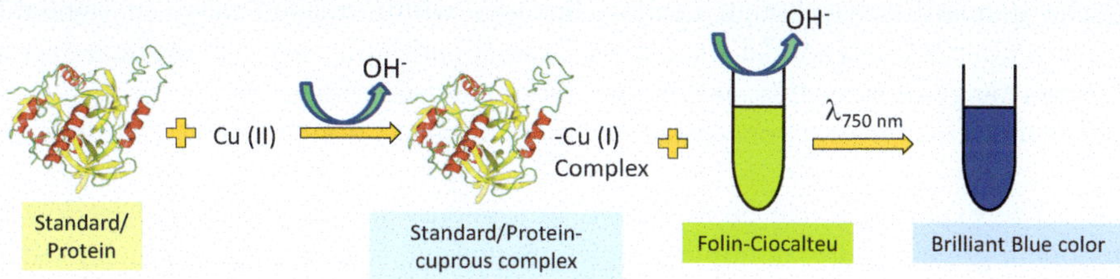

Fig. 11.2 Schematic representation of Lowry assay. The figure is self-explanatory and the details are provided in the protocol below

Protocol: There are several versions of the assay (each with subtle variations); the most popular method will be discussed below.

Reagents Required

Sodium carbonate (Na_2CO_3), sodium hydroxide (NaOH), copper sulphate ($CuSO_{4.}5H_2O$), sodium tartrate (Na Tartrate), Folin–Ciocalteu reagent, sodium dodecyl sulphate (SDS), and standard protein solution (BSA).

Solution Preparatory Steps

Prepare sodium-carbonate solution by adding 10 g of the salt in 50 ml of double-distilled (dd) water (*Solution* 1).

Prepare $CuSO_4 \cdot 5H_2O$-sodium tartarate solution in the ratio of 1:2 by adding 100 and 200 mg of the salts, respectively, in 50 ml dd water: $CuSO_4$ in Na-Tartarate (*Solution* 2).

Mix *Solutions* 1 and 2 slowly such that $CuSO_4 \cdot 5H_2O$ dissolves (*Solution* 3). It needs to be prepared fresh prior to the experiment.

Alkaline solution: Combine copper sulphate solution (*Solution* 3), 5% SDS (w/v) and 3.2% NaOH (w/v) in the ratio of 1:1:2 (*Solution* 4).

Folin–Ciocalteu reagent: Commercially available Folin Ciocalteu phenol reagent is diluted with dd water in the ratio 1:5, with a final volume of ~60 ml.

BSA solution: 100 mg BSA was dissolved in 100 ml dd water (1 mg/ml stock solution).

Experimental Procedure

- Take approximately 1 ml of protein of interest (unknown concentration) or standard (BSA) within the concentration range of 5–100 mg/ml.
- Add alkaline solution (*Solution* 4) to the protein solutions and incubate for 10–15 min.
- Mix 500 μl of Folin–Ciocalteu solution (as prepared above) to the above solution, vortex and incubate for 30 min.
- Combine the entire solution mixture by vortexing again post incubation.
- Measure absorbance at 750 nm.
- Prepare standard curve in the same way as mentioned for Bradford assay.

- The standard curve though exhibits linearity within a small range of standard concentration, polynomial, exponential, and logarithmic curves fit the data over a larger range of standards and provide more precise results.
- The unknown protein concentration is then obtained from the fit to the curve.

Note: The experiment can be replicated in a microwell plate format where lesser amount of protein and/or multiple samples are present. Moreover, several interfering agents might lead to erroneous concentration calculations and care should be taken to maintain the tolerance level of the assay in the buffer used. The BCA assay is more sensitive than Lowry and hence has become more popular over the years.

11.2.1.4 Bicinchoninic Acid (BCA) Assay

Principle

The BCA assay follows similar principle as Lowry (reduction of Cu^{2+} to Cu^{+}); however the Folin–Cioalteu's reagent is substituted with BCA. The chemical reaction involving BCA is more robust and tolerant towards interfering compounds, thus leading to more sensitivity and less variability [9]. The interaction of reduced cuprous ions (that are formed as a consequence of Cu^{2+}–protein reaction) with Bicinchoninic acid solution leads to formation of a bright purple coloured solution that can be spectroscopically measured at wavelengths between 500 and 570 nm [9] as shown in Fig. 11.3. Although the initial protein-Cupric ion interaction is dependent on types of amino acids as described earlier, the advantage of BCA assay is that it is temperature dependent. Therefore, sensitivity and tolerance towards additives can be modulated optimally by elevating the assay temperature [28] from 37 to 60 °C. Furthermore, the background noise and interference due to additives can be reduced by varying the protein sample-BCA ratio (typically between 8- and 20-fold of the latter is added to the sample or standard solution).

Protocol: (adapted from Ref. [1]).

Reagents

Sodium bicinchoninate, sodium carbonate (Na_2CO_3), sodium tartrate, sodium hydroxide (NaOH), sodium bicarbonate ($NaHCO_3$), $CuSO_4 \cdot 5H_2O$, BSA.

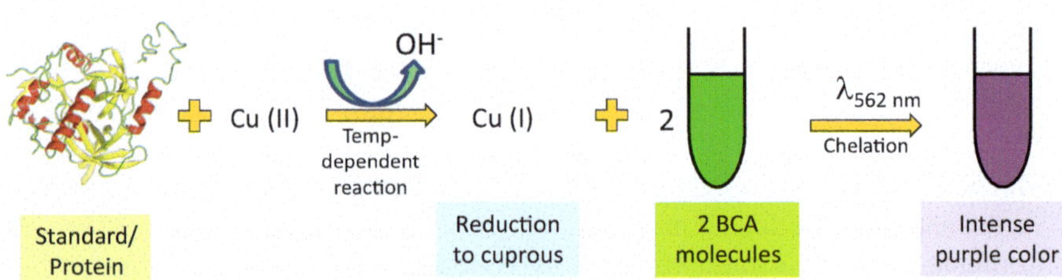

Fig. 11.3 Schematic representation of BCA assay. The figure is self-explanatory and the detail is provided in the protocol below

Solution Preparatory Steps

Reagent A: 1 g sodium bicinchoninate, 2 g Na_2CO_3, 0.16 g sodium tartrate, 0.4 g NaOH, and 0.95 g $NaHCO_3$ are added to DD H_2O and volume adjusted to 100 ml. The optimal pH for the solution is 11.25 that is adjusted with NaOH solution.

Reagent B: 0.4 g $CuSO_4 \cdot 5H_2O$ dissolved in 10 ml DD H_2O.

Both Reagents A and B are stable and can be stored at room temperature.

Working solution: Mix reagents A and B in the ratio of 50:1. A green coloured solution is formed that is stable for 5–7 days at room temperature.

Experimental Procedure

- For protein of interest or standard solution, a volume of 200 μl–1 ml for cuvette-based and ~25 μl for microplate-based assay is usually taken.
- Almost 20–60-fold BCA solution for the former and 8–10-fold for the microplate method are added.
- The mixture is incubated at either 37 or 60 °C (based on the amino acid composition and additives in buffer). A prior standardization step might help to find the temperature that works best for that particular protein.
- Incubate the solution for ~30 min. Longer incubation period can be selected if experiment is performed at room temperature.
- It is advisable to cover the micro plates especially if assay is performed at higher temperature to minimize sample loss due to evaporation.
- The signal at 562 nm is measured where the intensity of the purple coloured dye is maximum.
- Usually, the BCA assay produces a linear curve for a wider spectrum of concentration; however quadratic equation is also used for a better fit and precise analysis for a broader range of standard solutions.

Notes

This assay is sensitive to agents that reduce or chelate copper (e.g. DTT and EDTA, respectively). If a highly concentrated solution precipitates, the sample needs to be diluted to prevent the same.

11.3 Assays Involving Ultra Violet (UV) Absorption Spectroscopy

Principle

This is one of the popular and routine methods for determination of protein concentration in laboratories especially when protein is limited and needs to be recovered post quantitation analysis. This method relies on absorption measurement at 280 nm wavelength where predominantly aromatic amino acids, tyrosine, tryptophan and phenylalanine absorb as shown in Fig. 11.4. Although role of phenylalanine is not major, the presence of other two aromatic amino acids do play a pivotal role in the accuracy of this method [29]. Therefore, it might not be a suitable method for

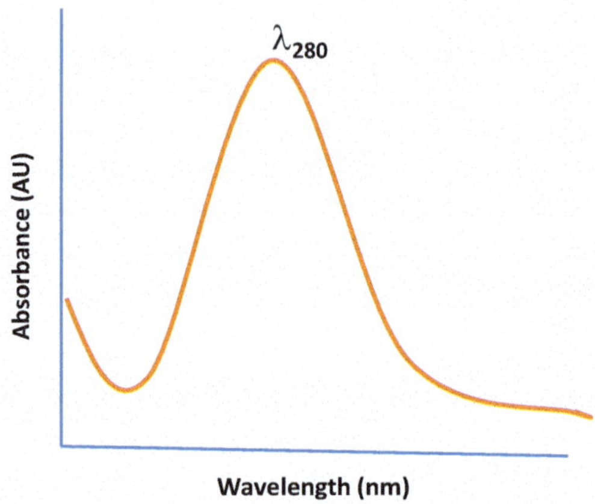

Fig. 11.4 Schematic representation of a typical blank-corrected protein spectrum with emission maximum at 280 nm. AU represents "absorbance units"

proteins with minimal or no aromatic amino acids. However, it is extremely popular due to its simplicity as standards are not required.

Protein concentration is calculated using Beer–Lambert law, the equation of which is provided below [30, 31]:

$$A = \varepsilon \times c \times l \tag{11.1}$$

where, "A" is measured absorbance at 280 nm; "ε" is the Extinction coefficient of the protein [2]; "c" is concentration of protein and "l" is the path length of the cuvette (usually 1 cm but might vary). The molar extinction coefficient of a protein with unit of M^{-1} cm^{-1} shows the absorbance affinity of light on a particular substance or chemical species at a particular wavelength.

Reagents: Autoclaved water, buffer, sample solution in the same buffer.

Protocol

- Turn on the spectrophotometer and keep the lamp on for 30 min and set the wavelength to 280 nm.
- Use an appropriate buffer in which the sample is dissolved for blank correction.
- The reading must be between the ranges of 0.2 and 1 (not more than 1); if the reading shows a value greater than 1, dilute the samples appropriately and take the reading again. This is due to the fact that Beer–Lambert equation does not follow linearity if the protein is too concentrated, i.e. absorbance reading goes above 1 [32, 33].
- For a protein with an unknown extinction coefficient (if sequence not known), then a linear standard curve can be generated with a standard protein solution (e.g. BSA) as described earlier. The protein concentration can then be determined from the fit of the graph.

Cautionary note: Quartz cuvette and not plastic cuvettes are to be used for measurement in the UV region as plastic cuvettes are not UV-transparent. Care should be taken to properly clean quartz cuvettes after every measurement to prevent cross-contamination.

Notes:

Buffer containing detergents (e.g. Triton X-100) should be avoided as they absorb UV light. Blank correction should be done with the same buffer in which the protein is present. The samples must be free from nucleic acid contamination as they absorb strongly at 280 nm [34]. If the protein has either none/minimal or several aromatic amino acids, this method might give incorrect results. Sample must be filtered using 0.2 μm filters or centrifuge at high speed for 20 min to avoid aggregates or other particles which absorb or scatter light. An alternative to 280 nm, other wavelengths such as 205, 214 and 220 nm can be used to measure proteins and peptide concentration where the absorbance from peptide bonds is calculated [35]. The molar extinction coefficient can be obtained from Expasy ProtParam tool (https://web.expasy.org/protparam/) [36]; however, the caveat is that the exact sequence of the protein of interest along with any attached tag should be known. Molar extinction coefficient can also be obtained mathematically using standard formula provided by [2], and its modified version [37].

11.3.1 UV Absorption Using Micro-Volume Spectroscopy

Principle

This micro-volume spectroscopy (Nano drop instruments) is widely used in all biochemical labs. The amount of sample needed is very less (typically 0.5 to 2.0 μl). These instruments allow very fast, accurate quantification of proteins. The principle includes the use of inherent surface tension of liquids analysed to create columns between ends of the optical sensor or fibres and the measurement of the formed optical path [3]

Reagents: Autoclaved water, buffer, sample in the same buffer.

Protocol

- Open the arm of the equipment, clean it using a tissue wipe. Add 1 μl of DD water to the sensor and bring down the arm.
- Open the respective Nano drop software and select "A_{280} or Protein Reading" option, followed by the "Water" option.
- Clean the sensor and add a buffer in which the protein is present and click "Blank" and wait for the calibration.
- Clean the arm and add 1 or 1.5 μl protein sample and click on "Measure".
- Take the reading at least three times and check for consistency in the values obtained (sometimes small air bubbles interfere with the readings).

- Check for the contaminants (cross-check A_{260}/A_{280} ratios: Pure nucleic acid samples would have an A_{260}/A_{280} ratio of ~1.8, while for protein, it would be ~0.6) [38, 39].
- Clean the arm with a tissue wipe and switch off the instrument.
- Protein concentration using Absorbance or Optical Density (O.D.) value is shown by Eq. ((11.2) below:

$$\frac{\text{O.D}}{\text{Molar extinction coefficient/Molecular weight in Dalton}} = \text{concentration in } \frac{\text{mg}}{\text{ml}} \tag{11.2}$$

Notes: The samples must be free from nucleic acid contamination as they absorb strongly at 280 nm. Sample must be filtered using 0.2 μm filters or centrifuged at high speed for 20 min to avoid aggregates or other particles, which absorb/scatter light.

11.4 Troubleshooting for Protein Quantitation

1	Can the standard curve be extrapolated if the data point of the sample is out of the standard curve range?	No. It is very important to ascertain that the sample data points fall within the standard curve range. The dilution of either or both sample and BSA should be adjusted.
2	Is BSA a good standard?	The most commonly used protein standard is BSA. It has limitations as it is difficult to get very highly purified BSA protein because of its high affinity for its binding partners. It also has high sensitivity in Bradford assay than other proteins, thus the concentration of the protein sample is likely underestimated [10]. Other proteins, such as immunoglobulin G, and lysozyme can also be used as protein standards.
3.	How to apply dilution factors?	Dilution factor is very important to consider when the samples are diluted than the standard protein or highly concentrated to be assayed by any kit with lesser assay range.
4.	How to use Microsoft excel to plot graph and apply standard curve?	Export your absorbance data to excel file and save with particular label. In excel sheet, first take average of the absorbance readings and do blank subtraction. Different formulae can be very easily used in Microsoft excel files. Select the standard protein concentration and blank-subtracted absorbance values to plot scatter graph. Then by clicking one of the points on graph,

(continued)

		open a dialogue box to add trend-line. Also select polynomial graph and display equation on chart. You will get equation in format of Y = mX + C, where X will be the blank subtracted absorbance of test sample, while Y will be the protein concentration of sample.
5.	The protein precipitates in sample.	Remove or reduce the level of detergents in your protein buffer by either diluting the sample or dialyzing the protein sample.

11.5 Purity Analysis of Proteins of Interest

Apart from knowing the exact concentration of a purified protein before it can be used for further in vitro characterization, it is also equally important to determine its purity as most of the proteins for structural and biophysical studies need to be >95% pure. One of the easiest, convenient and most widely used methods for identifying the impurities present in the purified protein sample is by running it on SDS-PAGE or Sodium Dodecyl Sulphate Poly Acrylamide Gel Electrophoresis [40, 41].

11.5.1 SDS PAGE

Principle

SDS PAGE is widely used for separating the mixture of proteins based on their size. SDS, an anionic detergent, denatures as well as evenly coats the entire length of the proteins with negative charge, so that the separation through electrophoresis is based solely on the size of the protein [41]. However, it is important to note that the monomeric size of the protein is determined using the method since SDS denatures any oligomeric form of a protein. The presence of β-mercaptoethanol (βME) reduces all disulphide bonds and helps in protein denaturation. It is also the best method for detecting other protein impurities present in the purified protein sample. It comprises a resolving gel at the bottom and a stacking gel on its top. The two types of gels help separation of proteins of different sizes on the gel efficiently. They differ in pH, pore size and in polyacrylamide content. While the stacking gel with very small pore size allows the protein samples in the wells to line up and enter the resolving gel at the same time; the resolving gel separates out protein bands based on their molecular weights (M.W.). Although most SDS PAGE gels are prepared with a Bisacrylamide: Acrylamide molar ratio of 1:29, which is sufficient to separate polypeptides that differ in size by ~3%, the ratio can be further optimized based on different requirements of separation. Bisacrylamide is a cross-linker that determines the pore size and hence mobility of the protein molecules passing through the gel [42]. Two other important reagents are Ammonium Per sulphate (APS), which is an oxidizing agent, and *N,N,N′N′*-Tetra methylene diamine (TEMED), which acts as

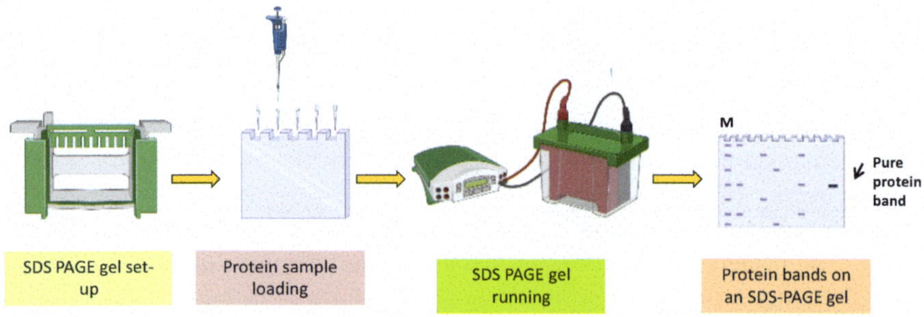

Fig. 11.5 Schematic representation of protein samples run on an SDS PAGE gel. The stained gel shows a marker (M) on the extreme left lane, followed by protein samples that have impurities along with them. The extreme right lane shows pure protein of interest, the molecular weight can be estimated from the known molecular weight marker on the left

a catalyst. Both these reagents help polymerization of the Bisacrylamide/acrylamide gel and make it ready for electrophoresis. The tracking dye that is added to the protein solution with its extremely high mobility runs faster than the protein molecules and ensures that the protein does not run out of the gel. The gels are then stained normally with Coomassie brilliant blue that forms dye:protein complex through interactions with charged amino acids. This is further stabilized by hydrogen bonding, van der Waals and hydrophobic interactions. A schematic representing proteins run on SDS PAGE gel is shown in Fig. 11.5.

Reagents

5× SDS sample loading buffer (10 ml)	
(250 mM Tris–HCl pH 6.8, 10% SDS, 0.5% BME, 30% glycerol).	
1 M Tris–HCl (pH 6.8)	2.5 ml from stock 1 M
SDS	10 g
Glycerol	3 ml
BME	0.5 ml from stock 14.3 M
Bromophenol blue	0.02 g
Milli Q Water	4 ml
30% acrylamide	
(29.2% Acrylamide, 0.8% *N′N′*-bis-methylene-acrylamide)	
Acrylamide	29.2 g (29.2%)
N′N′-bis-methylene-acrylamide	0.8 g (0.8%)
SDS-PAGE Running Buffer (1× SDS buffer) (for 1 L)	
(25 mM Tris, 192 mM glycine, pH 8.3, 1% SDS)	
Tris Base	3.02 g (25 mM)
Glycine	14.4 g (192 mM)
Milli Q water	1 L
No need to adjust pH	

(continued)

Staining/Destaining solution (for 1 L)	
(50% water, 40% methanol, 10% acetic acid, 0.1% Coomassie blue R-250)	
Methanol	400 ml
Acetic Acid (glacial)	100 ml
Milli Q	500 ml
Coomassie blue R-250	1 g
Destaining solution is the same, minus the Coomassie blue.	

11.5.2 SDS-Page Gel Preparation

Experimental Procedure

- Clean and assemble the plates, secure with clamp, add water to check for the leaks; if it leaks, then reassemble the plates.
- Remove water, make an appropriate percentage of resolving solution as mentioned in Table 11.3 and add resolving solution on top of the plates leaving 2 cm for stacking gel. Allow it to polymerize (~30 min). Add 100% ethanol if you see any air bubbles on the top.
- Prepare appropriate stacking gel as shown in Table 11.4, pour on the top of the resolving gel, insert comb and allow the gel to polymerize.
- Fill the tank with 1× SDS buffer and transfer the plates to the gel-running chamber. Remove the comb, fill the tank with 1× SDS buffer, load the samples and run under appropriate conditions (~90–100 volts for 60–90 min).
- After the run is over, stain the gel followed by destaining to visualize the proteins along with other impurities if any.

Table 11.3 Composition of resolving gel for SDS PAGE (Total Volume prepared: ~16 ml)

Percentage	6%	7%	10%	12%	15%
Distilled water (ml)	8.48	7.68	6.4	5.28	3.68
30% ACRYLAMIDE (ml)	3.2	4	4	6.4	8
1.5 M Tris–8.8 pH (ml)	4	4	4	4	4
10% SDS (μl)	160	160	160	160	160
10% APS (μl)	160	160	160	160	160
TEMED (μl)	16	14	8	8	8

Table 11.4 Composition of stacking gel for SDS PAGE (total volume ~9 ml)

Volume (ml)	For two gels
Distilled water(ml)	1.4
30% ACRYLAMIDE (ml)	3.2
1.5 M Tris–pH 6.8 (ml)	4
10% SDS (μl)	160
10% APS (μl)	160
TEMED (μl)	16

Notes:

For detecting low molecular weight proteins, Tris Tricine gel can be used, and for the proteins with low concentrations, silver staining can be employed to visualize them, which is 100 times more sensitive than Coomassie stain [43]. Suitable molecular weight markers (can be laboratory-made or purchased) need to be run side by side to have an estimate of the size of the protein. Analytical gel filtration (Chap. 8) can be used to analyse the homogeneity of the sample. To understand the post-translational modification present within the protein, various Mass Spectrometry [44] methods can be employed. If the purified protein is found to have more than 5% impurity, further purification is recommended prior to its characterization.

Acknowledgments The authors thank Advanced Centre for Treatment, Research and Education in Cancer (ACTREC) for providing necessary infrastructure and resources for successful completion of the chapter. The authors acknowledge Ms. Chanda Baisane, Bose Lab for formatting the manuscript.

Problems

Multiple Choice Questions

1. The statement that is false about the Bradford assay is:
   (a) Ethidium bromide is the dye used in the assay
   (b) The Bradford reagent dye colour changes from red to blue after binding with protein
   (c) The colour intensity is directly proportional to the bound proteins
   (d) The Bradford reagent has an absorption maximum of 470 nm in the absence of protein
2. The wavelength at which the absorption will be maximum after binding of Lowry reagent with protein is:
   (a) 295 nm
   (b) 595 nm
   (c) 750 nm
   (d) 562 nm
3. For UV absorption spectroscopy, which of the following assumptions is (are) correct?
   I. The method is suitable for proteins with aromatic amino acids.
   II. The absorption value should not be more than 1.
   III. The method relies on Beer–Lambert law.
   (a) I and II only
   (b) II and III only
   (c) I and III only
   (d) All of them

Subjective Questions

1. A researcher wants to check the concentration of the protein purified through affinity chromatography. To quantify the proteins in the test sample, a Bradford assay was carried out with 1 μl of test sample and BSA as standard (5 μl). The following absorbance values were recorded (A1, A2, and A3 being triplicates) at 595 nm. Calculate the protein concentration using the readings provided in the table below:

	A1	A2	A3
Blank	0.221	0.217	0.214
0.0625 mg/ml BSA	0.233	0.244	0.223
0.125 mg/ml BSA	0.297	0.25	0.283
0.25 mg/ml BSA	0.385	0.352	0.326
0.5 mg/ml BSA	0.482	0.467	0.45
1 mg/ml BSA	0.53	0.608	0.597
Test sample	0.473	0.504	0.525

2. A researcher purified a protein of interest using affinity chromatography and wanted to check the approximate concentration of the protein with Nano drop instrument. He took three absorbance readings of values 0.42, 0.5, and 0.48 at 280 nm. Molar extinction coefficient of the protein is 43,824 M^{-1} cm^{-1} and molecular weight is 66,400 Daltons. What will be the concentration of the protein?

References

1. Noble JE, Bailey MJA. Chapter 8: Quantitation of protein. In: Methods in enzymology. New York: Academic Press Inc.; 2009. p. 73–95.
2. Gill SC, von Hippel PH. Calculation of protein extinction coefficients from amino acid sequence data. Anal Biochem. 1989;182:319–26. https://doi.org/10.1016/0003-2697(89)90602-7.
3. Campion EM, Loughran ST, Walls D. Protein quantitation and analysis of purity. In: Methods in molecular biology. Totowa, NJ: Humana Press Inc.; 2017. p. 225–55.
4. Noble JE, Knight AE, Reason AJ, et al. A comparison of protein quantitation assays for biopharmaceutical applications. Mol Biotechnol. 2007;37:99–111. https://doi.org/10.1007/s12033-007-0038-9.
5. Johnson M. Protein quantitation. Mater Methods. 2012;2:115. https://doi.org/10.13070/mm.en.2.115.
6. Lebendiker M, Danieli T. Production of prone-to-aggregate proteins. FEBS Lett. 2014;588:236–46.
7. Bradford MM. A rapid and sensitive method for the quantitation of microgram quantities of protein utilizing the principle of protein-dye binding. Anal Biochem. 1976;72:248–54. https://doi.org/10.1016/0003-2697(76)90527-3.

8. Peterson GL. A simplification of the protein assay method of Lowry et al. which is more generally applicable. Anal Biochem. 1977;83:346–56. https://doi.org/10.1016/0003-2697(77)90043-4.
9. Smith PK, Krohn RI, Hermanson GT, et al. Measurement of protein using bicinchoninic acid. Anal Biochem. 1985;150:76–85. https://doi.org/10.1016/0003-2697(85)90442-7.
10. Stoscheck CM. Quantitation of protein. Methods Enzymol. 1990;182:50–68. https://doi.org/10.1016/0076-6879(90)82008-P.
11. Lovrien R, Matulis D. Assays for total protein. In: Current protocols in microbiology, vol. 3A. Hoboken, NJ: John Wiley & Sons, Inc.; 2005. p. 14.
12. Olson BJSC, Markwell J. Assays for determination of protein concentration. Curr Protoc Protein Sci. 2007;48:3.4.1–3.4.29. https://doi.org/10.1002/0471140864.ps0304s48.
13. Klein B. Standardization of serum protein analyses. Ann Clin Lab Sci. 1978;8:249–53.
14. Crowley TE, Kyte J. Experiments in the purification and characterization of enzymes: a laboratory manual. Netherland: Elsevier Inc.; 2014.
15. Yuan L, Zhang D, Jemal M, Aubry AF. Systematic evaluation of the root cause of non-linearity in liquid chromatography/tandem mass spectrometry bioanalytical assays and strategy to predict and extend the linear standard curve range. Rapid Commun Mass Spectrom. 2012;26:1465–74. https://doi.org/10.1002/rcm.6252.
16. Walker JM, Hammond JBW, Kruger NJ. The Bradford method for protein quantitation. In: New protein techniques. Totowa, NJ: Humana Press; 2003. p. 25–32.
17. López J, Imperial S, Valderrama R, Navarro S. An improved Bradford protein assay for collagen proteins. Clin Chim Acta. 1993;220:91–100. https://doi.org/10.1016/0009-8981(93)90009-S.
18. Schegg KM, Denslow ND, Andersen TT, et al. Quantitation and identification of proteins by amino acid analysis: ABRF-96AAA collaborative trial. Tech Protein Chem. 1997;8:207–16. https://doi.org/10.1016/S1080-8914(97)80023-3.
19. Kambhampati S, Li J, Evans BS, Allen DK. Accurate and efficient amino acid analysis for protein quantification using hydrophilic interaction chromatography coupled tandem mass spectrometry. Plant Methods. 2019;15:46. https://doi.org/10.1186/s13007-019-0430-z.
20. Sittampalam GS, Ellis RM, Miner DJ, et al. Evaluation of amino acid analysis as reference method to quantitate highly purified proteins. J Assoc Off Anal Chem. 1988;71:833–8. https://doi.org/10.1093/jaoac/71.4.833.
21. de Moreno MR, Smith JF, Smith RV. Mechanism studies of coomassie blue and silver staining of proteins. J Pharm Sci. 1986;75:907–11. https://doi.org/10.1002/jps.2600750919.
22. Fountoulakis M, Juranville JF, Manneberg M. Comparison of the Coomassie brilliant blue, bicinchoninic acid and Lowry quantitation assays, using non-glycosylated and glycosylated proteins. J Biochem Biophys Methods. 1992;24(3–4):265–74. https://doi.org/10.1016/0165-022X(94)90078-7.
23. Olson BJSC. Assays for determination of protein concentration. Curr Protoc Pharmacol. 2016;73:A.3A.1–A.3A.32. https://doi.org/10.1002/cpph.3.
24. Lowry OH, Rosebrough NJ, Farr AL, Randall RJ. Protein measurement with the Folin phenol reagent. J Biol Chem. 1951;193:265–75. https://doi.org/10.1016/s0021-9258(19)52451-6.
25. Herriott RM. Isolation, crystallization, and properties of pepsin inhibitor. J Gen Physiol. 1941;24:325–38. https://doi.org/10.1085/jgp.24.3.325.
26. Wu H. Studies on hemoglobin. J Biochem. 1922;2:173–80. https://doi.org/10.1093/oxfordjournals.jbchem.a125832.
27. Peterson GL. Review of the folin phenol protein quantitation method of Lowry, rosebrough, farr and Randall. Anal Biochem. 1979;100:201–20.
28. Wiechelman KJ, Braun RD, Fitzpatrick JD. Investigation of the bicinchoninic acid protein assay: identification of the groups responsible for color formation. Anal Biochem. 1988;175:231–7. https://doi.org/10.1016/0003-2697(88)90383-1.
29. Leach SJ, Scheraga HA. Effect of light scattering on ultraviolet difference spectra. J Am Chem Soc. 1960;82:4790–2. https://doi.org/10.1021/ja01503a008.

30. IUPAC. Beer–Lambert law (or Beer–Lambert–Bouguer law). In: IUPAC Compendium of Chemical Terminology. IUPAC. Oxford: Blackwell Scientific Publications; 2008.
31. Beer. Bestimmung der Absorption des rothen Lichts in farbigen Flüssigkeiten. Ann der Phys und Chemie. 1852;162:78–88. https://doi.org/10.1002/andp.18521620505.
32. Reule A. Testing spectrophotometer linearity. Appl Opt. 1968;7:1023. https://doi.org/10.1364/ao.7.001023.
33. Hawes RC. Technique for measuring photometric accuracy. Appl Opt. 1971;10:1246. https://doi.org/10.1364/ao.10.001246.
34. Groves WE, Davis FC, Sells BH. Spectrophotometric determination of microgram quantities of protein without nucleic acid interference. Anal Biochem. 1968;22:195–210. https://doi.org/10.1016/0003-2697(68)90307-2.
35. Scopes RK. Measurement of protein by spectrophotometry at 205 nm. Anal Biochem. 1974;59:277–82. https://doi.org/10.1016/0003-2697(74)90034-7.
36. Gasteiger E, Hoogland C, Gattiker A, et al. Protein identification and analysis tools on the ExPASy server. In: The proteomics protocols handbook. Totowa, NJ: Humana Press; 2005. p. 571–607.
37. Pace CN, Vajdos F, Fee L, et al. How to measure and predict the molar absorption coefficient of a protein. Protein Sci. 1995;4:2411–23. https://doi.org/10.1002/pro.5560041120.
38. Glasel JA. Validity of nucleic acid purities monitored by 260nm/280nm absorbance ratios. BioTechniques. 1995;18:62–3.
39. Goldfarb AR, Saidel LJ, Mosovich E. The ultraviolet absorption spectra of proteins. J Biol Chem. 1951;193:397–404. https://doi.org/10.1016/s0021-9258(19)52465-6.
40. Walker JM, Smith BJ. SDS polyacrylamide gel electrophoresis of proteins. In: Proteins. New Jersey: Humana Press; 2003. p. 41–56.
41. Laemmli UK. Cleavage of structural proteins during the assembly of the head of bacteriophage T4. Nature. 1970;227:680–5. https://doi.org/10.1038/227680a0.
42. Studier FW. Analysis of bacteriophage T7 early RNAs and proteins on slab gels. J Mol Biol. 1973;79:237–48. https://doi.org/10.1016/0022-2836(73)90003-X.
43. Schägger H. Tricine-SDS-PAGE. Nat Protoc. 2006;1:16–22. https://doi.org/10.1038/nprot.2006.4.
44. Mann M, Jensen ON. Proteomic analysis of post-translational modifications. Nat Biotechnol. 2003;21:255–61.

Answer Key

Chapter 2: Cloning and Gene Manipulation

Aasna L. Parui, Lalith K. Chaganti, Rucha Kulkarni, and Kakoli Bose

Multiple Choice Questions
1. (d), 2. (b), 3. (c)

Subjective Questions
Answer 1

After transforming the host *E. coli* with the ligation product of insert and the vector, four types of products are expected from the colonies obtained

1. The correctly ligated product of the vector and the insert
2. The circularized vector without the insert
3. Vector-vector ligation
4. Vector and wrongly oriented insert ligation product

Methods to identify the correctly cloned construct are as follows:

- Blue white colony screening
- Sanger sequencing
- Colony PCR
- Restriction digestion with the enzymes used for cloning the insert and observing insert release
- Using a positive selection system, e.g., use of antibiotic for selection of recombinants

Answer 2

To ease the process of blunt-ended DNA ligation, two methods can be used:

1. Use of Adaptors:
   Adaptors are preformed cohesive ends that are attached to a blunt-ended DNA molecule to ease the process of ligation.

K. Bose (ed.), *Textbook on Cloning, Expression and Purification of Recombinant Proteins*, https://doi.org/10.1007/978-981-16-4987-5

2. Use of Linkers:
   Linkers are DNA sequences that comprise a recognition site for sticky end producing enzyme. Thus, after cleavage with such an enzyme, it would produce cohesive ends at both the termini of the blunt-ended DNA molecule, thus making ligation easier.

Chapter 3: Selection of Cloning and Expression Plasmid Vectors

Rucha Kulkarni, Roshnee Bose, and Kakoli Bose

Multiple Choice Questions
1. (b), 2. (c), 3. (d)

Subjective Questions
Answer 1

In a situation where the expressed 6x-His tagged protein is unstable/insoluble, the gene can be cloned in a different vector system, such as pMAL or pGEX. The tags present in these systems, MBP and GST, respectively, act as chaperones, thus helping the unstable protein to fold correctly and preventing the formation of misfolded aggregates. Additionally, MBP tag also directs the protein to periplasmic space, preventing the formation of inclusion bodies. Thus, such a protein when expressed as a fusion protein with one of these tags is stabilized to some extent and can be purified.

Answer 2

A cosmid vector would be the most suitable vector for cloning this gene. It has a large insert capacity as compared to cloning vectors like pBR322 and pUC series of vectors. Since cosmids are plasmids with cos sites from lambda phage, the vector along with the cloned insert is packaged inside the phage head forming viral particles. These viral particles in turn infect the *E. coli* host, thus making the transformation process more successful. This is not the case with pBR and pUC series of plasmids since cloning such a large fragment in these vectors leads to a decrease in their copy number followed by loss of transformation efficiency.

Chapter 4: Transformation and Protein Expression

Shubham Deshmukh, Rucha Kulkarni, and Kakoli Bose

Multiple Choice Questions
1. (d), 2. (b), 3. (b)

Subjective Questions

Answer 1

There could be multiple reasons as to why colonies failed to appear, some of which are listed below.

1. DNA concentration used for transformation was not sufficient. A higher concentration ~50 ng/μl might be helpful.
2. Antibiotic concentration could be too high.
3. The DNA used for transformation might have residues of phenol or ethanol, which inhibit transformation.
4. The expressed protein might be toxic for *E. coli*. Alternative host strains and vectors should be tried in this case. Including glucose (0.2–2%) in the LBA plates also helps alleviate the problem.

Answer 2

Certain degree of modifications in the host or culture system is required to maximize protein production but at the same time it is crucial to maintain the quality and activity of the purified protein. Despite making these modifications and providing the optimal conditions for growth, sometimes the host fails to produce the desired protein. Some of the reasons for this are discussed below:

1. Some of the protein coding genes contain rare codons, the tRNAs for which are not present in the bacteria. Thus, while translating the protein, the bacterial machinery is not able to read the rare codon and thus is not able to incorporate the amino acid. Hence, protein production stops abruptly. In such a case instead of BL21 (DE3), Rosetta (DE3) pLysS cells should be preferred that contain the tRNAs for those rare codons.
2. The protein to be overexpressed might sometimes prove to be toxic to the host leading to cell death
3. Temperature at which the culture is kept at the time of induction also plays an important role. Keeping the temperature low (~18°C) at the time of induction might be of help since the protease enzymes in the bacterial cells would be inactive at a low temperature and hence degradation of the expressed protein would not occur.

Chapter 5: Introduction to Recombinant Protein Purification

Nitu Singh and Kakoli Bose

Multiple Choice Questions

1. (c), 2. (d), 3. (d)

Subjective Questions
Answer 1

Firstly, generate a standard plot by dividing "Ve" of standards by the "V_o" (V_e/V_o), and plot this value against the log of the molecular weights (log M). Fit the plot using a linear equation and determine the slope of the line. Using this fitted value, the molecular weight of the unknown protein with elution volume 88 mL is 25.12 kDa.

Answer 2

In anion-exchange chromatography, the most negatively charged protein will be most attracted to the stationary phase and will therefore elute last while the protein with the highest positive charge will elute first. Since AS (pI-4.6) will be most negative in buffer of pH-6.5, it will be eluted last, while BS (pI-5.0) will be second and CS (pI-7.0) will be first. In short, the order of elution is CS, BS, and AS.

Chapter 6: Protein Purification by Affinity Chromatography

Shubhankar Dutta and Kakoli Bose

Multiple Choice Questions
1. (c), 2. (a), 3. (b)

Subjective Questions
Answer 1

(A) White precipitate is a sign of protein aggregation and can be solved by adding an appropriate protease inhibitor in the lysis buffer. Additionally, protease inhibitors also reduce the problem of aggregation effectively when used in the eluates after purification. Maintaining target protein in an EDTA containing storage buffer can also avert this problem.
(B) To reduce non-specific protein interactions, a slightly higher concentration of Imidazole can be included in the binding buffer. A concentration of 20 mM or up to 50 mM Imidazole can be used if the protein does not elute at this concentration. For elution, a higher concentration of Imidazole can be tried. A gradient of 50–500 mM may help reduce impurities. Increasing the salt concentration also gives good results.
(C) To reduce the problem of co-eluted chaperones, the following can be tried. Addition of 5 mM ATP and 1 mM $MgCl_2$ in all buffers might help to get rid of chaperones. The chaperone-protein complexes can be disrupted by addition of glycerol (around 10%), sucrose (up to 500 mM) or by detergents. This could be done during loading and washing Ni-NTA column or as an additional

purification step using gel filtration column. If the protein is stable at low pH (~4.5–5.0), lowering the pH of the elution buffer also helps.

Answer 2

If proteins A and B are interacting partners, then in ideal scenario, the SDS-PAGE gel analysis should show bands at 48 kDa (protein B), 42 kDa (control) and 22 kDa (protein A) staring from top to bottom. However, presence of a band at 70 kDa (22 + 48 kDa) suggests formation of a complex between protein A and protein B. The complex might have formed due to covalent bonds, such as disulfide bonds among the free or exposed surface residues (mostly cysteines) of the two proteins.

To circumvent the formation of sticky complex between the two proteins during pull-down, reducing agents can be added to the elution buffer or in the SDS-PAGE solution. Application of reducing agents such as 1 mM Dithiothreitol (DTT) or β-mercaptoethanol can successfully result in the emergence of the interacting proteins as separate bands at designated positions of the gel.

Chapter 7: Protein Purification by Ion Exchange Chromatography

Ayon Chakraborty, Rashmi Puja, and Kakoli Bose

Multiple Choice Questions
1. (c), 2. (a), 3. (b)

Subjective Questions
Answer 1

The pH at which a protein's total charge is zero is known as its isoelectric point (pI). At pH > pI, protein is negatively charged, whereas when the pH < pI, protein is positively charged. If the pH of the buffer solution is lower than the pI of a protein (i.e., acidic solution), the protein will be positively charged by accepting H^+ from the acidic solution. On the contrary, when the pH of the buffer solution is higher than the pI of the protein (i.e., alkaline solution), it will be negatively charged by donating H^+ to the alkaline solution. Because the pH of the buffer is greater than the pI value of the protein, it will be negatively charged in this case.

Answer 2

(A) Proteins are negatively charged when the pH is greater than pI and positively charged when the pH is less than pI. Protein 1 has a pI value lesser than the pH of the buffer and hence adheres to the positively charged column matrix. At this pH, proteins 2, 3, and β-galactosidase will be positively charged and will flow through the column.

(B) The bound protein 1 can be eluted by the addition of increasing concentrations of salt (NaCl). Cl^- ions compete with the bound protein for attachment to the positively charged matrix. When the concentration of Cl^- ions are high, it replaces the protein. Thus, protein is eluted from the column.

(C) The cation exchange column should be first equilibrated at a pH that is greater than 5.3 but less than 6.8 (5.3 < pH < 6.8). At this pH, β-galactosidase will have a negative charge (pH > pI), whereas proteins 2 and 3 will both be positively charged (pH < pI). Thus, proteins 2 and 3 will attach to the negatively charged matrix of the cation exchange column, whereas β-galactosidase will be repelled by the negatively charged matrix and will be found in the flow-through.

Chapter 8: Gel Filtration Chromatography

Raghupathi Kummari and Kakoli Bose

Multiple Choice Questions

1. (b), 2. (b), 3. (d)

Subjective Question

Answer 1

1. K_{av} of the standards as shown below.

(A) K_{av} of BSA:

$$\mathbf{K_{av}} = \mathbf{V_e - V_0/(V_t - V_0)}$$
$$= 77 - 46.5/120 - 46.5$$
$$= 0.4149$$

(B) K_{av} of MBP:

$$= 86.5 - 46.5/120 - 46.5$$
$$= 0.5442$$

(C) K_{av} of Lysozyme:

$$= 111.5 - 46.5/120 - 46.5$$

$$= 0.8843$$

2. Now, to determine the K_{av} of "**PROTEIN X**," the following steps are followed:
   (A) Take the Log molecular weight of the standards.

S. no	Protein	Mol. wt.	Log Mol. Wt.	K_{av} values from step 1
1	BSA	66.5 kDa	1.822	0.4149
2	MBP	45 kDa	1.653	0.5442
3	Lysozyme	14 kDa	1.146	0.8843

(B) Plot the logarithm of corresponding molecular weight (on *X*-axis) versus K_{av} value (on *Y*-axis).

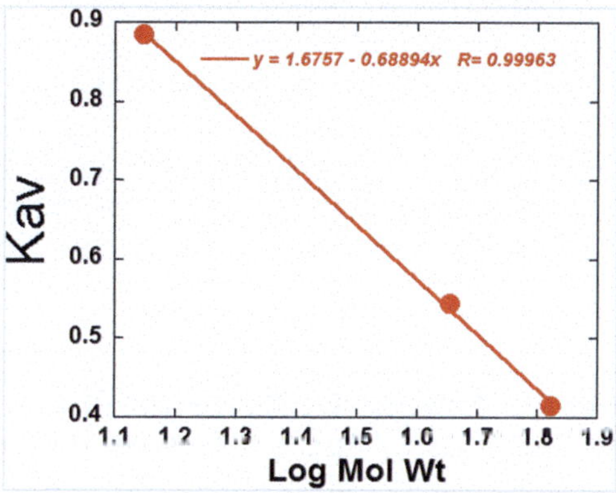

Use linear fitting to generate the straight line as shown in the plot above:

$$Y = -0.688(X) + 1.675.$$

(C) To calculate the unknown protein's molecular weight, we have to first determine K_{av} from the elution volumes.

Elution volume of "**PROTEIN X**" peak 1 = 64 ml

$$K_{av} = V_e - V_0/(V_t - V_0)$$

$$= 64 - 46.5/120 - 46.5$$

Therefore, $K_{av1} = 0.2380$
Substituting this K_{av1} value in the linear equation:

$$Y = -0.688(X) + 1.675$$

$$0.2380 = -0.688(X) + 1.675$$

$$0.2380 - 1.675 = -0.688(X)$$

$$-1.437 = -0.688(X)$$

$$X = -1.437/-0.688$$

$$X = 2.08 \text{ ('X' is the slope of the straight line)}$$

Anti-logarithmic value of 2.08 equals to **120 kDa** that corresponds to peak 1 (major peak)

Elution volume of "**PROTEIN X**" peak 2 = 87ml and K_{av2} is equal to 0.551. Substituting this K_{av2}value in the linear equation will give log molecular weight value of 1.633. Anti-logarithmic value is equal to **42 kDa** that corresponds to peak 2 (minor peak).

Since "**PROTEIN X**" gives two peaks that correspond to 42 kDa (minor) and 120 kDa (major), it can be concluded that "**PROTEIN X**" primarily maintains a trimeric state (~42X 3 kDa) that has been determined by the gel filtration experiment within experimental errors.

Chapter 9: Protein Purification by Reversed Phase Chromatography and Hydrophobic Interaction Chromatography

Rucha Kulkarni and Kakoli Bose

Multiple Choice Questions
1. (b), 2. (b), 3. (d)

Subjective Questions
Answer 1

To increase the yield of the hydrophobic protein the following parameters should be standardized:

Salt used: Using an alternate salt such as ammonium sulfate which is strong lyotropic salt that promotes tight binding between hydrophobic protein and the matrix. Moreover, it will be better to use a matrix for a moderately hydrophobic protein. This would enable maximum amount of protein to bind onto the resin. Phosphate buffer with sodium chloride can be used only when the protein is inherently very hydrophobic and does not require a strong lyotropic salt.

pH used: Acidic pH should be tried out while binding and eluting the protein. Acidic groups interact with the charged groups on proteins and prevent their interference in the hydrophobic interactions between the matrix and the protein.

Use of ligands: If ligand was already used, one should try using an alternative ligand with a longer alkyl chain as the length of the chain decides the hydrophobicity of the base matrix.

Answer 2

The following could be the probable solutions of the abovementioned problems:

1. Increasing the range of gradient of organic modifier: The stringency of washes must be increased by using a gradient of the organic modifier (here acetonitrile) to specifically elute the protein of interest. A gradient of 10% to 80% of acetonitrile can be tried out. This would enable the tightly bound protein to be eluted at a higher acetonitrile concentration.
2. For the second problem, certain organic solvents (e.g., methanol) that help to refine the polarity of the matrix should be tried out. The polarity of these solvents is such that they help in partially dissolving the less hydrophobic proteins so that they do not interfere as impurities later.

Chapter 10: Purification of Difficult Proteins

Saujanya Acharya and Kakoli Bose

Multiple Choice Questions

1. (d), 2. (b), 3. (c)

Subjective Questions

Answer 1

1. Using an MBP tag might be helpful. Change of expression host could be considered as well.
2. Optimizing culture conditions: lower temperatures, induction time, and IPTG concentration.
3. Incorporating a harsh detergent such as sarkosyl might help in solubilization.
4. If the protein has disulfide bonds, using BME might help.

Answer 2

Urea denaturation and refolding is the most promising strategy at this point.

Chapter 11: Protein Quantitation and Detection

Raghupathi Kummari, Rashmi Puja, and Kakoli Bose

Multiple Choice Questions
1. (d), 2. (b), 3. (c)

Subjective Questions
Answer 1

First find the average of three absorbance values and do blank correction from the absorbance. Then, multiply the concentration of BSA with five (to take care of the dilution factor). Plot the blank corrected absorbance values on *X* axis and final BSA protein concentration on *Y* axis (as highlighted below) to obtain the standard graph. In the standard graph, click on any point to add polynomial trend-line and select the equation that displays R^2 value. The R^2 value should be ~1 for best fitted curve. Now use the resulting equation to calculate protein concentration in sample.

	A1	A2	A3	Average
Blank	0.221	0.217	0.214	0.22
0.0625 mg/ml BSA	0.233	0.244	0.223	0.23
0.125 mg/ml BSA	0.297	0.25	0.283	0.28
0.25 mg/ml BSA	0.385	0.352	0.326	0.35
0.5 mg/ml BSA	0.482	0.467	0.45	0.47
1 mg/ml BSA	0.53	0.608	0.597	0.58
Test sample	0.473	0.504	0.525	0.50

	Absorbance (average)	Blank corrected absorbance	Final BSA protein conc.
Blank	0.22		
0.0625 mg/ml BSA	0.23	0.01	0.3125
0.125 mg/ml BSA	0.28	0.06	0.625
0.25 mg/ml BSA	0.35	0.13	1.25
0.5 mg/ml BSA	0.47	0.25	2.5
1 mg/ml BSA	0.58	0.36	5
Test sample	0.50	0.28	

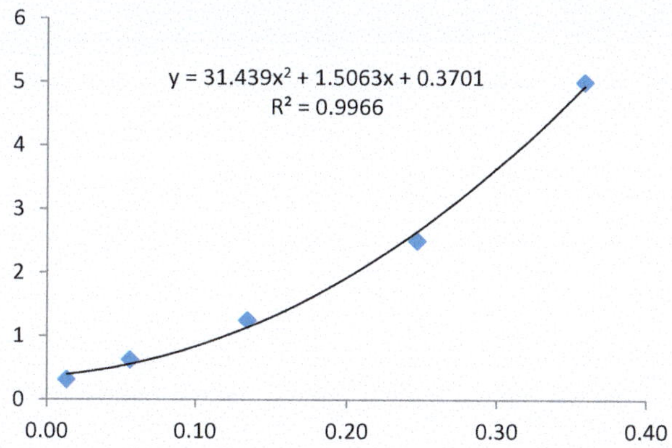

For determining the protein concentration in sample, put the blank corrected absorbance of sample as the value of X in standard equation.

$$\begin{aligned}\text{Protein concentration in sample (mg/ml)}\\ &= (31.43 \text{ X } 0.28 \text{ X } 0.28) + (1.506 \text{ X } 0.28) + 0.370\\ &= 2.46 + 0.42 + 0.37\\ &= 3.25 \text{ mg/ml}\end{aligned}$$

Answer 2

To calculate the protein concentration, we can use abovementioned formula, i.e.

$$\frac{\text{O.D.}}{\text{Molar extinction coefficient/Molecular weight in D}} = \text{concentration in mg/ml}$$

(Use the average value of absorbance taken in triplicate as O.D.)

$$\text{concentration (C)in} \frac{\text{mg}}{\text{ml}} = \frac{0.47}{43824\ 66400}$$

$$C = \frac{0.47}{0.66}$$

$$C = 0.71 \text{ mg/ml}$$

The manufacturer's authorised representative in the EU is Springer Nature Customer Service Centre GmbH, Europaplatz 3, 69115 Heidelberg, Germany. If you have any concerns regarding our products, please contact ProductSafety@springernature.com

Printed and bound by CPI Group (UK) Ltd, Croydon, CR0 4YY
30/09/2026
02225240-0003